AF594950

Cognitive Buildings

Cognitive Buildings

Editors

Lavinia Chiara Tagliabue
Ibrahim Yitmen

MDPI • Basel • Beijing • Wuhan • Barcelona • Belgrade • Manchester • Tokyo • Cluj • Tianjin

Editors
Lavinia Chiara Tagliabue
Computer Science Department
University of Turin
Turin
Italy

Ibrahim Yitmen
Department of Construction
Engineering and Lighting
Science
Jönköping University
Jönköping
Sweden

Editorial Office
MDPI
St. Alban-Anlage 66
4052 Basel, Switzerland

This is a reprint of articles from the Special Issue published online in the open access journal *Applied Sciences* (ISSN 2076-3417) (available at: www.mdpi.com/journal/applsci/special_issues/cognitive_buildings).

For citation purposes, cite each article independently as indicated on the article page online and as indicated below:

LastName, A.A.; LastName, B.B.; LastName, C.C. Article Title. *Journal Name* **Year**, *Volume Number*, Page Range.

ISBN 978-3-0365-3952-2 (Hbk)
ISBN 978-3-0365-3951-5 (PDF)

Contents

About the Editors

Lavinia Chiara Tagliabue

Dr. Lavinia Chiara Tagliabue, Associate Professor, of Computer Science Department, University of Turin, Turin, Italy.

Her research interests are related to sustainability and sustainable buildings; environmental and energy protocols; energy saving; renewable energies; energy retrofit and NZEB; BIM to BEM (building information modeling to building energy modeling) interoperability; cognitive buildings; behavioral design; probabilistic models; building management systems (BMS), Digital Twins.

Ibrahim Yitmen

Dr. Ibrahim Yitmen, Associate Professor, Department of Construction Engineering and Lighting Science, School of Engineering, Jönköping University, Jönköping, Sweden.

His research interests are related to digital transformation in the AEC industry; digital twin-based smart built environment; augmented reality/mixed reality for cognitive buildings; integration of digital twins and deep learning for smart planning and construction; blockchain technology in construction supply chains; cyber-physical systems for construction 4.0; integration of unmanned aerial vehicles and BIM for construction safety planning and monitoring.

Preface to "Cognitive Buildings"

The book *Cognitive Buildings* proposes pioneering scientific works addressing the paradigm shift toward a novel concept of buildings where spaces and users are connected and supported by a technological environment with the aim of creating a synergic system that learns from the users' feedback and external variables to improve energy efficiency and comfort promoting users' capability and supporting social progress and the best future built environment.

The editors of this volume collected the valuable and brilliant scientific research of the international community from colleagues involved in the field, and they were able to depict a comprehensive panorama of the proposed concept that will be our common future built environment. The editors acknowledge all the authors and the editorial assistance that have made it possible to successfully compile this volume. The book is addressed to researchers and professionals that will shape our future common evolving environment.

Lavinia Chiara Tagliabue and Ibrahim Yitmen
Editors

Editorial

Special Issue Cognitive Buildings

Lavinia Chiara Tagliabue [1,*] and Ibrahim Yitmen [2]

1 Computer Science Department, University of Turin, Corso Svizzera 185, 10149 Torino, Italy
2 Department of Construction Engineering and Lighting Science, School of Engineering, Jönköping University, Gjuterigatan 5, Box 1026, 551 11 Jönköping, Sweden; ibrahim.yitmen@ju.se
* Correspondence: laviniachiara.tagliabue@unito.it; Tel.: +39-335-590-2507

1. Introduction

Cognitive building is a pioneering topic to envision the future of our built environment. The concept of "cognitive" steps towards a paradigm shift, from the static concept of the building as a container of human activities, is nearer to the modernist vision of "machine à habiter" of Le Corbusier, where the technological content adds the capability of learning from users' behavior and environmental variables to adapt itself to achieve main goals such as users' comfort, energy-saving, flexible functionality, high durability, and good maintainability. The concept is based on digital frameworks and IoT networks towards the smart city concept. A BIM-based approach to cognitive buildings and assets is widely represented in the evolving research, oriented to different fields of application throughout the buildings' lifecycle that will shape the future of the built environment, where the users could extend and improve their daily experience towards a better life.

So far, 11 papers have been published in the Special Issue. The next sections provide a brief summary of each of the papers published.

2. Explainable Post-Occupancy Evaluation Using a Humanoid Robot

Bonomolo et al. [1] propose a new methodological approach for evaluating the comfort condition using the concept of explainable post-occupancy to make the user aware of the environmental state in which (s)he works. Such an approach was implemented on a humanoid robot with social capabilities that aims to enforce human engagement to follow recommendations. The humanoid robot helps the user to position the sensors correctly to acquire environmental measures corresponding to the temperature, humidity, noise level, and illuminance. The distribution of the last parameter due to its high variability is also retrieved by the simulation software Dialux. Using the post-occupancy evaluation method, the robot also proposes a questionnaire to the user for collecting his/her preferences and sensations. In the end, the robot explains to the user the difference between the suggested values by the technical standards and the real measures, comparing the results with his/her preferences and perceptions. Finally, it provides a new classification into four clusters: true positive, true negative, false positive, and false negative. This study shows that the user is able to improve her/his condition based on the explanation given by the robot.

3. An openBIM Approach to IoT Integration with Incomplete As-Built Data

Moretti et al. [2] discuss how Digital Twins (DT) are powerful tools to support asset managers in the operation and maintenance of cognitive buildings. Building Information Models (BIM) are critical for Asset Management (AM), especially when used in conjunction with Internet of Things (IoT) and other asset data collected throughout a building's lifecycle. However, information contained within BIM models is usually outdated, inaccurate, and incomplete as a result of unclear geometric and semantic data modeling procedures during the building life cycle. The aim of this paper is to develop an openBIM methodology to support dynamic AM applications with limited as-built information availability. The

Citation: Tagliabue, L.C.; Yitmen, I. Special Issue Cognitive Buildings. *Appl. Sci.* **2022**, *12*, 2460. https://doi.org/10.3390/app12052460

Received: 7 December 2021
Accepted: 19 December 2021
Published: 26 February 2022

Publisher's Note: MDPI stays neutral with regard to jurisdictional claims in published maps and institutional affiliations.

workflow is based on the use of the IfcSharedFacilitiesElements schema for processing the geometric and semantic information of both existing and newly created Industry Foundation Classes (IFC) objects, supporting real-time data integration. The methodology is validated using the West Cambridge DT Research Facility data, demonstrating good potential in supporting an asset anomaly detection application. The proposed workflow increases the automation of the digital AM processes, thanks to the adoption of BIM-IoT integration tools and methods within the context of the development of a building DT.

4. Building Information Modeling and Energy Simulation for Architecture Design

Bonomolo et al. [3] deal with Building Information Modeling (BIM) that, over the years, has undergone a significant increase, both in terms of functions and use. This tool can almost completely manage the entire process of the design, construction, and management of a building internally. However, it is not able to fully integrate the functions and especially the information needed to conduct a complex energy analysis. Indeed, even if the energy analysis has been integrated into the BIM environment, it still fails to make the most of all of the potential offered by building information modeling. The main goals of this study are the analysis of the interaction between BIM and energy simulation through a review of the main existing commercial tools (available and user-friendly), and the identification and application of a methodology in a BIM environment by using Graphisoft's BIM software Archicad and the plug-in for dynamic energy simulation EcoDesigner STAR. The application on a case study gave the possibility to explore the advantages and the limits of these commercial tools and, consequently, to provide some possible improvements. The results of the analysis, satisfactory from a quantitative and qualitative point of view, validated the methodology proposed in this study and highlighted some limitations of the tools used; in particular, for the aspects concerning the personalization of heating systems.

5. Building Information Modeling and Internet of Things Integration for Facility Management—Literature Review and Future Needs

Mannino et al. [4] show how the digitization of the built environment is seen as a significant factor for innovation in the architecture, engineering, construction and operation sector. However, a lack of data and information in as-built digital models considerably limits the potential of building information modeling in facility management. Therefore, the optimization of data collection and management is needed, all the more so now that Industry 4.0 has widened the use of sensors into buildings and infrastructures. A literature review on the two main pillars of digitalization in construction, building information modeling and Internet of things, is presented, along with a bibliographic analysis of two citations and abstracts databases focusing on the operations stage. The bibliographic research has been carried out using Web of Science and Scopus databases. The article is aimed at providing a detailed analysis of BIM–IoT integration for Facility Management (FM) process improvements. Issues, opportunities, and areas where further research efforts are required are outlined. Finally, four key areas of further research development in FM management have been proposed, focusing on optimizing data collection and management.

6. Towards an Occupancy-Oriented Digital Twin for Facility Management: Test Campaign and Sensors Assessment

Seghezzi et al. [5] focus on calibration and test campaigns of an IoT camera-based sensor system to monitor occupancy as part of an ongoing research project aiming at defining a Building Management System (BMS) for facility management based on an occupancy-oriented Digital Twin (DT). The research project aims to facilitate the optimization of the building operational stage through advanced monitoring techniques and data analytics. The quality of the collected data, which are the input for analyses and simulations on the DT virtual entity, is critical in order to ensure the quality of the results. Therefore, calibration and test campaigns are essential to ensure the data quality and efficiency of the IoT sensor system. The paper describes the general methodology for the BMS definition, and the method and results of the first stages of the research. The preliminary analyses includes Indicative Post-Occupancy Evaluations (POEs) supported by Building Information Modeling (BIM) to optimize sensor system planning. Test campaigns are then performed to evaluate the collected data quality and system efficiency. The method was applied on a Department of Politecnico di Milano. The period of the year in which tests are performed was critical for lighting conditions. In addition, spaces' geometric features and user behavior caused major issues and faults in the system. Incorrect boundary definition: areas that are not covered by boundaries; thus, they are not monitored.

7. An Adapted Model of Cognitive Digital Twins for Building Lifecycle Management

Yitmen et al. [6] present the digital transformation era in the Architecture, Engineering, and Construction (AEC) industry, where Cognitive Digital Twins (CDT) are introduced as part of the next level of process automation and control towards Construction 4.0. CDT incorporates cognitive abilities in order to detect complex and unpredictable actions, and reasons about dynamic process optimization strategies to support decision-making in Building Lifecycle Management (BLM) are provided. Nevertheless, there is a lack of understanding of the real impact of CDT integration, Machine Learning (ML), Cyber–Physical Systems (CPS), Big Data, Artificial Intelligence (AI), and Internet of Things (IoT), all connected to self-learning hybrid models with proactive cognitive capabilities for different phases of the building asset lifecycle. This study investigates the applicability, interoperability, and integrability of an adapted model of CDT for BLM to identify and close this gap. Surveys of industry experts were performed, focusing on life cycle-centric applicability, interoperability, and the CDT model's integration in practice, besides decision support capabilities and AEC industry insights. The evaluation of the adapted model of the CDT model supports approaching the development of CDT for process optimization and decision-making purposes, as well as integrability enablers confirm a progression towards Construction 4.0.

8. The Implementation of Visual Comfort Evaluation in the Evidence-Based Design Process Using Lighting Simulation

Davoodi et al. [7] work on a validation of the EBD-SIM (evidence-based design simulation) framework, a conceptual framework developed to integrate the use of a lighting simulation in the EBD process, which suggested that EBD's Post-Occupancy Evaluation (POE) should be conducted more frequently. A follow-up field study was designed for a subjective–objective results implementation in the EBD process using lighting simulation tools. In this real-time case study, the visual comfort of the occupants was evaluated. The visual comfort analysis data were collected via simulations and questionnaires for subjective visual comfort perceptions. The follow-up study, conducted in June, confirmed the results of the original study, conducted in October, but additionally found correlations with annual performance metrics. This study shows that, at least for the variables related to daylight, a POE needs to be conducted at different times of the year to obtain a more comprehensive insight into the users' perception of the lit environment.

9. IoT Open-Source Architecture for the Maintenance of Building Facilities

Villa et al. [8] debate about the introduction of the Internet of Things (IoT) in the construction industry, which is evolving Facility Maintenance (FM) towards predictive maintenance development. The predictive maintenance of building facilities requires continuously updated data on construction components to be acquired through integrated sensors. The main challenges in developing predictive maintenance tools for building facilities is IoT integration, IoT data visualization on the building 3D model, and the implementation of a maintenance management system on the IoT and Building Information Modeling (BIM). The current 3D building models do not fully interact with IoT building facilities data. Data integration in BIM is challenging. The research aims to integrate IoT alert systems with BIM models to monitor building facilities during the operational phase and to visualize building facilities' conditions virtually. To provide efficient maintenance services for building facilities, this research proposes an integration of a digital framework based on IoT and BIM platforms. Sensors applied in the building systems and IoT technology on a cloud platform with open source tools and standards enable the monitoring of a real-time operation and the detection of different kinds of faults in the case of malfunction or failure, therefore sending alerts to facility managers and operators. The proposed preventive maintenance methodology applied on a proof-of-concept Heating, Ventilation, and Air Conditioning (HVAC) plant adopts open source IoT sensor networks. The results show that the integrated IoT and BIM dashboard framework and implemented building structures preventive maintenance methodology are applicable and promising. The automated system architecture of building facilities is intended to provide a reliable and practical tool for real-time data acquisition. An analysis and 3D visualization to support the intelligent monitoring of the indoor condition in buildings will enable the facility managers to make faster and better decisions and to improve building facilities' real time monitoring with fallouts on the maintenance timeliness.

10. BIM-Based Research Framework for Sustainable Building Projects: A Strategy for Mitigating BIM Implementation Barriers

Manzoor et al. [9] underline that although Building Information Modeling (BIM) can enhance the efficiency of sustainable building projects, its adoption is still plagued with barriers. In order to incorporate BIM more efficiently, it is important to consider and mitigate these barriers. The aim of this study is to explore and develop strategies to alleviate barriers in developing countries, such as Malaysia, in order to broaden the implementation of BIM with the aid of quantitative and qualitative approaches. To achieve this aim, a comprehensive literature review was carried out to identify the barriers, and a questionnaire survey was conducted with construction projects' stakeholders. The ranking analysis results revealed the top five critical barriers to be the "unavailability of standards and guidelines", "lack of BIM training", "lack of expertise", "high cost", and "lack of research and BIM implementation". Comparative study findings showed that the "lack of research and BIM implementation" is the least important barrier in other countries, such as China, United Kingdom, Nigeria, and Pakistan. Furthermore, a qualitative analysis revealed the strategies to mitigate the BIM implementation barriers to enhance sustainable goals. The final outcome of this study is the establishment of a framework incorporated with BIM implementation barriers and strategies, namely, the "BIM-based research framework", which can assist project managers and policymakers towards effective sustainable construction.

11. Transforming Building Criteria to Evidence Index

Fischl and Johansson [10] emphasize how there is increasing pressure from developers toward architects and engineers to deliver scientifically sound proposals for often complex and cost-intensive construction products. An increase in digitalization within the construction industry, and the availability of intelligently built assets and overall sustainability, make it possible to customize a construction product. This servitization of construction products is assumed to perform much more preferably in satisfying stakeholders' physical, psychological, and social needs. The degree to which these products are performing can be evaluated through an evidence index. This article aims to introduce a conceptual model of an evidence index and to test it in the programming stage of a case study. The investigation follows the evidence-based design approach and renders evidence through key performance indicators in the programming stage of the building process. For testing the concept, a case study investigation was performed by simulating a novice research assistant, and the amount of evidence was collected and appraised for the evidence index. The case study showed that key performance indicators of a servitized project could be evaluated on a four-point scale. The quality of the evidence index generation depended on the level of expertise the evaluator has in the research and the skillful use of scientific databases.

12. Heuristic and Numerical Geometrical Methods for Estimating the Elevation and Slope at Points Using Level Curves—Application for Embankments

Deaconu and Deaconu [11] highlight that both the calculation of ground slopes at points on the map and the elevation estimation for these points bear significant importance and also have applications in various domains, such as civil engineering and road and railway design. The paper presents two methods that use level curves: one that is fast and approximate, and another that is slower but more precise. The running speed of the two proposed methods and their results are compared by performing 100 million experiments. The paper also presents how these methods can be applied to optimize embankments. An accurate method to calculate the horizontal plane of the excavation/filling when building a new house is also presented.

Author Contributions: Conceptualization, L.C.T.; methodology, L.C.T. and I.Y.; validation, L.C.T. and I.Y. All authors have read and agreed to the published version of the manuscript.

Funding: This research received no external funding.

Institutional Review Board Statement: Not applicable.

Informed Consent Statement: Not applicable.

Data Availability Statement: Not applicable.

Acknowledgments: Thanks are due to all of the authors and peer reviewers for their valuable contributions to this Special Issue. The MDPI management and staff are also to be congratulated for their untiring editorial support for the success of this project.

Conflicts of Interest: The authors declare no conflict of interest.

References

1. Bonomolo, M.; Ribino, P.; Vitale, G. Explainable Post-Occupancy Evaluation Using a Humanoid Robot. *Appl. Sci.* **2020**, *10*, 7906. [CrossRef]
2. Moretti, N.; Xie, X.; Merino, J.; Brazauskas, J.; Parlikad, A. An openBIM Approach to IoT Integration with Incomplete As-Built Data. *Appl. Sci.* **2020**, *10*, 8287. [CrossRef]
3. Bonomolo, M.; Di Lisi, S.; Leone, G. Building Information Modelling and Energy Simulation for Architecture Design. *Appl. Sci.* **2021**, *11*, 2252. [CrossRef]
4. Mannino, A.; Dejaco, M.; Re Cecconi, F. Building Information Modelling and Internet of Things Integration for Facility Management—Literature Review and Future Needs. *Appl. Sci.* **2021**, *11*, 3062. [CrossRef]
5. Seghezzi, E.; Locatelli, M.; Pellegrini, L.; Pattini, G.; Di Giuda, G.; Tagliabue, L.; Boella, G. Towards an Occupancy-Oriented Digital Twin for Facility Management: Test Campaign and Sensors Assessment. *Appl. Sci.* **2021**, *11*, 3108. [CrossRef]
6. Yitmen, I.; Alizadehsalehi, S.; Akıner, İ.; Akıner, M. An Adapted Model of Cognitive Digital Twins for Building Lifecycle Management. *Appl. Sci.* **2021**, *11*, 4276. [CrossRef]
7. Davoodi, A.; Johansson, P.; Aries, M. The Implementation of Visual Comfort Evaluation in the Evidence-Based Design Process Using Lighting Simulation. *Appl. Sci.* **2021**, *11*, 4982. [CrossRef]
8. Villa, V.; Naticchia, B.; Bruno, G.; Aliev, K.; Piantanida, P.; Antonelli, D. IoT Open-Source Architecture for the Maintenance of Building Facilities. *Appl. Sci.* **2021**, *11*, 5374. [CrossRef]
9. Manzoor, B.; Othman, I.; Gardezi, S.; Altan, H.; Abdalla, S. BIM-Based Research Framework for Sustainable Building Projects: A Strategy for Mitigating BIM Implementation Barriers. *Appl. Sci.* **2021**, *11*, 5397. [CrossRef]
10. Fischl, G.; Johansson, P. Transforming Building Criteria to Evidence Index. *Appl. Sci.* **2021**, *11*, 5894. [CrossRef]
11. Deaconu, A.; Deaconu, O. Heuristic and Numerical Geometrical Methods for Estimating the Elevation and Slope at Points Using Level Curves. Application for Embankments. *Appl. Sci.* **2021**, *11*, 6176. [CrossRef]

Article

Heuristic and Numerical Geometrical Methods for Estimating the Elevation and Slope at Points Using Level Curves. Application for Embankments

Adrian Marius Deaconu [1,*] and Ovidiu Deaconu [2]

1 Department of Mathematics and Computer Science, Transilvania University of Brașov, B-dul Eroilor 29, 500036 Brasov, Romania
2 Department of Civil Engineering, Transilvania University of Brașov, Str. Turnului, 500036 Brasov, Romania; ovideaconu@unitbv.ro
* Correspondence: a.deaconu@unitbv.ro; Tel.: +40-744-378-881

Abstract: Both the calculation of ground slopes at points on the map and the elevation estimation for these points bear significant importance and also have applications in various domains, such as civil engineering, road and railway design. The paper presents two methods that use level curves: one that is fast and approximate and another which is slower, but more precise. The running speed of the two proposed methods and their results are compared by performing 100 million experiments. The paper also presents how these methods can be applied to optimize embankments. An accurate method to calculate the horizontal plane of the excavation/filling when building a new house is also presented.

Keywords: level curves; ground slopes; embankments; road and rail design; optimization

Citation: Deaconu, A.M.; Deaconu, O. Heuristic and Numerical Geometrical Methods for Estimating the Elevation and Slope at Points Using Level Curves. Application for Embankments. *Appl. Sci.* **2021**, *11*, 6176. https://doi.org/10.3390/app11136176

Academic Editor: Igal M. Shohet

Received: 22 May 2021
Accepted: 29 June 2021
Published: 2 July 2021

1. Introduction

The possibility of calculating the slope and elevation using level curves has practical applicability in several fields of activity, such as civil engineering [1], architecture [2,3], railway [4], road and bridge engineering [5–8], topography [9], etc. Although the results found can be used in various areas, this article is limited to solving two problems related to civil engineering and to railways, roads, and bridges, both ultimately related to the embankments optimization. The problem is well known, but it was not closely examined by design engineers, who preferred an intuitive, approximate solution. The current article presents a rigorous and mathematically accurate solution to find the optimum between excavations and fillings, resulting in a reduction of the embankments related costs. The article presents in detail the solutions of two concrete problems in two different fields of activity.

The first problem investigated is in the field of civil engineering and consists of determining the optimal elevation corresponding to the finished floor of the ground level, when the building has to be positioned on a slope [1–3]. The second problem investigated is in the design of a new railway or road when there are several fixed points on the topographic map through which the new road [10–13] or the new railway [4,11] will pass.

The designer has to know the slope and elevation in each of these points as accurately as possible. Using the methods proposed in this paper, the two parameters (slope and elevation) can be determined fast and with a high level of accuracy using the level curves of the map, without the need for direct measurement on the spot. Furthermore, when designing roads or railways, the risk of failure, the frictional resistance, the safety factor against sliding or failure, the translational fill failure, or the cut volume and the road fill per meter require knowing the slopes (side slope, cut slope, or fill slope).

The paper is organized as follows: an algorithm to calculate the ground slope at a given point, using level curves, is deduced in Section 2. This problem is reduced to estimating the shortest segment that connects two lines through a point. Two methods are proposed, a fast and approximate one and a more exact one, respectively. The two methods are compared, and it is shown that the errors of the first method are low. In Section 3, the optimum excavation/filling plane is calculated when a new house is built. Two methods are proposed and discussed in Section 4, where they are compared. To finalize the paper, conclusions will be drawn based on the arguments presented.

2. Materials and Methods

A method of calculating the soil slope through a given point is being introduced. The input data are: the level curves, in Figure 1, the two level curves located at elevation 610 and, respectively, 620, and the 2D coordinates of a point Q on the topographic map. The direction of the Ox axis is considered from west to east, the direction of the Oy axis is south-north. The output is the slope calculated at point Q. The slope can be determined using two other points collinear with Q for which the elevation is known. The points on the map for which the elevation is known are located on level curves. Therefore, it is enough to find two points collinear with Q located on two different level curves. Of course, for accuracy reasons, it is preferable that Q is located between these 2 points and the distance between the chosen 2 points is minimum. Consequently, the problem can be accurately solved if we determine the closest two level curves to Q (Q is located in between these curves) and two points P_1 and P_2, each located on one of the level curves, and the distance between P_1 and P_2 is minimum (see Figure 1).

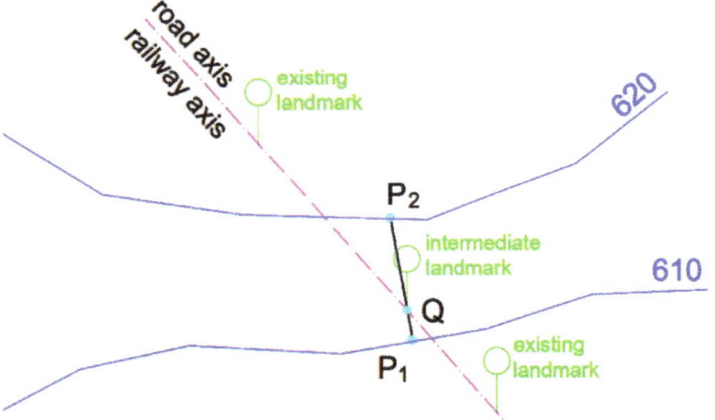

Figure 1. Finding the slope at a point using level curves.

Since a level curve is approximated with consecutive coplanar segments of straight lines, our problem can be reduced to finding the shortest length segment $|P_1P_2|$ that connects two coplanar line segments located on two consecutive level curves. When the two 2D points $P_1(x_1,y_1)$ and $P_2(x_2,y_2)$ are determined, they are transformed into 3D points by adding the elevations z_1 and, respectively, z_2 of the level curves on which each of the points are located. Using the 3D coordinates of these 2 points, the slope m = tan(α) at point Q (see Figure 2) can be calculated as follows:

$$\begin{cases} m = \frac{|z_2 - z_1|}{\sqrt{(x_2-x_1)^2+(y_2-y_1)^2}} \\ \alpha = \operatorname{atan2}\left(|z_2 - z_1|, \sqrt{(x_2-x_1)^2+(y_2-y_1)^2}\right) \end{cases} \quad (1)$$

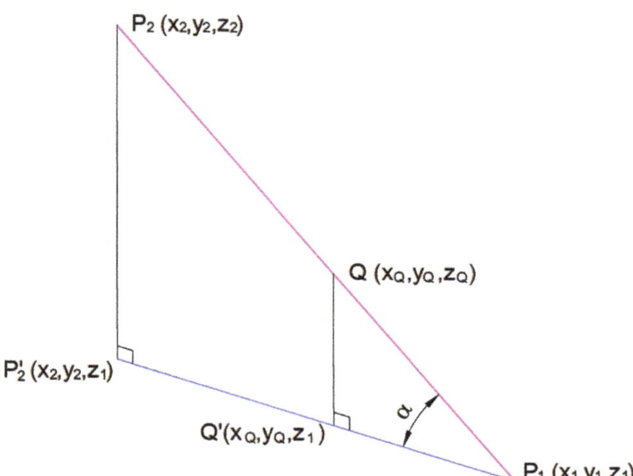

Figure 2. Determining the slope at Q using P_1 and P_2.

The first problem (denoted IPS) intended to be solved is to identify the pair of segments (located on two level curves) to which the finding of the shortest length segment should be applied. Of course, for each of the two level curves, the candidates are the closest segments to point Q. Let us denote by S_i the set of candidates (segments) for curve i (i = 1, 2). For each pair of segments $s_1 \in S_1$ and $s_2 \in S_2$, respectively, the corresponding shortest length segment $[P_1P_2]$ between s_1 and s_2 is calculated. If s_1 contains P_1 and s_2 contains P_2, then the pair (s_1, s_2) is added to the set (denoted C) of solution candidates for IPS (Figure 3).

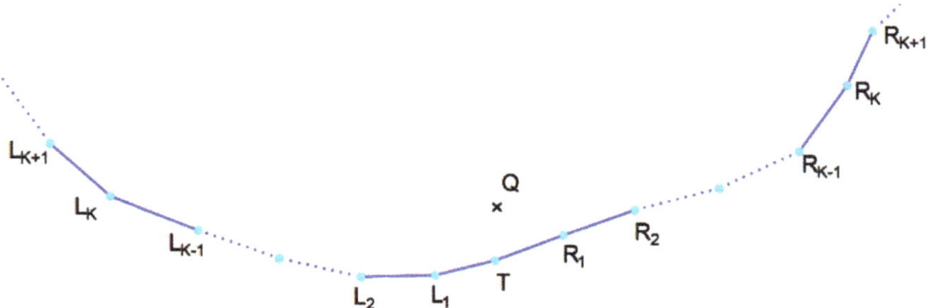

Figure 3. Finding the closest line segments to Q.

The set S_i (i = 1, 2) can be built as follows: first, we find the point T belonging to the level curve i that is closest to Q. We consider then $k \geq 1$ points $(L_1, L_2, \ldots, L_k)$ to the left of T on the level curve and k points $(R_1, R_2, \ldots, R_k)$ located to right (see Figure 4). Finally, the set S_i consists of 2k segments of straight lines given by the consecutive points on the level curve i: $L_k, L_{k-1}, \ldots, L_1, T, R_1, R_2, \ldots, R_k$, i.e.,:

$$S_i = \{[L_{h+1}L_h] \mid h = 1,2,\ldots,k-1\} \cup \{[L_1T],[TR_1]\} \cup \{[R_hR_{h+1}] \mid h = 1,2,\ldots,k-1\} \quad (2)$$

Now, let us see how k (giving the number of points in the vicinity of T) is considered. Usually (in real tests), k is enough to be 1. In few cases k is found to be greater. So, the calculations are started with k = 1 and, if necessary, it is increased.

The Algorithm 1 for solving IPS (denoted AIPS) is:

Algorithm 1 Solving IPS (denoted AIPS)

```
AIPS:
Input: the points of the closest two level curves to Q
Output: the shortest line segment [P1P2] that connects the given two level curves
finished = false;
While not finished do
      C = ϕ (empty set);
      Build the sets of line segments S1 and S2 (see Equation (2));
      For each segment s1 from S1 do
            For each segment s2 from S2 do
                  Calculate the shortest length segment P1P2 that connects s1 and s2 and passes through Q;
                  If P1 ∈ s1 and P2 ∈ s2 then
                        Add [P1P2] to C;
                  End if;
            End for;
      End for;
      If C not is empty then
            finished = true;
      else
            k = k+1;
      End if;
End while;
```

The shortest line segment that connects both level curves is the shortest segment $[P_1P_2]$ from C.

After applying AIPS, the problem is reduced to calculating the shortest segment that passes through a given point Q and connects two coplanar lines d_1 and d_2 (Figure 4). We shall denote this problem as SSTPC2L (shortest segment through a point that connects 2 lines). In Figure 4 the points A_{11} and A_{12} determine the line d_1, and the points A_{21} and A_{22} determine the line d_2.

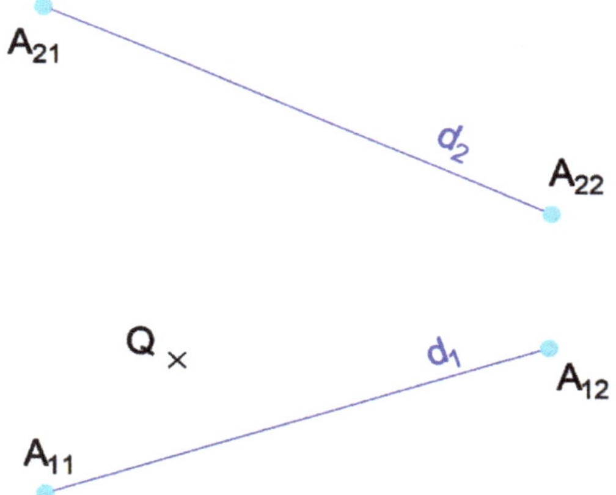

Figure 4. Connecting the point Q to the lines d_1 and d_2.

We present two methods of solving SSTPC2L: an approximate method and an exact mathematical method. These two methods (presented in Sections 2.1 and 2.2) were partially described in [7].

2.1. Heuristic (Approximate) Method for SSTPC2L

From the topographic map, the point Q is known and two segments $A_{11}A_{12}$ and $A_{21}A_{22}$ are determined belonging to the level curves from Figure 1 ($A_{11}A_{12}$ on curve 610 and $A_{21}A_{22}$ on curve 620). These two segments are found according to Figure 3. The method consists of drawing the perpendiculars from the given point Q on the two given lines d_1 and, respectively, d_2 (Figure 5). We construct a line passing through Q which is parallel to the line determined by the feet of the two perpendiculars. The obtained line is denoted by d_3, and its intersection with the lines d_1 and d_2 is the searched segment $[P_1P_2]$ (approximate solution of SSTPC2L).

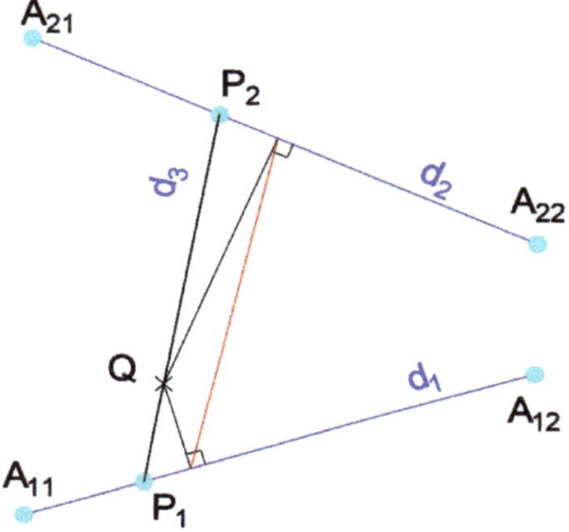

Figure 5. Approximate method of finding points P_1 and P_2.

Now, let's present how the points P_1 and P_2 are calculated.

The equation of the line d_i (i = 1, 2) starting from the points A_{i1} and A_{i2} is:

$$y = m_i x + n_i, \text{ where } m_i = \frac{y_{i2} - y_{i1}}{x_{i2} - x_{i1}} \text{ and } n_i = y_{i1} - m_i x_{i1} \tag{3}$$

The perpendicular from Q on d_i (i = 1, 2) is:

$$y = -\frac{1}{m_i}x + n'_i, \text{ where } n'_i = y_Q + \frac{1}{m_i}x_Q \tag{4}$$

The foot $Q_i\left(x_i^Q, y_i^Q\right)$ (i = 1, 2) of the perpendicular on d_i is obtained by solving the system of Equations (3) and (4).

The slope of the line Q_1Q_2 is:

$$m_{12} = \frac{y_2^Q - y_1^Q}{x_2^Q - x_1^Q} \tag{5}$$

Since m_{12} is the slope of Q_1Q_2, then the parallel through the point $Q\left(x_Q, y_Q\right)$ is:

$$y = y_Q + m_{12}(x - x_Q) \tag{6}$$

Finally, the points P_i (i = 1, 2) are obtained by solving the system of Equations (3) and (6).

The following Algorithm 2 (denoted A1SSTPC2L) is obtained to solve the problem SSTPC2L:

Algorithm 2 Solving the problem SSTPC2L

A1SSTPC2L:
Input: Q and the points of the closest two level curves to Q
Output: The slope at point Q
Apply AIPS to find A_{11}, A_{12}, A_{21}, A_{22};
Calculate the point P_i by solving system of Equations (3) and (6) (i = 1, 2).
Calculate the slope at point Q using Equation (1).

2.2. The Exact (Mathematical) Method for SSTPC2L

For this method, in order to simplify calculus, we apply some initial transformations (rotations and translations) to the points A_{11}, A_{12}, A_{21}, A_{22}, and Q so that the bisector of the angle between the lines d_1 and d_2 becomes parallel to Oy axis and the point Q is in the origin O, i.e., Q = O (0,0). To do so, we first calculate the intersection I of d_1 and d_2:

$$d_1 \cap d_2 = \{I\} \tag{7}$$

by solving the system of linear equations:

$$\begin{cases} d_1 : \frac{x - x_{11}}{x_{12} - x_{11}} = \frac{y - y_{11}}{y_{12} - y_{11}} \\ d_2 : \frac{x - x_{21}}{x_{22} - x_{21}} = \frac{y - y_{21}}{y_{22} - y_{21}} \end{cases}$$

The following transformations to A_{11}, A_{12}, A_{21}, A_{22} and Q are applied:

1. Translation with $(-x_I, -y_I)$ (I is moved into origin O)
2. Rotation with sin(−u) = −sin(u) and cos(−u) = cos(u), where:

$$\begin{cases} r = \sqrt{(x_{12} - x_{11})^2 + (y_{12} - y_{11})^2} \\ \cos(u) = (x_{12} - x_{11})/r \\ \sin(u) = (y_{12} - y_{11})/r \end{cases} \tag{8}$$

After this rotation, the line d_1 is over Ox axis of the map's system of coordinates.

3. Rotation with sin(v) and cos(v), where:

$$\begin{cases} w = \operatorname{atan2}(y_{22} - y_{21}, x_{22} - x_{21}) \\ \cos(v) = \cos(\pi/2 - w/2) \\ \sin(v) = \sin(\pi/2 - w/2) \end{cases} \tag{9}$$

After this rotation, the bisector of the angle given by lines d_1 and d_2 is over Oy.

4. Translation with $(-x_Q, -y_Q)$ (Q is moved into origin O)

The following 2 systems of equations have to be solved in order to find the points $P_i(x_i,y_i)$ (i = 1,2) where the lines d_i (i = 1,2) intersect the line d:

$$\begin{cases} y = a_i x + b_i \\ y = ax \end{cases} \tag{10}$$

Since A_{i1}, A_{i2} define the line d_i (i = 1,2) in (10) we have:

$$a_i = \frac{(y_{i2} - y_{i1})}{(x_{i2} - x_{i1})} \tag{11}$$

$$b_i = y_{i1} - a_{i1} x_{i1} \tag{12}$$

Solving the 2 systems of equations, the points $P_i(x_i,y_i)$ (i = 1,2) are found, where:

$$x_i = \frac{b_i}{a - a_i} \tag{13}$$

$$y_i = \frac{ab_i}{a - a_i} \tag{14}$$

The distance between P_1 and P_2 must be minimized, i.e.,:

$$\min_a f(a) \tag{15}$$

where:

$$f(a) = (y_2 - y_1)^2 + (x_2 - x_1)^2 \tag{16}$$

By replacing (13) and (14) in (16) the following formula for f(a) is obtained:

$$f(a) = \left(\frac{a_2b_2}{a - a_2} - \frac{a_1b_1}{a - a_1} + b_2 - b_1\right)^2 + \left(\frac{b_2}{a - a_2} - \frac{b_1}{a - a_1}\right)^2 \tag{17}$$

We make the following notations:

$$a_L = \min\left\{\frac{-1}{a_1}, \frac{-1}{a_2}\right\} \tag{18}$$

$$a_R = \max\left\{\frac{-1}{a_1}, \frac{-1}{a_2}\right\} \tag{19}$$

It is easy to see that the function f is decreasing for $a \le a_L$, and is increasing for $a \ge a_R$, since f is continuous, a_L and a_R are the slopes of the perpendiculars QQ_1 and QQ_2 from Q on the lines d_1 and, respectively, d_2, and "a" is the slope of the line d_3. Consequently, there is a minimum of the function f denoted a_{min} in the interval $[a_L, a_R]$. So, the value a_{min} is one of the solutions of the equation:

$$f'(a) = 0 \tag{20}$$

on the interval $[a_L, a_R]$, where:

$$f'(a) = 2\left(\frac{a_2b_2}{a - a_2} - \frac{a_1b_1}{a - a_1} + b_2 - b_1\right)\left[\frac{-a_2b_2}{(a - a_2)^2} + \frac{a_1b_1}{(a - a_1)^2}\right] + 2\left(\frac{b_2}{a - a_2} - \frac{b_1}{a - a_1}\right)\left[\frac{-b_2}{(a - a_2)^2} + \frac{b_1}{(a - a_1)^2}\right] \tag{21}$$

Remark 1. *Since the bisector of the angle between d_1 and d_2 is parallel to Oy axis, it is not difficult to see that*:

$$a_L = -a_R \tag{22}$$

In order to obtain the solution a_{min} of the Equation (20), a numerical method [14] can be used such as bisection method [15], or tangent method. Since $f'(a_L) < 0$ and $f'(a_R) > 0 \Rightarrow f'(a_L) \cdot f'(a_R) < 0$, the bisection method is suitable to be applied on the interval:

$$a \in [a_L, a_R] \tag{23}$$

After a_{min} is calculated, the points P_i (i = 1, 2) are obtained using (13) and (14) and the distance between these two points is:

$$\sqrt{f(a_{min})} \tag{24}$$

In order to go back to the initial system of coordinates, the inverse initial transformations must be applied in reverse order to the points P_1 and P_2, i.e.,:

1. Translation with $(-x_Q, -y_Q)$
2. Rotation with sin(u) and cos(u) (were calculated in (8))
3. Rotation with sin(-v) = sin(v) and cos(-v) = cos(v) (calculated in (9))

4. Translation with (x_I, y_I).

The following Algorithm 3 (denoted A2SSTPC2L) is obtained to solve SSTPC2L:

Algorithm 3 Solving SSTPC2L

A2SSTPC2L:
Input: Q and the points of the closest two level curves to Q
Output: The slope at point Q
Apply AIPS to find $A_{11}, A_{12}, A_{21}, A_{22}$;
Transform the points $A_{11}, A_{12}, A_{21}, A_{22}$ and Q so that the bisector of the angle between the lines d_1 and d_2 is parallel to Oy axis and the point Q is in the origin O;
Find the solution a_{min} of the equation $f'(a) = 0$ using bisection method on the interval (a_L, a_R) (see Equations (19) and (20));
Calculate the coordinates of the points P_i (i = 1,2) using (13) and (14), where $a = a_{min}$;
Apply the inverse initial transformations in reverse order to the points P_1 and P_2;
Calculate the slope at point Q using (1).

We implemented the above algorithm in Visual C++ and in the bisection method we set the error err = 0.0001 for finding the value a_{min}. In Figure 6, a graphical output of our program is presented.

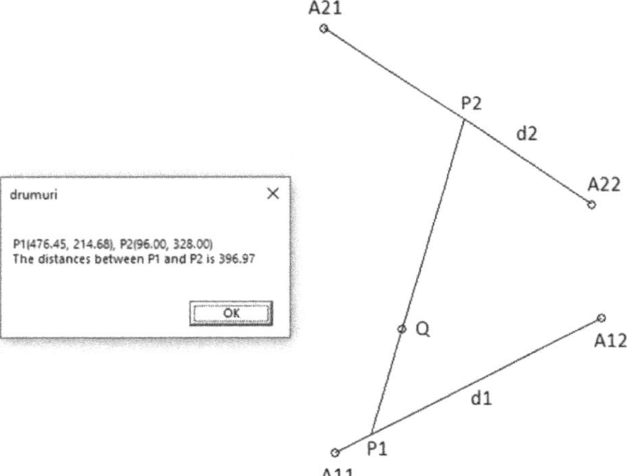

Figure 6. Output of the Visual C++ program.

3. Results

Besides estimation of the ground slope for a given point, other two possible problems can be solved since we have now obtained a method to calculate the points P_1, and P_2 on the closest two level curves to a given point Q so that P_1, Q, and P_2 are collinear and the distance between P_1 and P_2 is minimum.

3.1. Elevation Estimation at a Point

The challenge is to estimate as well as possible the elevation z of a point Q. Assuming that $z_1 < z_2$ (see Figure 3) and since QQ′ is parallel to $P_2P'_2$, in the triangle $P_1P_2P'_2$ we have:

$$\frac{z_Q - z_1}{z_2 - z_1} = \frac{\text{dist}(P_1, Q')}{\text{dist}(P_1, P'_2)} \quad (25)$$

From (25) we obtain the elevation of Q:

$$z_Q = z_1 + \sqrt{\frac{(x_Q - x_1)^2 + (y_Q - y_1)^2}{(x_2 - x_1)^2 + (y_2 - y_1)^2}}(z_2 - z_1)$$

It is immediate that, without the assumption of $z_1 < z_2$, the following formula can be used to estimate the elevation of Q:

$$z_Q = \min\{z_1, z_2\} + \sqrt{\frac{(x_Q - x_1)^2 + (y_Q - y_1)^2}{(x_2 - x_1)^2 + (y_2 - y_1)^2}}|z_2 - z_1| \quad (26)$$

3.2. The Slope between Two Points

The problem of estimating the slope between two points is significant, seeing that when designing roads, the slope between any two points on a road cannot exceed a given maximum slope, e.g., for highways this value has to be less than 6% or 7%. Therefore, the chosen points from the future road can be tested for this eligibility. To do that, we consider the elevations of the points Q_1 and Q_2 calculated using formula (26) and we can estimate the slope $m(Q_1,Q_2)$ of the road passing through $Q_1(x_1,y_1,z_1)$ and $Q_2(x_2,y_2,z_2)$ using the following formula (see Figure 7):

$$m(Q_1, Q_2) = \tan(\beta) = \frac{|z_2 - z_1|}{\mathrm{dist}(Q_1, Q_2')} = \frac{|z_2 - z_1|}{\sqrt{(x_2 - x_1)^2 + (y_2 - y_1)^2}} \quad (27)$$

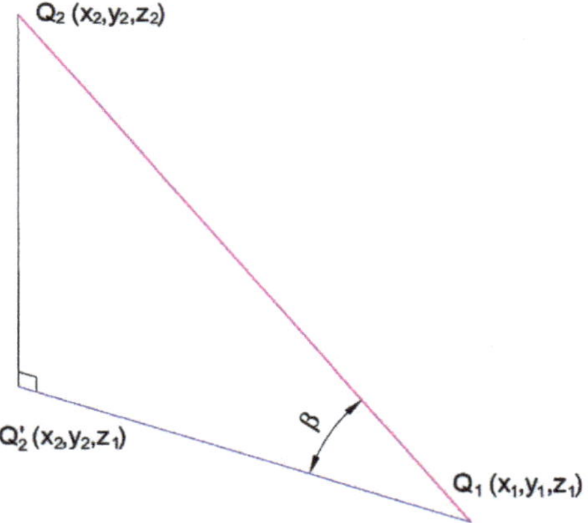

Figure 7. The slope of the road passing through Q_1 and Q_2.

3.3. Optimum Excavation/Filling

When a house is built on sloping land and/or with elevations and depressions, the elevation of a horizontal plane must be calculated so that the excavations over the plane would fill the space below the plane resulting in a horizontal platform on which a house can be built. In such a manner, the optimal solution between excavations and fillings is calculated, resulting in a reduction in the costs of embankments. We shall follow up with a presentation of a method to accurately calculate the elevation of this plane. The well-known "flood fill" algorithm [16] from computer graphics is adapted to deal with this problem. Since the algorithm runs in a discrete space, the coordinates will be transformed into their discrete counterparts.

We start with the level curves from the vicinity of the house (Figure 8). Then a discretization is applied (Figure 9), meaning that the plane is divided into equidistant 2D points, e.g., in Figure 9, the distance of 40 cm between points was considered. The points of the contour of the house are transformed into their discrete counterparts (for each such a point, the closest discrete point is determined). Using Bresenham's algorithm [17], the discrete points of the contour are found (green points in Figure 9). Using a flood fill algorithm [16] in the discrete plane of the house, the discrete points inside the contour are determined (red points from Figure 9). The elevation of each red or green point is calculated using the level curves. Using these elevations, the optimal elevation of the excavation/filling plane is computed. So, the following algorithm denoted AOEF is obtained.

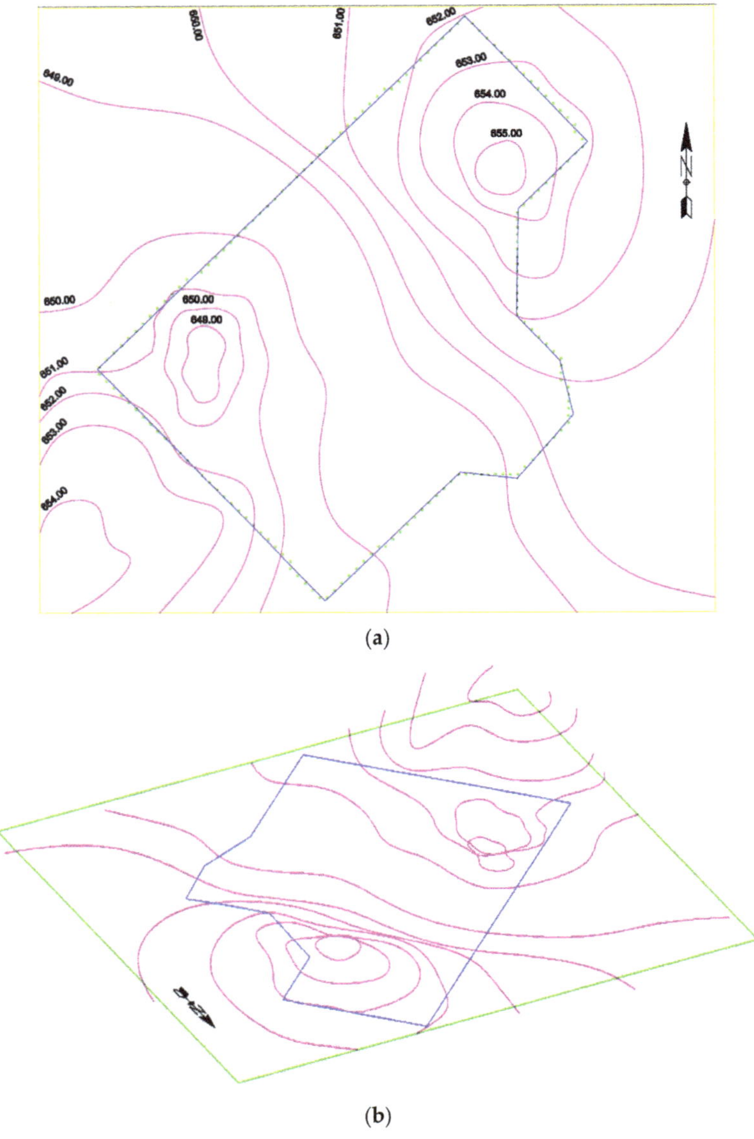

(a)

(b)

Figure 8. Level curves (pink lines) in the vicinity of the house (blue lines): (**a**) 2D view; (**b**) 3D view.

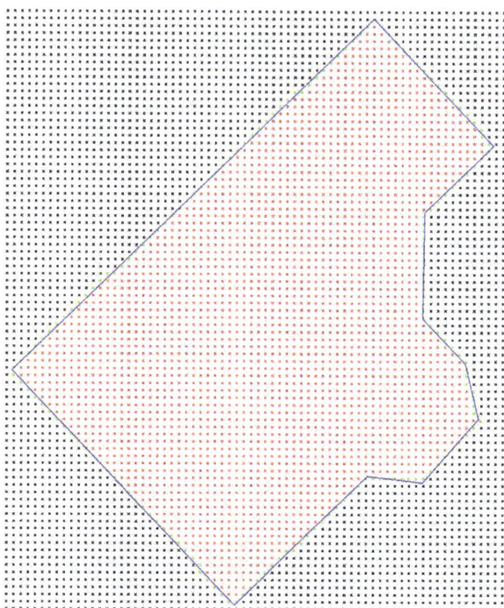

Figure 9. Discretization of the plane of the house. Black points are outside the house, green points give the contour of the house, red points are inside.

Algorithm 4 Optimum Excavation/Filling (AOEF):

Algorithm 4 Optimum Excavation/Filling (AOEF)

Input: the points P_i (i = 1, 2, ..., n) of the contour;
for each point P_i **do**
    Find the closed discrete point P'_i;
end for;
for i = 1 **to** n − 1 **do**
    Find the set of points C_i using Bresenham's algorithm from point P_i to point P_{i+1};
end for;
Using Bresenham's algorithm from point P_n to point P_1, find the set of points C_n;
Find the points Q_i (i = 1, 2, . . . , m) inside the contour given by $C_1 \cup C_1 \cup \ldots \cup C_n$;
for each point Q_i **do**
    Apply **A2SSTPC2L** to find the elevation z_i of Q_i;
end for;

Using the elevations z_i, calculate the elevation of the optimal excavation/filling plane (see AHEFP).

Applying the elevations z_i of the discrete points Q_i inside the contour of the house, the elevation of the optimal excavation/filling plane can be determined as follows. The elevations are sorted ascendingly. A horizontal plane $\pi(h)$ (h is the height/elevation of the plane) is placed consecutively starting with the first (the lowest) elevation, continuing with the second, and ending with the last one (the highest). The following algorithm is obtained:

Algorithm 5 Height of Excavation/Filling Plane (AHEFP):

Algorithm 5 Height of Excavation/Filling Plane (AHEFP)

```
Input: vector of elevations z = (z_i)_{i = 1, 2, ..., m};
Sort ascending the vector z;
S = 0;
for i = 2 to n do
        S = S + z_i − z_1;
  end for;
S_1 = 0;
S_2 = S;
for i = 2 to n − 1 do
        S_1 = S_1 + (i − 1) · (z_i − z_{i-1});
        S_2 = S_2 − (n − i + 1) · (z_i − z_{i-1});
        if S_2 ≤ S_1 then
                opt = i;
                break;
        end if;
end for;
```

At each iteration of the last "for" loop from AFHEFP, the height of the plane is considered equal to z_i, S_1 is the sum of the distances between the elevation of z_i the plane and the elevations below the plane, and S_2 is the sum of the distances between the elevations over the plane and the elevation z_i of the plane, i.e.,:

$$S_1 = \sum_{k=1}^{i-1} (z_i - z_k) \tag{28}$$

$$S_2 = \sum_{k=i+1}^{n} (z_k - z_i) \tag{29}$$

when starting a new iteration, S_1 is the sum of i-1 components of z and was calculated in the previous iteration. The new sum S_1 can be obtained from the previous by adding $(i-1) \cdot (z_i - z_{i-1})$.

At the beginning of each iteration, S_2 is the sum of n-i+1 components of z and was calculated in the previous iteration. Thus, using the value of the previous S_2, the new sum can be obtained from the previous by subtracting $(n - i + 1) \cdot (z_i - z_{i-1})$.

Since S_1 is increasing from 0 to S and S_2 is decreasing from S to 0, the optimum is reached when S_2 becomes less or equal to S_1. More exactly, the optimum height of the plane is inside the interval $[z_{opt-1}, z_{opt}]$. If the values z_{opt-1} and z_{opt} are close enough, i.e., $z_{opt} - z_{opt-1} < \varepsilon$, where $\varepsilon > 0$ is the fixed maximum distance to the optimum, e.g., $\varepsilon = 5$ cm, then any of the values z_{opt-1} and z_{opt} are good approximations of the optimum. If $z_{opt} - z_{opt-1} \geq \varepsilon$, then a divide and conquer strategy is applied to get closer to optimum. To do that, two initial planes $\pi(z_{opt-1})$ and $\pi(z_{opt})$ are considered. A new plane is placed in the middle. If the distance between S_2 calculated for $\pi((z_{opt-1}+z_{opt})/2)$ and S_1 for $\pi(z_{opt-1})$ is less than the distance between S_2 for $\pi(z_{opt})$ and S_1 for $\pi((z_{opt-1}+z_{opt})/2)$, then the optimum is further calculated in the interval $[z_{opt-1}, (z_{opt-1}+z_{opt})/2]$. Otherwise, the optimum is calculated in the interval $[(z_{opt-1}+z_{opt})/2, z_{opt}]$ and so on.

Algorithm 6 divide and conquer for optimum excavation/filling plane (ADCOEFP):

Algorithm 6 Divide and conquer for optimum excavation/filling plane (ADCOEFP)

```
a = z_opt-1;
b = z_opt;
S_a,1 = S_1 − (opt − 1) · (z_opt − z_opt-1);
S_a,2 = S_2 + (n − opt + 1) · (z_opt − z_opt-1);
S_b,1 = S_1;
S_b,2 = S_2;
while b − a ≥ ε do
    c = (a+b)/2;
    S_c,1 = S_a,1 + (opt − 1) · (c − a);
    S_c,2 = S_a,2 − (n − opt + 1) · (c − a);
    if | S_c,2 − S_a,1 | < | S_b,2 − S_c,1 | then
        a = c;
        S_a,1 = S_c,1;
    else
        b = c;
        S_b,2 = S_c,2;
    end if;
end while;
S_1 = S_a,1;
S_2 = S_a,2;
```

At the end of the while loop, any of the values, a or b are good approximations of the optimum. We considered a as the optimal solution. The approximate excavation volume is $V_e = S_2 \cdot d_2$, and the filling volume is $V_f = S_1 \cdot d_2$. Consequently, the difference between the two volumes is:

$$|V_e - V_f| \;=\; |S_2 - S_1|d^2 \leq \varepsilon S_{house} \approx \varepsilon md^2 \tag{30}$$

So, the two volumes are very similar.

3.4. Numerical Example

Figure 10 shows a real topographic map for a plot of land on which a house is to be built. The topographic map of the land was drawn up by a topometric engineer, with specific high-precision equipment, called in specialized terms "station." For small areas of land, the station is successively placed and GPS 3D coordinates are measured relative to a fixed position (in our situation it is a terminal in the Black Sea area located at a distance of approximately 717 km). The first chosen points of the topographic survey are those that describe the contour of the plot of land. In our case, the terrain contour is described by the 15 points (numbered in black). These points are inventoried and noted in a table (coordinate inventory) located on the topographic map. The following points will describe the contour of the level elevations, the positioning of the access road, the position of the neighboring houses (they are marked on the map in red together with the elevation), etc. In the case of lands with a high slope, such as the one exemplified, the measured points must be thickened for a greater accuracy of tracing the level curves (in total we have over 100 such points).

The designer receives in electronic format a topographic map similar to the one in Figure 10. On this topographic support, the designer will place the designed building on the scale, having to respect a series of rigors imposed by law, among which: the minimum distance from the property limit and/or from the road axis as in Figure 11. Once the construction is located, the new contour will appear on the topographic map described by a few new points (minimum 4, in the example, for the description of the contour, 6 points are used). A situation plan is drawn up in which, according to the legislation, the 2D coordinates of the corners of the construction must be specified (see Table 1) together with the elevation of the horizontal excavation/filling plane. In Figure 11, this elevation is denoted by CTA and is estimated at 707 m by the designer. A good determination of this elevation is not easy, especially if the terrain has a high slope.

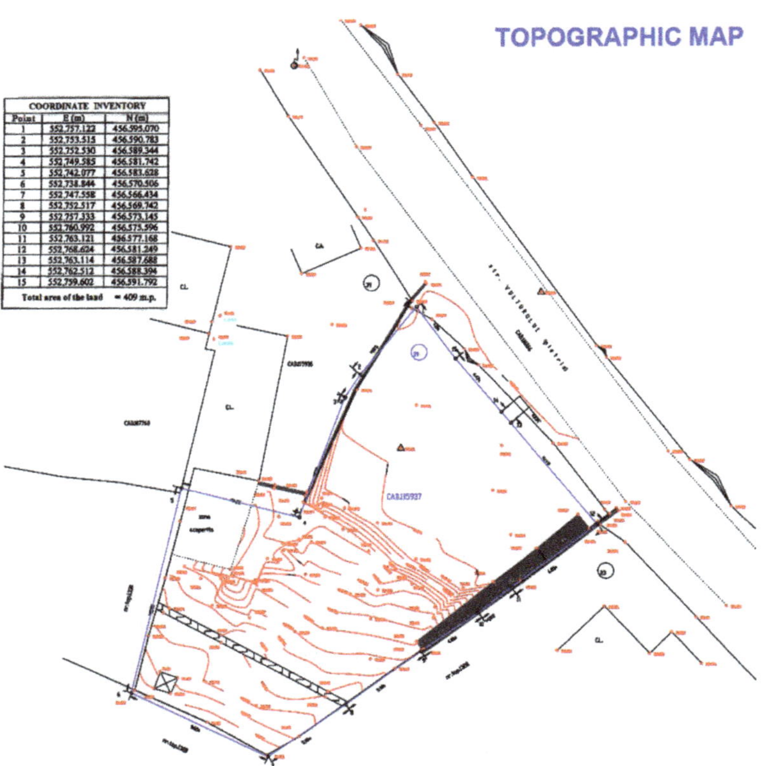

Figure 10. Topographic map.

Table 1. The x and y coordinates of the Q points.

Points	E (m)	N (m)
Q_1	552,749.248	456,578.380
Q_2	552,754.270	456,571.835
Q_3	552,766.210	456,580.997
Q_4	552,760.244	456,588.772
Q_5	552,754.733	456,584.542
Q_6	552,753.385	456,581.554

A detail of Figure 11 is presented in Figure 12 illustrating the points A^2_{11}, A^2_{12}, A^2_{21}, and A^2_{22} found using the algorithm AIPS.

Table 2 presents the coordinates of the points A^k_{11}, A^k_{12}, A^k_{21}, and A^k_{22} calculated for each point Q_k (k = 1, 2, . . . , 6) by applying AIPS. Using A2SSTPC2L and (26), the elevation for each point Q_k (k = 1, 2, . . . , 6) is calculated.

Usually, for the dimensions of the project to be in accordance with the real construction, before commencing all the construction works, a plot is made on the field of the future building. For this purpose, the ground tracing is performed. The corners of the house are very precisely marked (the Q points). In our example, the provided measured elevations of the 6 points are presented in Table 3.

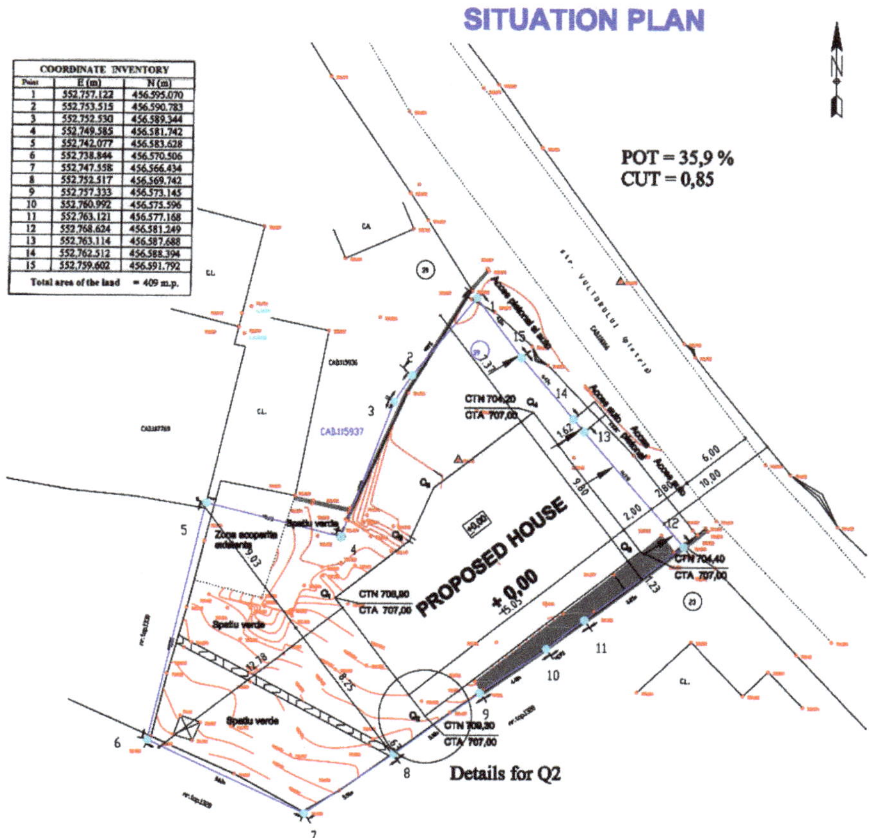

COORDINATE INVENTORY		
Point	E (m)	N (m)
1	552,757.122	456,595.070
2	552,753.515	456,590.783
3	552,752.530	456,589.344
4	552,749.585	456,581.742
5	552,742.077	456,583.628
6	552,738.844	456,570.506
7	552,747.558	456,566.434
8	552,752.517	456,569.742
9	552,757.333	456,573.145
10	552,760.992	456,575.596
11	552,763.121	456,577.168
12	552,768.624	456,581.249
13	552,763.114	456,587.688
14	552,762.512	456,588.394
15	552,759.602	456,591.792
Total area of the land	= 409 m.p.	

Figure 11. Situation plan.

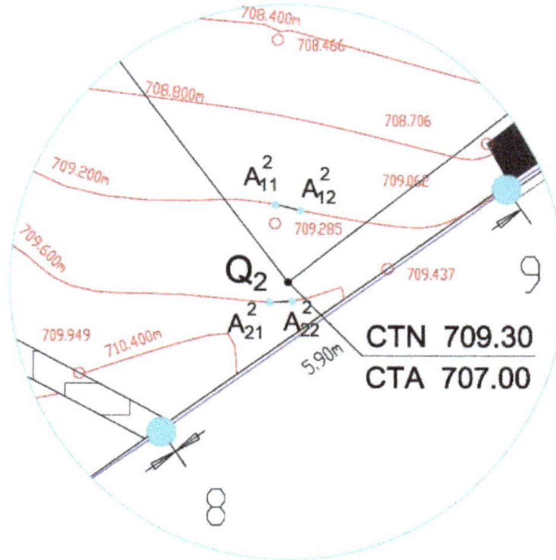

Figure 12. Detail of Figure 11 for Q_2.

Table 2. 3D coordinates of the Q points.

Points		E (m)	N (m)	Z(m)	Calculated Z(m)
Q_1		552,749.2478	456,578.3797		707.8851
	A^1_{11}	552,749.7024	456,578.7371	707.6000	
	A^1_{12}	552,749.4725	456,578.9520	707.6000	
	A^1_{21}	552,748.8265	456,578.0230	708.0000	
	A^1_{22}	552,749.2769	456,577.9832	708.0000	
Q_2		552,754.2700	456.,571.8350		709.3279
	A^2_{11}	552,754.0980	456,572.9276	709.2000	
	A^2_{12}	552,754.4526	456,572.8445	709.2000	
	A^2_{21}	552,754.0176	456,571.5626	709.6000	
	A^2_{22}	552,754.3366	456,571.5767	709.6000	
Q_3		552,766.2100	456,580.9970		704.2017
	A^3_{11}	552,768.2179	456,583.3340	704.0000	
	A^3_{12}	552,768.7357	456,582.6679	704.0000	
	A^3_{21}	552,760.8755	456,578.3463	704.4000	
	A^3_{22}	552,760.9848	456,577.8984	704.4000	
Q_4		552,760.2440	456,588.7720		704.2330
	A^4_{11}	552,762.5489	456,590.7208	704.0000	
	A^4_{12}	552,762.9178	456,590.3418	704.0000	
	A^4_{21}	552,755.2345	456,585.1132	704.4000	
	A^4_{22}	552,754.9360	456,585.3872	704.4000	
Q_5		552,754.7330	456,584.5420		704.7360
	A^5_{11}	552,755.2345	456,585.1132	704.4000	
	A^5_{12}	552,755.1438	456,584.4878	704.4000	
	A^5_{21}	552,752.7819	456,582.6377	704.8000	
	A^5_{22}	552,752.3362	456,582.9349	704.8000	
Q_6		552,753.3850	456,581.5540		705.0855
	A^6_{11}	552,753.3671	456,581.7908	704.8000	
	A^6_{12}	552,753.7000	456,581.5900	704.8000	
	A^6_{21}	552,753.2125	456,581.6078	705.2000	
	A^6_{22}	552,753.5536	456,581.3981	705.2000	

Table 3. The estimated and the measured z-coordinates of the Q points.

Points	Estimated Z (m)	Measured Z (m)
Q_1	707.8921	707.8990
Q_2	709.3279	709.3260
Q_3	704.2017	704.2140
Q_4	704.2330	704.2210
Q_5	704.7360	704.7500
Q_6	705.0855	705.0970

As seen in Table 3, the estimated elevation and the measured one for each Q point are very close. The difference between them is less than 2 cm.

Using the coordinates of the points Q_k (k = 1, 2, . . . , 6) from Table 1, Bresenham, and, then flood fill algorithms were applied resulting in 803 points located inside the perimeter of the house. The elevations of these points were calculated using A2SSTPC2L and (26). The algorithm AHEFP was applied for these 803 elevations and the height of the horizontal excavation/filling plane was accurately estimated as 706.2571 m, instead of the value 707 m proposed by the designer. Out of the 803 points, 412 were located below the horizontal plane, and 391 had the elevation over this plane.

4. Discussion

In order to compare the two proposed methods (from Sections 2.1 and 2.2) for determining the points P_1 and P_2 100 million experiments were performed. For each experiment, two random lines d_1 and d_2 were selected and the point Q was also randomly chosen.

The tests were performed using an ASUS ROG GL752VW-T4015D laptop with Intel® Core™ i7-6700HQ 2.60 GHz processor, and 8 GB of RAM.

First, we analyzed the speed of the proposed methods. The total running time of the first method was 5.82 s. The total running time of the second method was 40.14 s. So, the first method is almost 7 times faster than the second one (see Figure 13). The first method is considerably faster because it requires elementary calculations, while the second one runs an iterative algorithm to solve the Equation (18).

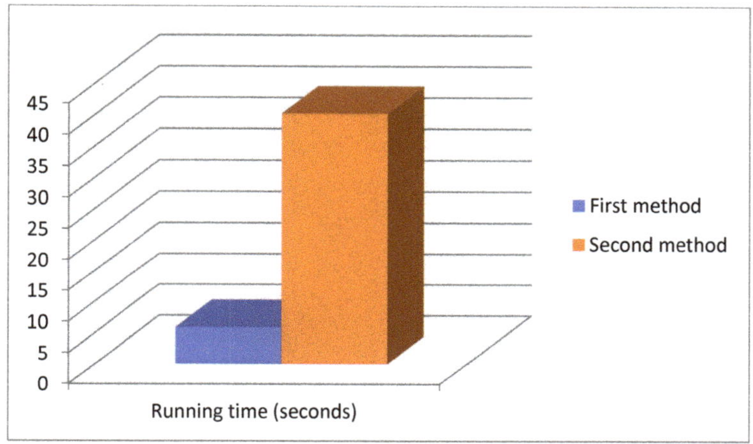

Figure 13. Running time comparison of the two methods (100 million experiments).

We also calculated the accuracy of the first method against the second method since the error for the second method can be set as low as desired. An average error of 0.61% was obtained, the lowest error being 0 and the highest 1.17% (Table 4). So, we concluded that the accuracy of the approximate first method is good enough.

Table 4. Errors of the first method obtained from 100 million experiments.

	Errors (First Method against Second Method)
Lowest error	0%
Average error	0.61%
Highest error	1.17%

The steps of positioning the house will be next discussed, and the utility of the methods proposed in this paper will be shown. The designer receives a topographic map in electronic format. On this topographic support, the designer will place the designed building on the scale, having to comply with a series of rigors imposed by law, among which: the minimum distance from the property limit and/or from the road axis, etc. Once the construction is located, the outline of the house described by several new points will appear on the topographic map (minimum 4). At these new points, it is necessary to specify: the coordinates on the three dimensions according to which the elevation is specified, and the distances of the construction location from the property limits. The determination of the coordinates in the plan does not raise any concerns, but there are difficulties in obtaining the real vertical elevation especially if the terrain has a high slope. The article proposes a method for determining the exact elevation of the land at the corners of the building, and in fact for the entire contour. Depending on the contour dimensions, the designer establishes a position of the horizontal plane, a plane that corresponds to the finished floor of the ground level. The decision to establish this quota should be contingent on the volumes of embankments that will be made. The optimum is obtained when the excavation volume is approximately equal to the filling volume. The article solves this problem in Section 3.3.

For the design of a road, a primary route of the road is made on a topographic map and road recognition is performed. A final design of the road follows both along its entire length and on its cross-sections. Figure 14 shows with a red line the optimal trajectory of a road depending on the elevation of the natural land. Because the considered road is a highway, the rules impose stricter restrictions on the slopes. If the allure of the natural terrain in certain areas is followed, the maximum allowed inclination of the road will inevitably be exceeded. As it can be seen in Figure 14, in these areas, measures must be taken to straighten the slope, such as fillings, excavations or the provision of bridges and viaducts, tunnels respectively. Through detailed awareness of the elevations and the inclination of the land in the points near the road, an optimal vertical tracing of the road can be established so that we have minimum excavations and fillings for an imposed road slope, or we can optimize the lengths of the bridges, respectively tunnels [12].

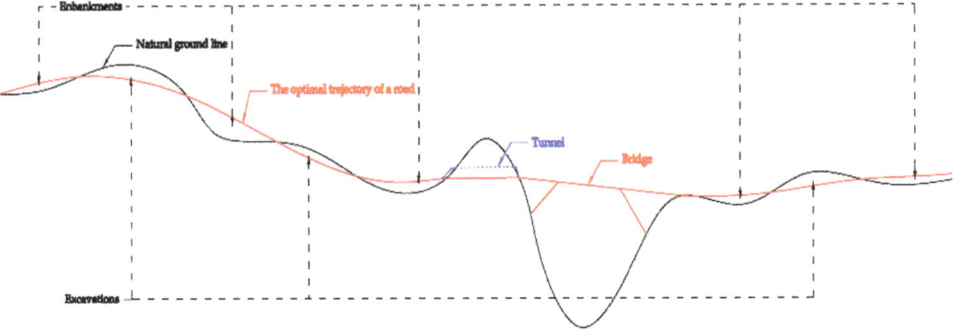

Figure 14. The optimal trajectory of a road depending on the elevation of the natural land.

When designing a new road or railway, a number of initial design factors must be taken into account, such as: road category, design speed, width of a traffic lane, number of lanes, traffic frequency, equipment size, etc. Depending on the route imposed on the road, there are also other specific aspects that must be taken into account:

1. Avoid high erosion hazard sites, particularly where mass failure is a possibility.
2. Utilize natural terrain features such as stable benches, ridgetops, and low gradient slopes to minimize the area of road disturbance.
3. If necessary, include short road segments with steeper gradients to avoid problem areas or to utilize natural terrain features.
4. Avoid midslope locations on long, steep, or unstable slopes.
5. Locate roads on well-drained soils and rock formations which dip into slopes rather than areas characterized by seeps, highly plastic clays, concave slopes hummocky topography, cracked soil and rock strata dipping parallel to the slope.
6. For logging road, utilize natural log landing areas (flatter, benched, well-drained land) to reduce soil disturbance associated with log landings and skid roads.
7. Avoid undercutting unstable, moist toe slopes when locating roads in or near a valley bottom.
8. Roll or vary road grades where possible to dissipate flow in road drainage ditches and culverts and to reduce surface erosion.
9. Select drainage crossings to minimize channel disturbance during construction and to minimize approach cuts and fills.
10. Locate roads far enough above streams to provide an adequate buffer, or provide structure or objects to intercept sediment moving down slope below the road.

11. If an unstable area such as a headwall must be crossed, consider end hauling excavated material rather than using sidecast methods. Avoid deep fills and compact all fills to accepted engineering standards. Design for close culvert and cross drain spacing to effectively remove water from ditches and provide for adequate energy dissipators below culvert outlets. Horizontal drains or interceptor drains may be necessary to drain excess groundwater [13].

As seen in Figure 15, the angle of inclination of the terrain is an important parameter to know in order to avoid erosion and then landslides [18,19].

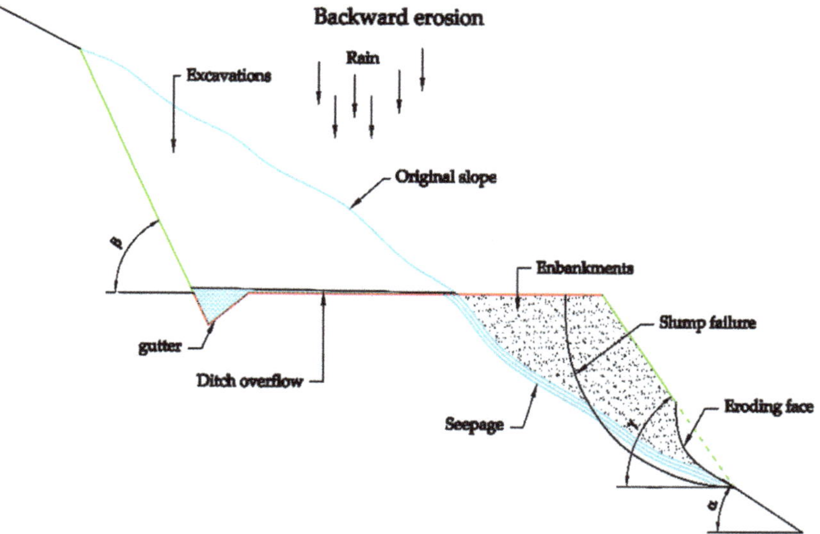

Figure 15. Draining excess water to prevent landslides.

From the point of view of the cross-sections, the position of the road in relation to the terrain slope must take into account the above constraints. By the same token, knowing the slope of the land is essential for establishing the position both horizontally and especially vertically of the road, and for achieving a low cost related to the embankment works. This position is established according to the Full Bench Road Prism, which contains the elements in Figure 16.

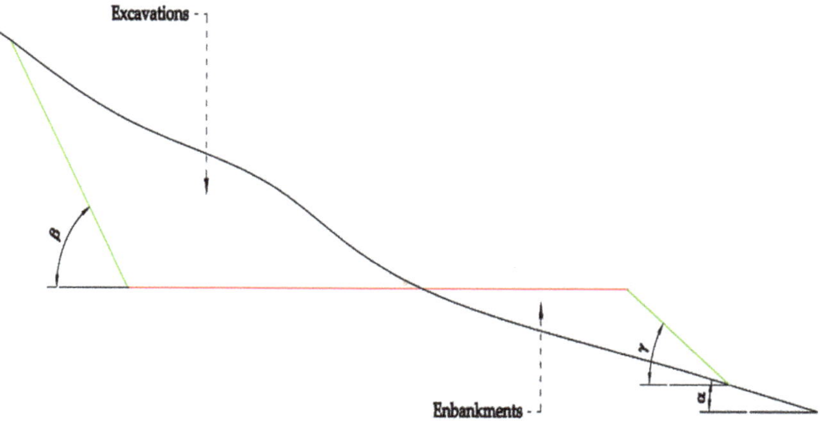

Figure 16. Elements of road prism geometry.

For very steep road areas such as the one in Figure 17, knowing the slope is imperative because substantial savings in embankment volumes can be made, resulting in decreasing the quantities of gabions or excavations [13].

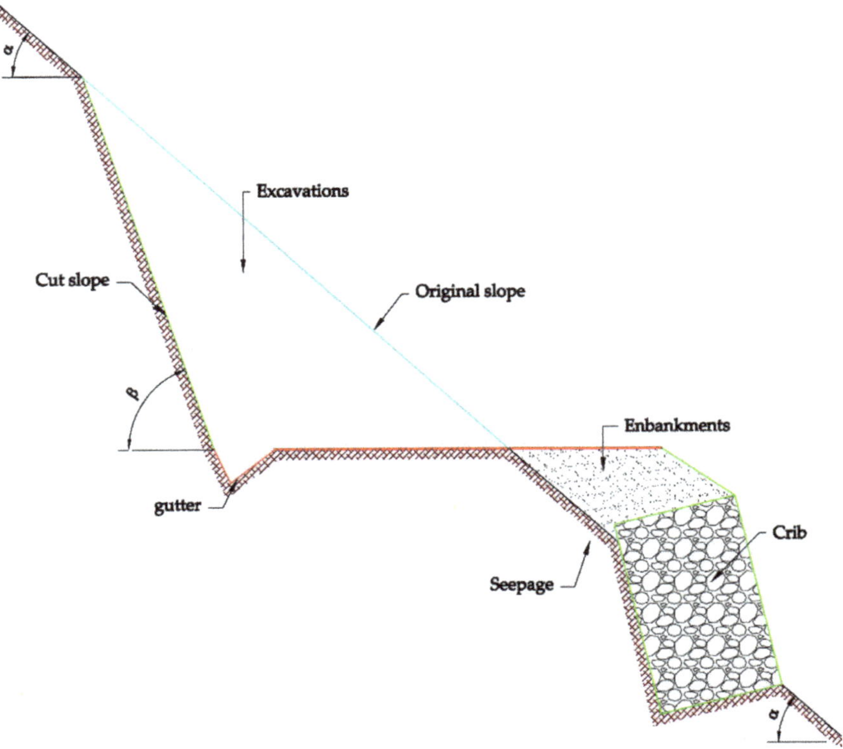

Figure 17. Reduction in excavation made possible on a steep slope by the use of cribbing.

The problem related to determining the slope of natural land at a point on the topographic map is necessary both in the longitudinal section and in the cross-section of roads or railways and have the same solution described in Sections 3.1 and 3.2.

The proposed methods can also be applied in environment protection, and hydrotechnical engineering [18,19]. The evaluation of the slope is also necessary for river levees. Slope stability analyses are conducted with rising water levels until certain failure is reached. Discharge occurs at a water height of around 7 cm above the crown, which means the slope would actually fail before reaching the hydraulic heads used for slope stability calculation during the overflow [20].

5. Conclusions

To conclude, the correct estimation of elevation at points has a practical application to calculate the optimum excavation/filling plane when a new house is built. The article presented a method to estimate accurately the elevation of this plane which reduces the costs of embankments.

When designing a new road or railway, designers need to know the ground slope at some points of the future road or rail as exactly as possible, as well as the elevation of each considered point, and, also, the slope of each consecutive two points. Instead of sending workers on the field to perform measurements, these values can be calculated with very good accuracy using the methods described in Section 2.1, and Section 2.2. Two algorithms to solve SSTPC2L are presented, a fast and approximate one and an exact but slower one. The error of the first method is low (see Table 1). It has the advantage of being very fast and accurate enough. Thus, if there are many points for which SSTPC2L is applied, the first method can be preferred. However, if we look for a more accurate solution, the second method is more suitable.

Author Contributions: Conceptualization, A.M.D. and O.D.; methodology, A.M.D. and O.D.; software, A.M.D.; validation, O.D.; formal analysis, A.M.D.; investigation, O.D.; resources, A.M.D. and O.D. data curation, A.M.D. and O.D.; writing—original draft preparation, A.M.D. and O.D.; writing—review and editing, A.M.D. and O.D.; supervision, A.M.D.; funding acquisition, A.M.D. and O.D. Both authors have read and agreed to the published version of the manuscript.

Funding: This research was funded by University Transilvania of Brașov.

Institutional Review Board Statement: Not applicable.

Informed Consent Statement: Not applicable.

Data Availability Statement: Not applicable.

Conflicts of Interest: The authors declare no conflict of interest.

References

1. Levin, A.H. *Hillside Building: Design and Construction*, US 2nd ed.; Builder's Book Ink: Los Angeles, CA, USA, 1999.
2. Tarbatt, J.; Tarbatt, C.S. *The Urban Block—A Guide for Urban Designers, Architects and Town Planners*; RIBA Publishing: London, UK, 2020; ISBN 9781859468746.
3. Kicklighter, C.E.; Thomas, W.S. *Architecture: Residential Drafting and Design*; G-W Publisher: Tinley Park, IL, USA, 2018.
4. Pyrgidis, C.N. *Railway Transportation Systems: Design, Construction and Operation*; CRC Press: Boca Raton, FL, USA, 2019; ISBN 9780367027971.
5. Ellis, P. *Guide to Road Design Part 1: Introduction to Road Design*; Austroads Ltd.: Sydney, Australia, 2015.
6. Fannin, R.J.; Lorbach, J. *Guide to Forest Road Engineering in Mountainous Terrain*; Rome (Italy) FAO: Rome, Italy, 2007.
7. Deaconu, O.; Deaconu, A. Numerical methods to find the slope at a point on map using level curves. In *IOP Conf. Series: Materials Science and Engineering, Proceedings of the Application in Road Designing, International Conference CIBv2019 Civil Engineering and Building Services, Brașov, Romania, 1–2 November 2019*; IOP Publishing: Philadelphia, PA, USA, 2020; p. 789.
8. Aruga, K.; Sessions, J.; Akay, A.E.; Ching, W. Forest Road Deisign and Construction in Mountain Terrain. In Proceedings of the Conference of IUFRO, Vancouver, BC, Canada, 13–16 June 2004.
9. Sutton, R. *An Introduction to Topography*; Larsen and Keller Education: New York, NY, USA, 2017; ISBN 13 978-1635492774.
10. Suresh, N.; Akhilesh, K.M.; Avijit, M.; Prasanta, S.; Praveen, E. Review of Alignment design consistency in Mountainous/Hilly Roads. In Proceedings of the 3rd Conference of Transportation Research Group of India, Kolkata, India, 17–20 December 2015.
11. Jin, X.; Wei, L.; Yiming, S. New design method for horizontal alignment of complex mountain highways based on "trajectory–speed" collaborative decision. *Adv. Mech. Eng.* **2017**, *9*, 1–18.
12. Siviero, L. Roads and Verticality Strategy and Design in Mountain Landscape. Ph.D. Thesis, Faculty of Engineering, University of Trento, Trento, Italy, 2012.
13. Sheng, T.C. *Watershed Management Field Manual, Food and Agriculture Organization of the United Nations*; Rome (Italy) FAO: Rome, Italy, 1998.
14. Quarteroni, A.; Sacco, R.; Saleri, F. *Numerical Mathematics*; Springer: Berlin/Heidelberg, Germany, 2007.
15. Burden, R.L.; Faires, J.D. *2.1 The Bisection Algorithm, Numerical Analysis*, 3rd ed.; PWS Publishers: Boston, MA, USA, 1985; ISBN 0-87150-857-5.
16. Smith, A.R. Tint Fill. In *SIGGRAPH '79: Proceedings of the 6th Annual Conference on Computer Graphics and Interactive Techniques*; Association for Computing Machinery: New York, NY, USA, 1979; pp. 276–283. ISBN 978-0-89791-004-0.
17. Bresenham, J.E. Algorithm for computer control of a digital plotter. *IBM Syst. J.* **1965**, *4*, 25–30. [CrossRef]
18. Jadid, R.; Montoya, B.M.; Gabr, M.A. Strain-based Approach for Stability Analysis of Earthen Embankments. In Proceedings of the Dam Safety Conference, Online, Palm Springs, CA, USA, 21 September 2020–31 December 2020.
19. Jiang, S.; Liu, X.; Huang, J. Non-intrusive reliability analysis of unsaturated embankment slopes accounting for spatial variabilities of soil hydraulic and shear strength parameters. *Eng. Comput.* **2020**, *333*, 1–14. [CrossRef]
20. Rossi, N.; Bačić, M.; Kovačević, M.S.; Librić, L. Development of Fragility Curves for Piping and Slope Stability of River Levees. *Water* **2021**, *13*, 738. [CrossRef]

Article

Transforming Building Criteria to Evidence Index

Géza Fischl * and Peter Johansson

Department of Construction Engineering and Lighting Science, School of Engineering, Jönköping University, 553 18 Jönköping, Sweden; peter.johansson@ju.se
* Correspondence: geza.fischl@ju.se

Abstract: There is increasing pressure from developers toward architects and engineers to deliver scientifically sound proposals for often complex and cost-intensive construction products. An increase in digitalization within the construction industry and the availability of intelligently built assets and overall sustainability make it possible to customize a construction product. This servitization of construction products is assumed to perform much preferably in satisfying stakeholders' physical, psychological, and social needs. The degree to which these products are performing can be evaluated through an evidence index. This article aims to introduce a conceptual model of an evidence index and test it in the programming stage of a case study. The investigation follows the evidence-based design approach and renders evidence through key performance indicators in the programming stage of the building process. For testing the concept, a case study investigation was performed by simulating a novice research assistant, and the amount of evidence was collected and appraised for evidence index. The case study showed that key performance indicators of a servitized project could be evaluated on a four-point scale. The quality of the evidence index generation depended on the level of expertise the evaluator has in research and the skilful use of scientific databases.

Keywords: construction product; servitization; evidence-based design; level of evidence; cognitive buildings

Citation: Fischl, G.; Johansson, P. Transforming Building Criteria to Evidence Index. *Appl. Sci.* **2021**, *11*, 5894. https://doi.org/10.3390/app11135894

Academic Editor: Lavinia Chiara Tagliabue

Received: 16 May 2021
Accepted: 17 June 2021
Published: 24 June 2021

1. Introduction

A typically assisted workflow by Building Information Modeling (BIM) process is the building performance evaluation [1], for instance, for energy consumption [2] and lighting [3], and air quality through simulations [4]. The optimal building performance can be achieved with technical considerations and a close fit between the building and its users' needs, providing comfort, health, and safety [5]. In terms of a computer-aided approach, development toward an easy-to-use data input is emerging for human behaviour regarding the programming and design phase. For improving design, a variety of quantitative approaches surfaced, like the probabilistic method [6], which reflects variation in the energy consumption models and the agent-based model (see, e.g., in [2,7,8]) that is investigating complex systems composed of interacting agents. In connection to building performance evaluation, a knowledge-oriented value generation process [1,7] in which stakeholders find satisfactory proofs, concerning key performance indicators (KPIs), treated like evidence for reasoning their needs and activities had surfaced. Key performance indicators are instrumental for optimizing the goals of the organization. The organizational goals are usually higher; meanwhile, the KPIs operationalise these goals and make them measurable, understandable, and actionable [8,9]. Therefore, KPIs in the programming and design phase of the building performance evaluation are often connected to the developer's detailed list of building criteria or physical attributes that are expected to be incorporated in the project. This detailed list of building criteria usually emerges through a long-term collaboration between the industry partners to ensure technical and functional service quality. One of this list is the Swedish Program for Technical Standard [8,10], used in the case study below, a database for optimizing healthcare facilities' construction and design. The design criteria are set up with stakeholders, including the construction industry, to deliver a technical service solution to the users. This kind of servitization [11,12] is a key for delivering appropriate performance to stakeholders. In a traditional method, a building developer would provide a facility list that satisfies the needs in a technical/engineering approach, meanwhile, a servitized concept offers KPIs matching the stakeholders' personal needs. Today's technological possibilities allow building upon a new set of servitized construction products that are more efficient and less resource-intensive, connected through smart products and systems, and provide self-learning abilities that deliver an evidence-based optimization. For instance, lighting in an office is set to 200 lux as the general illuminance, but for well-being supportive lighting should include a glare-free setting that helps the individual. Furthermore, considering the daily intake of lighting energy for appropriate circadian rhythm functioning, the KPI should be supporting a human circadian rhythm. The critical issue here is how to assign the appropriate KPI for measuring the intended outcome. That is why researchers had turned their attention to evidence-based design (EBD) as a design method from the field of evidence-based medicine [13]. EBD for the built environment can be defined as the process of design decisions on credible research and lessons learned from previous design experiences as evidence to achieve the best possible outcomes [14,15]. EBD is the antecedent of the building performance evaluation [16,17] and recent research has shown how EBD can be connected to digital tools, like lighting simulations [18]. This research has identified the importance of selecting KPIs (or metrics) for identifying evidence [19] during the evidence-based optimization (EBO). EBO and the current development of cognitive abilities of the building system could lead to better servitization of construction products (Figure 1).

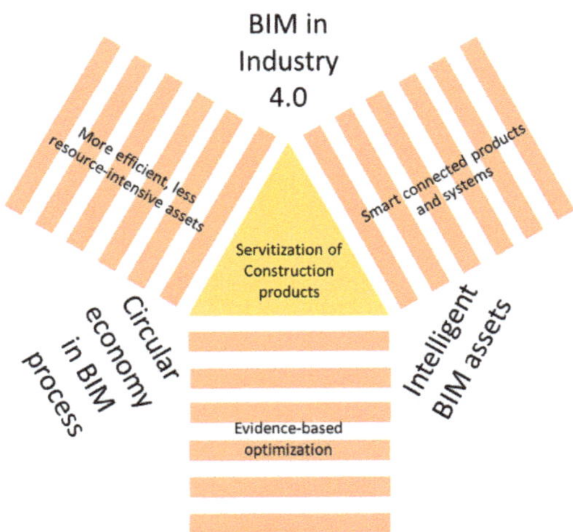

Figure 1. Servitisation of construction product in the context of BIM in Industry 4.0, Circular economy, and Intelligent Building Information Modeling (BIM) assets (Adapted from CPA, [20]; p 17).

Promoters of EBO may be representatives of developers, architecture designers, clients, users, researchers, and facility managers for improving an individual's physical, psychological, and social qualities together with environmental sustainability. From this perspective, the attainment of stakeholders' values would ensure the successful implementation of resources to the predefined goal. A possible measure of reaching the predefined goal is to assess an evidence index for all physical attributes or components listed in the building criteria; henceforth, early in the programming stage, the expected evidence index can indicate how well the predefined goals in the building criteria are in line with a scientific level of evidence.

Consequently, this article aims to introduce and, through a case study, test an evidence index capable of describing the level of evidence at the programming stage of a building process. It is assumed that a novice investigator can generate an evidence index, but the results need to be further scrutinized.

2. Methodology

In this section, the first conceptual method of an evidence index is described, and an enhancement of the EBD process with a cognitive building concept is suggested. Consequently, a case study focusing on an administrative office is tested with the suggested evidence index framework.

2.1. Establishing Evidence Index

The concept of an evidence index (EI) in a room or a building is complementing a post-occupancy evaluation method (POE) (see, e.g., in [21]). POEs have been around for several decades [16], and the depth of investigations is generally made on three distinct levels [22] (indicative, investigative, and diagnostic). A post-occupancy evaluation system arose for feedbacking stakeholders in building projects regarding how well the building performs after its inauguration in terms of user's satisfaction, energy performance, indoor environmental quality, and sustainability. An EI's conceptual development is originated in the EBD process [23]. The intention with a single value on a room/building project is to inform the stakeholders about the verified level of evidence that the building criteria is setting. As building criteria may arise from a previous POE and most likely a new project organization would introduce these additional values for the programming stage, this act of organizational learning [24] was the forerunner of a cognitive building. Cognition in terms of information processing requires working memory (POE) and long-term memory (building criteria, KPIs) in order to appraise information (evidence) for appropriate response selection [25] (project outcome). Consequently, the state-of-the-art understanding of the cognitive building solution is a sustainable building system that automatically integrates, analyses, and learns from the IoT-generated data [26]. The EI would be benefitting the cognitive building concept through the interconnectedness of scientific databases and machine learning of evidences in scientific publications to find and appraise a project-specific evidence index. For this to become a reality, more knowledge about EBD and EI is needed, and the work described here is a contribution to that.

2.1.1. Evidence-Based Design Process for Cognitive Buildings

POE is an integral part of the EBD process [3,18], which distinguishes between eight phases that refer to a continuous workflow stepwise progression. The modified EBD process fitted to a servitized construction product for cognitive building solution may include the following:

1. Defining the key goals and objectives: A vision is developed for the intentions, directions, and goals for the project. The multidisciplinary project team and a cognitive building solution articulate the goals and objectives to be reached. This process includes similar project-specific digitalized POE results and KPIs that proved to be appropriate and accurate for the type of the planned project.
2. Finding of sources for relevant evidence: A relevant evidence is obtained mainly from various digitalized scientific and expert testimonial databases to identify research results that may serve as evidence. This process requires robust digitalization and machine learning to locate evidence.
3. Critical interpretation of relevant evidence: The validity and reliability of the evidence need to be established by review. The automatized process of finding and evaluating the level of evidence in every project related sources is due to the fine-tuned algorithms capable of understanding and interpreting scientific results and closing the gap between evaluators. Informing the design phase and creating hypotheses for value generation also starts here.

4. Creation and innovation of evidence-based design concepts: The relevant evidence is translated into design guidelines and statements. Designers use guidelines for aesthetic, functional, or compositional decisions and incorporate digital tools to visualize the project. The cognitive building concept helps to prioritize among the possible design solutions due to its iterative design capabilities. The possible alternatives are tested in a parametric environment.
5. Development of hypotheses: Design hypotheses are generated and tested by various means. The parametrized design solutions are being evaluated mathematically and/or visually in order to set the subjective and objective method for hypothesis testing.
6. Collection of baseline performance measures: The building criteria with the embedded level of evidence are identified and assessed. The project values and requirements are translated into parametric design criteria expressed in terms of performance metrics and simulation results.
7. Monitoring of implementation of design and construction: With the help of cognitive building solution, the construction is monitored, and the project team makes sure that the design strategies are executed and delivered. In terms of deviation from the planned action, the cognitive building solution is waiting for human approval of the deviations. At the end of construction, the project team and the cognitive building solution verifies that the project is ready for post-occupancy research.
8. Measurement of post-occupancy performance measure: The project-specific KPIs are being analysed in situ or virtually. The necessary adjustments are made in the physical environment to match the stakeholders' expectations.

The above-suggested EBD process for cognitive building solution is still ahead of the present reality. However, for the realization of the EI in a non-cognitive building solution, the first three stages of the original EBD process are considered in this article. Starting with the definition of goals and objectives that describes the planned building project's intentions and direction. A team of decision-makers articulate project goals in terms of their desired outcomes. In the case of a well-known building typology to be delivered, the project goals and objectives are revisited from previous successful projects using POE. These building criteria may be industry standards and recommendations. In case of an innovative solution, building criteria are being set intuitively according to the team's experience. When this preliminary programming of domain-specific values is set, the next stage is finding sources for relevant evidence.

2.1.2. Finding Evidence

Relevant evidence is gathered mainly from scientific literature to identify gaps in knowledge and determine what relevant research has already been performed and which needs to be researched. Peavey and Vander Wyst [27] differentiate between evidence that incorrectly refers to a proof of a design decision. This misconception is caused by the difference between the commonly used definition of evidence as proof and the scientific interpretation of evidence. The latter divide evidence into several levels. Another shortcoming of using evidence as a proof is reported by Cama [28] when the practitioner indistictively using the evidence for any kind of built environment. To overcome such a misinterpretation of evidence, a critical interpretation is needed.

The methodological framework for ranking evidence is combined from a series of research design methods that gradually decrease the need for scientific rigour, validity, and reliability. Therefore, the evaluation of the level of evidence prerequisites a qualified person to interpret the specific evidence in accordance with an EBD guideline. This guideline was moulded from Pati [29], Stetler [30], and Stichler [31] recommendations for healthcare design settings (Table 1).

Table 1. Levels of evidence as it is originated from healthcare design.

Ranking	Evidence-Based Design
1	Meta-analysis and systematic reviews of randomized controlled trials or experimental studies
2	Single experimental study (randomized, controlled)
3	Single quasi-experimental study (randomized, concurrent, or historical controls)
4	Systematic, interpretive, or integrative review of multiple studies of observational or qualitative research
5	Single non-experimental study, correlational, descriptive, mixed methods, and qualitative research
6	Published evaluation data (e.g., facility evaluations, mock-ups) that were systematically collected and were verifiable
7	Consensus opinion of authorities (e.g., a nationally known guideline group with strong peer review)
8	Opinions of recognized experts, case studies

Note. Adapted from Pati [29], Stetler [30], and Stichler [32]. These levels should be used in conjunction with a critical appraisal of quality at each level.

In a building project, ranking of evidence in the early project phase is imperative because it has a significant outcome for the programming stage that will impact the stakeholders' physical, and psychological and social wellbeing. At the critical interpretation of relevant evidence, awareness about potentially misinterpreted evidence can still be resolved in time before the project is suffering from serious financial expenses. However, the level of evidence may require qualified personnel in research methodology who can rank the scientific evidence and still give credits for opinions and individual observations. A categorization of such a comprehensive source of information should be guided by a value-generation process that gives meaning to complicated interpretations of the scientifically produced evidence. Marquardt and Motzek [33] suggested a helpful algorithm for architects and designers to critically appraise the quality of evidence in EBD. By adopting a four-level scale instead of a six-level, as Marquardt and Motzek [34] suggested, the investigation of the quality of evidence may take less time to perform with less trained personnel. To appraise the level of evidence into a four-point scale, Evans (2003, p. 82) published a hierarchy to an indication of the validity and trustworthiness of different types of research. This process assists in the selection of the evidence to guide evidence-based clinical practice. However, a building delivery process is not seen directly as a comparable field of study to clinical practice, yet its systematic research-based approach to identification of evidence makes it possible to apply the principles of research to designers and engineering practitioners. Henceforth, the proposed four-point rating scale is the first attempt to measure the level of evidence in servitized construction delivery using EBD. The highest rating is *excellent*, when the evidence provides the strongest scientific base for the practice. This evidence level is at the least risk of error, therefore it is optimal for the development of practical design guidelines and recommendations. The next highest level is *good*. This rating provides a sound basis for practical cases and is at low risk of error. However, as it may have been generated by single studies, it also highlights areas where replication of research is needed. A less prefered rating is *fair*, which includes varying degrees of risk for error, and it does not provide a strong evidence for the practice. These studies usually represent exploration of interventions. The rationale behind this level is to accept a greater risk of error in the evidence, yet allow further identification of potentially beneficial KPIs that require additional investigation and evaluation. The least preferred and most common level of evidence can be ranked as *poor*, when there is a weak basis for practical use and is at serious risk of error or bias. The four-point scale rating has an advantage on the usability side as an evaluator is forced to avoid central tendencies and needs to be making a decision based on the criteria at hand [35]. The drawback of this four-point scale is tangible when the accuracy of the level of evidence is in question. In the EBD process, the scientific evaluations should be synchronized to laypersons or design experts too. A four-point category suggested on quality of evidence is summarized and retains the major scientific category differences and the contents (Table 2).

Table 2. Modified levels of evidence for quantitative research and EBD project.

Level	Determining the Level of Evidence
Excellent	Meta-analysis and systematic reviews of randomized controlled trials or experimental studies; Single experimental study (randomized, controlled)
Good	Single quasi-experimental study (randomized, concurrent, or historical controls); Systematic, interpretive, or integrative review of multiple studies of observational or qualitative research
Fair	Single non-experimental study, correlational, descriptive, mixed methods, and qualitative research; Published evaluation data (e.g., facility evaluations, mock-ups) that were systematically collected and were verifiable
Poor	Consensus opinion of authorities (e.g., a nationally known guideline group with strong peer review); Opinions of recognized experts, case studies

Note: Adapted from Pati [29], Stetler [30], and Stichler [32]. These levels should be used in conjunction with critical appraisal of quality at each level.

2.1.3. Critical Review of LOE

Two flow chart diagrams visualize the decision-making procedure evaluating the level of evidence (LOE) with a quantitative or qualitative study by Marquardt and Motzek [33]. These algorithms for rating the evidence distinguishes among six-levels LOE according to Stichler [32]. The step-by-step procedure follows "yes" and "no" options for the main methodological junctions. The answers on these methodological alternatives will eventually lead the rater to various level of evidence. A four-level LOE category is presented in relation to the major study types in Figure 2. When a qualitative or case study is investigated in an interpretative way, the rater evaluates if the study has a literature review, a framework, a clear method reported, and the diversity of views are represented, then the LOE might be reaching a *fair* rating. When these aspects are not addressed in the study, it is assigned as *poor*. For instance, a case study describing the renovation of a building is classified as *poor*, but if the study features several buildings with the same typology, including stakeholder interviews, and has a matching methodology, it is assigned to *fair*. The quantitative study employs statistical analyses and measures outcomes, therefore they belong to the observational study category. These can be a panel, cohort, case-control, and cross-sectional studies. A sample is followed over a period in a panel study, and the effects of exposures are examined. All observational studies are assigned to *fair*, considering a set of samples compared to each other in a methodologically appropriate way. In an experimental study, when the participants are randomly allocated to at least two randomized selected groups and compared under two or more conditions, the study can be called a randomized controlled trial (RCT). In this type of study, one group receives treatment, while the other group does not. The measurements taken in both cases are before and after the treatment.

When the groups are not randomized but grouped due to specific characteristics, the study is considered as quasi-experimental. In some quasi-experimental studies using within-subject tests, the measurements are taken before and after the intervention. If an experimental or quasi-experimental study is well-conducted, it is classified as *good*, otherwise as *fair*. These types of studies are well-conducted if (1) there are two separate groups of participants; (2) there is a low gradual reduction rate, under 20%; (3) the outcomes are analyzed according to initial treatment assignment; and (4) there are reliable outcomes with low dispersion. Additionally, an experimental RCT study is well conducted and can be assigned as *good* if there is a low ascertainment bias at sampling and the study maintain a high quality of blinding of participants and the researcher.

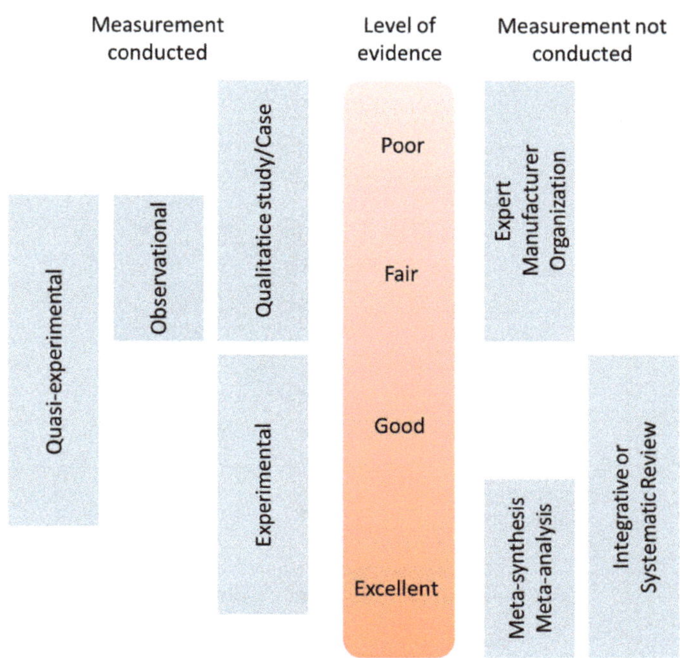

Figure 2. Study types in relation to the four-point scale of the level of evidence (LOE). Modified after Marquardt and Motzek [33].

The right side of the diagram (Figure 2), deals with nonsystematic, systematic, and mixed methodology papers. The nonsystematic refers to studies from manufacturers or consultants, including experts' opinions and guidelines of professional organizations and standards. The scientific robustness of the papers is lacking and the critical approach to the investigations is missing. Furthermore, these studies might have financial interest bias. These nonsystematic papers are all rated as *poor*. In contrast to this, the systematic reviews and meta-analyses identify, evaluate, and summarize objective and accurate approaches rated as *excellent* evidence. Meanwhile, in the systematic review and the integrative review with lower quality of study design, the studies summarize merely empirical or theoretical views, therefore they reach a *fair* level.

3. Case Study Application

The Real Estate (Regionfastigheter) organisation of Jönköping's County Council in Sweden has developed an IT-based management system for controlling and supporting its building process, called Program of Technical Standard (PTS). PTS is a knowledge database containing best practice and specific knowledge about how the building of premises for healthcare should be carried out. PTS is a widely accepted building criteria recommendation in 20 of the 21 County Councils in Sweden [36]. Among other things, PTS contains standard room requirements for various interior amenities and functions.

The case study involved an administrative office (approximately 10 m^2; but at least 4.2 m $\times$ 2.1 m, Figure 3) from PTS. The detailed list of building criteria for the single person occupancy administrative office was obtained through Regionfastigheter Jönköping, Sweden. The building criteria in the PTS are not explained or categorized in any specific way, making it rather difficult for formulating a prioritization about it. Therefore, the investigation of the list of building criteria was first categorized and later searched to find the value and possible evidence related to the particular criteria.

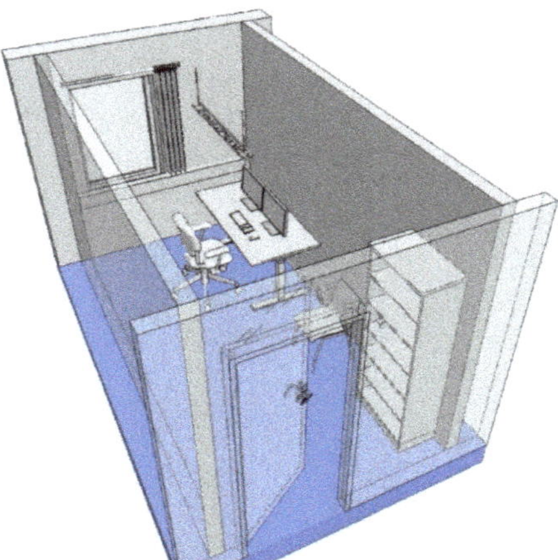

Figure 3. BIM model of administrative office typology visualized after the (After Program of Technical Standard) PTS building criteria. This visualization is done by ArchiCAD 22 Educational edition, Jönköping university.

The case study aimed to investigate the evidence index for this single office occupancy in the programming phase of the design process.

3.1. Procedure and Analysis

The case study procedure followed Table 3 steps to generate the EI. The manual steps are summarized which are set to make an EI for a servitized construction product. After identifying stakeholder as administrative personnel, the EBD process started. The building typology was set for a healthcare facility, and the building criteria were extracted from the PTS. The evidence and the domain of the evidence were described in each finding. The project relevance was chosen to be on a four-point scale: poor (1), fair (2), good (3), and excellent (4). The applicability of this four-point scale was earlier described. The priority of evidence to be used in the EI calculation was set between low (1), medium (2), or high (3). The assessment process did not aim to collect as many evidence as possible within one building criterion, instead as a general approach it aimed to provide at least one evidence for each criterion. The decision behind this category scale is the ease of use for the rater to set up a quick cognitive process. The LOE was appraised using the diagram of Figure 2 and the reference for the related evidence was indicated. The LOE was set between poor (1), fair (2), good (3), and excellent (4) depending on the scientific approach.

Table 3. Procedure to generate the Evidence index for a servitized construction product.

Procedure to Generate the Evidence Index	
1. Identifying Stakeholder	To whom?
2. EBD process	1–8 stages
2.1 Category	Building typology
2.2 Building criteria (KPI)	Physical attributes
2.3 Values	What domain?
2.4 Evidence	What exactly
2.5 Relevance	1, 2, 3, 4 (Poor, Fair, Good, Excellent)
2.6 Priority	1, 2, 3 (Low, Medium, High)
3. Level of evidence (LOE)	1, 2, 3, 4 (Poor, Fair, Good, Excellent)
4. EI	(Relevance × Priority × LOE)/2
5. Reference	Indicative

The appraisal of the EI resembled a novice research assistant searching strategy, implying that the person first uses Google or Google scholar engine and, if it is not successful, then uses a university library access for scientific literature.

3.2. Results and Discussion

The results are presented in Table 4. Altogether, 42 building criteria were taken into consideration when identifying 30 evidence. Among the evidence, best practice indicated a not identified evidence, therefore it was assumed that the building criteria is based on a practical need existing in the construction and use phases. The values for these items were not calculated and counted into the EI. Consequently, the final value on EI was 1.49, which is slightly better than a *poor* level but not reaching *fair*. Due to the high number of *best practice* designations, the novice research assistant had difficulties identifying the scientific evidence describing why specific building criteria exist. As a consequence of this finding, the level of expertise in evaluating healthcare buildings needs to be higher. Regarding the usability of the four-point scale LOE appraisal, it put a high demand on the evaluator to clearly identify the strength of evidence. However, when using internet forums or opinions for the search, the diagram could not be considered for appraising the evidence level. These building criteria were treated as best practices.

Table 4. Transformation of building criteria into evidence index.

Programming	Category	Building Criteria	Values	Evidence	Reference	Relevance	Priority	LOE	EI
Room is set up for administrative office work	Amenities	Cloth hangers	comfort	Best practice					
		Window blinds	electricity saving	1. obtained electricity savings for lights (with the window system and controllable highly reflective venetian blinds plus light dimming) reaching 76% on overcast days and 92% on clear days	[37]	1	2	2	0.33
			best utilization of blinds	2. of daylight is achieved with horizontal slats, because this evens out the big differences in luminances between the window zone and the rear wall zone	[38]	4	3	3	3.00
			threshold value for action	3. a threshold value of 2000 cd/m^2 was used, based on the assumptions that the primary task involved a LCD computer monitor with an average luminance of 200 cd/m^2. The window was within the occupant's peripheral field of view so that a maximum luminance ratio of 10:1 between window and task was just acceptable, and that the average background luminance was 50–100 cd/m^2. It was also based on subjective survey results that found that there was a 50% probability that blinds would be lowered when the average window luminance was 2100 cd/m^2	[39]	3	3	2	1.50
		Electrical wiring in walls, beside the door		Best practice					
		Electrical outlet		Best practice					
		General lighting: up and down	well-being	1. General lighting for writing task: Em = 500 lx;	[40]	4	3	1	1.00
			efficiency	2. Indirect light is more expensive to install, but 20% more energy efficient as indirect alone, light wall and ceiling color is needed	[41]	3	3	2	1.50
		Combined outlets (telephone, data)	comfort	Best practice					
	Work support	Vertically adjustable desk	awareness	1. Desks may be an important remedy in this endeavor, particularly in office settings, while ergonomics awareness may be able to contribute to further changes in sedentary behavior if enhanced and if supported by the work organization.	[42]	4	3	3	3.00
			well-being	2. The ability to alter one's position by sitting, standing, and walking is healthier than sitting continuously with 90 degree angles in knees and hips	[43]	4	3	1	1.00
		Bookshelf (L = 800 B = 420 H = 1700)	comfort	Best practice					
		Visitor's chair	comfort	Best practice					
		Curtain rod	control	Best practice					
		Curtain hanger	control	Best practice					

Table 4. *Cont.*

Programming	Category	Building Criteria	Values	Evidence	Reference	Relevance	Priority	LOE	EI
	Activity	PC with 2 screens	job satisfaction	Best practice					
	Functionality	Door, free size 840 mm	accessibility	Swedish standard for accessibility	[44]	4	3	2	2.00
		Adjustable room temp. 23 °C ± 1.5 °C	control, comfort	1. the need for temperature control is 4 K	[45]	4	3	2	2.00
			accuracy	2. 1 K is enough for accuracy in adjustment	[46]	4	3	2	2.00
		Min temp. 21 °C	stress, comfort	avoiding SBS, Recommended temperature 21 + −2	[47]	4	3	2	2.00
		Min. filtering F7	well-being, ozon indoors, bad odour removal	the particle size removal efficiency of the air filters for 0.4 mm particles were 14% (F5), 22% (F6), 65% (F7) and 82% (F8)	[48]	4	3	2	2.00
		Min airflows (L/s person) 15 L/s	well-being, staff turnover	avoiding SBS, min airflow 10 L/s	[47]	4	3	2	2.00
		Pressure conditions to other rooms: Balanced	well-being, staff turnover						
		General lighting	well-being, job satisfaction	1. General lighting for writing task: Em = 500 lx;	[40]	4	3	1	1.00
		Lighting Strength. Lighting power according to industry recommendation for this type of room. Normal	well-being, job satisfaction	1. General lighting for writing task: Em = 500 lx;	[40]	4	3	1	1.00
		Color rendering index Ra > 80 Normal	well-being, job satisfaction	More accurate perception of color	[49]	4	3	1	1.00
		Color temperature 4000 K Neutral color temperature	circadian rhythm	stronger melatonin supression at 18.7 lux	[50]	4	3	2	2.00
			cognitive performance	general recommended light	[51]	4	3	1	1.00
			job performance	highest relaxation	[52]	4	3	2	2.00
		Glare-free lighting Normal	visual cofort, alertness and mood, pleasant view	greater tolerance under daylight, positive glare ratings, more sensitive to glare the less relaxed,	[53]	3	2	2	1.00
		Lighting control-switch. Normal requirements (manual control). Switches	well-being, job satisfaction	Best practice					
		Lighting control-dimmer. Light control-manual control via dimmer Dimmer	well-being, job satisfaction	Best practice					
		Lighting control-absence controlled. Manual ignition with absence-controlled extinction Absence	well-being, job satisfaction	Best practice					
		Sound-proofing. Room with requirements for adequate sound insulation during conversations with moderate voice strength and spaces for rest and sleep.	well-being, job satisfaction	Best practice					
		Expeditions located in administration unit 44 dB 48 dB R'w	Privacy at moderate voice strength	Sound class "B";	[54]	4	3	1	1.00
		Step sound level Highest step sound level L'n, Tw (dB). 64 dB			[54]	4	3	1	1.00
		Room acoustics. Longest reverberation time (s) 0.6 s			[54]	4	3	1	1.00

Table 4. *Cont.*

Programming	Category	Building Criteria	Values	Evidence	Reference	Relevance	Priority	LOE	EI
		Noise from installations. Rooms with moderate requirements for sound levels. 35 dB (A)			[54]	4	3	1	1.00
		Daylight Requirements are required	visual cofort, alertness and mood	1. Reduced discomfort; 2. Improves circadian rhythm, 3. Max. visual performance, 4. Mood changes	[55,56]	4	3	2	2.00
		Power lighting. In% connected lighting ÖL 50% VL 50%	well-being, job satisfaction	Best practice					
		Power take-off 230 V. In% connected outlets or loads ÖL 50% VL 50%	well-being, job satisfaction	Best practice					
		Power take-off 230 V data. Very Important Last MVL 100%	well-being, job satisfaction	Best practice					
		Medical spaces. According to SS 436 40 00 ch. 710 Group 0	well-being, job satisfaction	Electrical installation rules	[57] ch. 710 Group 0	4	3	1	1
		Shooting signal. Indicator tab for busy marking. Switching on and off at the door	well-being, job satisfaction	Best practice					
		View towards greenery	Stress reduction	Short-term visits to urban nature areas have positive effects on stress relief.	[58,59]	4	3	2	2.00
			Cortisol reduction	The salivary cortisol concentration decreases in all urban environments.	[58]	4	3	2	2.00
			Noise reducer	Accessibility may not reduce noise annoyance	[60]	2	1	2	0.33
		Hygiene class 2	Hygiene	Surface layers on walls must withstand cleaning agents and point disinfection	[61]	4	3	1	1.00
		Locking Mechanical lock	well-being, job satisfaction	Best practice					
	Finishes	Flooring Carpet upholstered against the wall	safety	Best practice					
	Finishes	Wall. Painted, gloss value 20	well-being, job satisfaction	Best practice					
			Mean Value for the Evidence Index:						1.49

4. Discussion

Gedda [62] published an inspirational article on evidence index that is related to evidence-based medicine, in which the author refers to the evidence index as the "factual components on which the main decision-making is based" (p. 1) during a treatment. As evidence-based medicine gave rise to evidence-based design, the promoters of EBD rely on objective scientific data combined with stakeholders' perspective and expertise in the building project. Considering a complex construction project where building criteria are detailed and consensus-based among the partners, an evidence index could be validating the objectivity of the criteria set to fulfil the stakeholders' needs. However, the often servitized and multifaceted criteria in the era of digitalization can be the source of confusion with regards to prioritization between the building criteria to fulfil the project goals. Therefore, this study aimed to test an evidence index during the programming stage of a building process in order to understand the level of objective scientific data involved in the decision making. The study assumed that even a novice investigator could generate an evidence index.

The methodological development for the EI is fundamentally striving for a quantifiable measurement for the stakeholders' interests. The building performance evaluation (BPE) had always been a building process-oriented approach, and on a larger scale, it incorporates the quantitative research characteristics and the EBD process model. The EBD process model is a combination of quantitative research and a building project process model. Therefore, the use of EBD as the primary process model for EI generation seemed viable (Table 5).

Table 5. The main steps for quantitative research (after Polit and Beck [63] and Stichler [31]), evidence-based design (EBD) [23], and building performance evaluation (BPE) [17] are shown.

Steps	Quantitative Research	BPE	EBD
1	Identify problem, research question, or hypothesis (es)	Market/Needs analysis	Define evidence-based goals and objectives (with client and interdisciplinary team
2	Perform literature review	Program review	Find sources for relevant evidence
3	Use of a theoretical framework to explain the relationships among variables		Critically interpret relevant evidence, assess evidence applicability, quality and strength
4	Select an appropriate research design to test the hypothesis	Design review	Create and innovate EBD concepts
5	Identify measurements to quantify variables	Effectiveness review	Develop a hypothesis
6	Select the sample		Collect baseline performance measures
7	Data collection and analysis	Commissioning	Monitor implementation of design and construction
8	Statistical data analysis	POE	POE
9	Disseminate results in publications and presentations		

Note: POE = Post-occupancy evaluation.

The development of EI in this paper mainly focuses on the programming stage, which is an early stage of the building process, but this is the strategically important stage, where servitized construction products review scientific evidence on how well they can support the predefined goals. Theoretically, the EI could be extended throughout the entire EBD or BPE process and inform the stakeholders about the whole building process. The cognitive building solution for delivering EI for a building project is a challenging task. Today, the initial stages within the EBD process had been made by manual effort and resulted in a case study quality. However, the results showed that it was possible to generate an EI value between 1 and 4 on the rating scale and indicate the room EI, the process was time-consuming and often assigning best practice for reasoning for the building criteria. The frequent occurrence of best practices indicates that the level of evicence of the KPI used in a construction project is not measurable. This is troublesome in the present development where KPI based management of construction projects are promoted, mainly due to the new opportunities given by digitalization [1,9].

One of the main characteristics of this EI is the measurement scale on which it measures the scientific evidence. Literature used six to eight-level differentiation between the evidence while the current EI is reduced to four level in order to facilitate a quicker appraisal of the LOE and in the same time better correspond to the 8-point scale of LOE. Furthermore, the four-point scale measurement technic is supporting the evaluator for learning the basic differences when an LOE is appraised. In terms of a design project aiming to deliver a public building, the four-point scale seems appropriate for grasping the array of choices. What might be debatable is that the first level of evidence includes opinions of recognized experts and the use of case studies. In a business where all the experts are proud of their years of experiences, it may generate tension between the stakeholders, depending on who is more trusted in the process. The development of both EI and LOE, described above, has its theoretical background in research theory in general and having the fact that research theory is internationally applicable indicates that the framework could also be internationally applicable. However, more research on different types of buildings and their contexts is needed to evaluate the general applicability of the proposed framework. Discussion of results from a case study exercise is somewhat a straightforward activity now. As any case study, results bear a low-quality level of evidence. In this investigation, it is also shown that the level of evidence cannot exceed *fair*. However, the experience for the single investigator had been meaningful, as going through a number of building criteria without finding appropriate scientific relevance triggered the curiosity for criteria that cannot be easily found. Regarding the generalizability of the given case study, the findings should be carefully examined. The case study outcome would suggest that a more comprehensive investigation should take place with different background of the investigators and preferably in a randomized manner. As for the PTS, the building criteria is a country-specific knowledge that requires a culturally appropriate building tradition. Employing the same PTS criteria outside of Sweden would not mean failure, but adaptations of the criteria must be considered. With regards to using the building criteria in another building typology, such as, a culture center [64] it can indicate that the builing critera are similar for other building types.

In the future research, an expert pool evaluation of the building criteria may shed light on the various best practice designated findings in the search for evidence. Furthermore, as a concern for the industry regarding cost efficiency, if fresh graduates on first- and second cycle could contribute to the evaluation of evidence, it would generate a more economically feasible way to extend the EI related research.

5. Conclusions

This study presents the first steps of an EI for a built environment. The study conceptualizes on the basis of the EBD stages a cognitive building solution that is capable of automatizing a series of repetitive and research related tasks regarding evidence appraisal and evaluation. The procedure to establish an EI for the built environment was tested through a case study, in which a novice research assistant approach to evidence appraisal was assessed. The 4-point rating scale, together with a diagram of which studies may fulfil the level of evidence requirements, was used to assess the building criteria. The concept of a cognitive building solution is preferable due to the strenuous job a person needs to perform when evaluating building criteria. The limitation of this study entailed a single evaluator for the transition process of the building criteria into LOE and later to EI. The EI is an initial step for establishing the cognitive building solution for EBD, in which the technological solution is rendered to serve a servitized construction product.

Author Contributions: Conceptualization, G.F. and P.J.; methodology, G.F.; formal analysis, G.F.; investigation, G.F.; resources, P.J.; data curation, G.F.; writing—original draft preparation, G.F.; writing—review and editing, G.F. and P.J.; visualization, G.F.; All authors have read and agreed to the published version of the manuscript.

Funding: This research received no external funding.

Institutional Review Board Statement: Not applicable.

Informed Consent Statement: Not applicable.

Data Availability Statement: Not applicable.

Acknowledgments: The authors would like to thank Kaj Granath for his insight and comments on this paper's first version. Furthermore, the authors are grateful for Regionfastigheter Jönköping for sharing their PTS documentation for research purpose.

Conflicts of Interest: The authors declare no conflict of interest.

References

1. Fischer, M. *Integrating Project Delivery*; Wiley: Hoboken, NJ, USA, 2017.
2. Hafezparast Moadab, N.; Olsson, T.; Fischl, G.; Aries, M. Smart versus conventional lighting in apartments-Electric lighting energy consumption simulation for three different households. *Energy Build.* **2021**, *244*, 111009. [CrossRef]
3. Davoodi, A.; Johansson, P.; Henricson, M.; Aries, M. A conceptual framework for integration of evidence-based design with lighting simulation tools. *Buildings* **2017**, *7*, 82. [CrossRef]
4. Li, H.; Hong, T.; Lee, S.H.; Sofos, M. System-level key performance indicators for building performance evaluation. *Energy Build.* **2020**, *209*, 109703. [CrossRef]
5. Micolier, A.; Taillandier, F.; Taillandier, P.; Bos, F. Li-BIM, an agent-based approach to simulate occupant-building interaction from the Building-Information Modelling. *Eng. Appl. Artif. Intell.* **2019**, *82*, 44–59. [CrossRef]
6. Jang, H.; Kang, J. A stochastic model of integrating occupant behaviour into energy simulation with respect to actual energy consumption in high-rise apartment buildings. *Energy Build.* **2016**, *121*, 205–216. [CrossRef]
7. Johansson, P.; Fischl, G.; Granath, K. Towards Evidence-Based BIM. In *The Advances in ICT in Design, Construction and Management in Architecture, Engineering, Construction and Operations (AECO): Proceedings of the 36th CIB W78 2019 Conference*; University of Northumbria: Newcastle-upon-Tyne, UK, 2019.
8. Perjons, E.; Johannesson, P. A Value and Model Driven Method for Patient Oriented KPI Design in Health Care. In *Biomedical Engineering Systems and Technologies*; Springer: Berlin/Heidelberg, Germany, 2011; Volume 2011, pp. 123–137.
9. Won, J.; Lee, G. How to tell if a BIM project is successful: A goal-driven approach. *Autom. Constr.* **2016**, *69*, 34–43. [CrossRef]
10. Arkitektur, C.F.V. Administrativa Arbetsplatser inom Vården och dess Förvaltningar. 2015. Available online: https://www.ptsforum.se/media/1082/2015-02-05-admarbplatser.pdf (accessed on 2 May 2021).
11. Baines, T.; Lightfoot, H.W. Servitization of the manufacturing firm. *Int. J. Oper. Prod. Manag.* **2014**, *34*, 2–35. [CrossRef]
12. Doni, F.; Corvino, A.; Bianchi Martini, S. Servitization and sustainability actions. Evidence from European manufacturing companies. *J. Environ. Manag.* **2019**, *234*, 367–378. [CrossRef] [PubMed]
13. Ulrich, R.; Zimring, C.; Zhu, X.; Dubose, J.; Seo, H.; Choi, Y.; Quan, X.; Joseph, A. A Review of the Research Literature on Evidence-Based Healthcare Design. *HERD Health Environ. Res. Des. J.* **2008**, *1*, 61–125. [CrossRef]
14. Hamilton, D.K.; Watkins, D.H. *Evidence-Based Design for Multiple Building Types*; John Wiley & Sons: Hoboken, NJ, USA, 2009.
15. Levin, D.J. Defining Evidence-Based Design. *Healthc. Des. Mag.* **2008**, *8*, 8.
16. Preiser, W.F. Building performance assessment—from POE to BPE, a personal perspective. *Archit. Sci. Rev.* **2005**, *48*, 201–204. [CrossRef]
17. Preiser, W.; Vischer, J. *Assessing Building Performance*; Routledge: London, UK, 2006.
18. Davoodi, A.; Johansson, P.; Aries, M. The use of lighting simulation in the evidence-based design process: A case study approach using visual comfort analysis in offices. *Build. Simul.* **2020**, *13*, 141–153. [CrossRef]
19. Davoodi, A.; Johansson, P.; Aries, M. The Implementation of Visual Comfort Evaluation in the Evidence-Based Design Process Using Lighting Simulation. *Appl. Sci.* **2021**, *11*, 4982. [CrossRef]
20. Association, C.P. *The Future for Construction Product Manufacturing, Digitalisation, Industry 4.0 and the Circular Economy*; Construction Products Association: London, UK, 2016.
21. Li, P.; Froese, T.M.; Brager, G. Post-occupancy evaluation: State-of-the-art analysis and state-of-the-practice review. *Build. Environ.* **2018**, *133*, 187–202. [CrossRef]
22. Preiser, W.F.; White, E.; Rabinowitz, H. *Post-Occupancy Evaluation (Routledge Revivals)*; Routledge: London, UK, 2015.
23. Malone, E.; Nanda, U.; Harmsen, C.; Reno, K.; Edelstein, E.; Hamilton, D.K.; Salvatore, A.; Mann-Dooks, J.R.; Oland, C. *An Introduction to Evidence-Based Design: Exploring Healthcare and Design (EDAC Study Guides, Volume 1)*; The Center for Health Design: Concord, CA, USA, 2008.
24. Zimring, C.; Rosenheck, T. Post-Occupancy Evaluations and Organizational Learning1. In *Learning from Our Buildings: A State-of-the-Practice Summary of Post-Occupancy Evaluation*; National Academic Press: Washington, DC, USA, 2001; pp. 42–53.
25. Wickens, C.D.; Hollands, J.G.; Banbury, S.; Parasuraman, R. *Engineering Psychology and Human Performance*; Psychology Press: London, UK, 2015.
26. Ploennigs, J.; Ba, A.; Barry, M. Materializing the Promises of Cognitive IoT: How Cognitive Buildings Are Shaping the Way. *IEEE Internet Things J.* **2018**, *5*, 2367–2374. [CrossRef]
27. Peavey, E.; Vander Wyst, K.B. Evidence-Based Design and Research-Informed Design: What's the Difference? Conceptual Definitions and Comparative Analysis. *Herd Health Environ. Res. Des. J.* **2017**, *10*, 143–156. [CrossRef]
28. Cama, R. *Evidence-Based Healthcare Design*; John Wiley & Sons: Hoboken, NJ, USA, 2009.
29. Pati, D. A Framework for Evaluating Evidence in Evidence-Based Design. *Herd Health Environ. Res. Des. J.* **2011**, *4*, 50–71. [CrossRef] [PubMed]
30. Stetler, C. *Evidence-Based Practice and the Use of Research: A Synopsis of Basic Concepts & Strategies to Improve Care*; NOVA Foundation: Washington, DC, USA, 2002.
31. Stichler, J.F. Research or evidence-based design: Which process should we be using? *SAGE J.* **2010**, *4*, 4. [CrossRef] [PubMed]
32. Stichler, J.F. Weighing the evidence. *SAGE J.* **2010**, *3*, 3. [CrossRef]
33. Marquardt, G.; Motzek, T. How to rate the quality of a research paper: Introducing a helpful algorithm for architects and designers. *Herd Health Environ. Res. Des. J.* **2013**, *6*, 119–127. [CrossRef]

34. Evans, D. Hierarchy of evidence: A framework for ranking evidence evaluating healthcare interventions. *J. Clin. Nurs.* **2003**, *12*, 77–84. [CrossRef]
35. Chyung, S.Y.; Roberts, K.; Swanson, I.; Hankinson, A. Evidence-based survey design: The use of a midpoint on the Likert scale. *Perform. Improv.* **2017**, *56*, 15–23. [CrossRef]
36. Regionfastigheter, J.L. Program för Tekniskt Standard (Program for Technical Standard). Available online: https://www.ptsforum.se/ (accessed on 8 June 2021).
37. Athienitis, A.K.; Tzempelikos, A. A methodology for simulation of daylight room illuminance distribution and light dimming for a room with a controlled shading device. *Sol. Energy* **2002**, *72*, 271–281. [CrossRef]
38. Christoffersen, E.P.J.; Johnsen, K. An experimental evaluation of daylight systems and lighting control. *Right Light.* **1997**, *15*, 245–254.
39. Eleanor, S.L.; Glenn, D.H.; Robert, D.C.; Luis, L.F.; Sila, K.; Mary Ann, P.; Francis, M.R.; Stephen, E.S. *Daylighting the New York Times Headquarters Building: Final Report: Commissioning Daylighting Systems and Estimation of Demand Response*; Commissioning Daylighting Systems and Estimation of Demand Response: Berkeley, CA, USA, 2007.
40. Lighting, Z. *The Lighting Handbook*; Zumtobel Lighting: Dornbirn, Austria, 2013; p. A1-1.
41. Benya, J.R. *Lighting for Schools*; Educational Resources Information Center (ED): Washington, DC, USA, 2001.
42. Straker, L.; Abbott, R.A.; Heiden, M.; Mathiassen, S.E.; Toomingas, A. Sit–stand desks in call centres: Associations of use and ergonomics awareness with sedentary behavior. *Appl. Ergon.* **2013**, *44*, 517–522. [CrossRef] [PubMed]
43. Chandra, A.; Chandna, P.; Deswal, S.; Kumar, R. Ergonomics in the office environment: A review. In Proceedings of the International Conference on Energy and Environment, Chandigarh, Haryana, India, 23–24 March 2009.
44. Svensson, E. *Bygg Ikapp Handikapp: Att Bygga för Ökad Tillgänglighet och Användbarhet för Personer med Funktionshinder: Kommentarer till Boverkets Byggregler, BBR*; Svensk Byggtjänst: Eskilstuna, Sweden, 2006.
45. Karjalainen, S. Thermal comfort and use of thermostats in Finnish homes and offices. *Build. Environ.* **2009**, *44*, 1237–1245. [CrossRef]
46. Karjalainen, S. Usability guidelines for room temperature controls. *Intell. Build. Int.* **2010**, *2*, 85–97. [CrossRef]
47. Jaakkola, J.J.; Reinikainen, L.M.; Heinonen, O.P.; Majanen, A.; Seppänen, O. Indoor air quality requirements for healthy office buildings: Recommendations based on an epidemiologic study. *Environ. Int.* **1991**, *17*, 371–378. [CrossRef]
48. Hyttinen, M.; Pasanen, P.; Salo, J.; Björkroth, M.; Vartiainen, M.; Kalliokoski, P. Reactions of Ozone on Ventilation Filters. *Indoor Built Environ.* **2003**, *12*, 151–158. [CrossRef]
49. Ohno, Y. CIE fundamentals for color measurements. In *Proceedings of the NIP & Digital Fabrication Conference*; Society for Imaging ScienceTechnology: Springfieldu, VA, USA, 2000; pp. 540–545.
50. Figueiro, M.; Nagare, R.; Price, L. Non-visual effects of light: How to use light to promote circadian entrainment and elicit alertness. *Light. Res. Technol.* **2018**, *50*, 38–62. [CrossRef] [PubMed]
51. Morrow, B.; Kanakri, S. *The Effect of LED and Fluorescent Lighting on Children in the Classroom*; EDRA: Oklahoma City, OK, USA, 2018.
52. Wang, M.L.; Luo, M.R. Effects of LED lighting on office work performance. In Proceedings of the 2016 13th China International Forum on Solid State Lighting (SSLChina), Beijing, China, 15–17 November 2016; pp. 119–122.
53. Borisuit, A.; Linhart, F.; Scartezzini, J.-L.; Münch, M. Effects of realistic office daylighting and electric lighting conditions on visual comfort, alertness and mood. *Light. Res. Technol.* **2015**, *47*, 192–209. [CrossRef]
54. Svensk Standard, S. *25268: 2007. Byggakustik–Ljud. Av Utrymmen I Byggn. –VårdlokalerUndervis. Dag-Och FritidshemKontor Och Hotel*; Swedish Institute for Standards: Stockholm, Sweden, 2007.
55. Boyce, P.; Hunter, C.; Howlett, O. *The Benefits of Daylight through Windows*; Rensselaer Polytechnic Institute: Troy, NY, USA, 2003.
56. Löfberg, H.A.; Bäck, B.; Glaas, F. *Räkna Med Dagsljus*; Statens Institut för Byggnadsforskning: Gävle, Sweden, 1987.
57. Elstandard, S.S. *Elinstallationsreglerna SS 436 40 00, utgåva 3, med kommentarer*; SEK Svensk Elstandard: Stockholm, Sweden, 2017.
58. Tyrväinen, L.; Ojala, A.; Korpela, K.; Lanki, T.; Tsunetsugu, Y.; Kagawa, T. The influence of urban green environments on stress relief measures: A field experiment. *J. Environ. Psychol.* **2014**, *38*, 1–9. [CrossRef]
59. Farley, K.M.; Veitch, J.A. *A Room with a View: A Review of the Effects of Windows on Work and Well-Being*; Citeseer: Princeton, NJ, USA, 2001.
60. Li, H.; Chau, C.K.; Tang, S.K. Can surrounding greenery reduce noise annoyance at home? *Sci. Total Environ.* **2010**, *408*, 4376–4384. [CrossRef]
61. för Vårdhygien, S.F. Byggenskap och vårdhygien. In *Vårdhygieniska Aspekter vid ny-och Ombyggnation samt Renovering av vårdlokaler*; Arbetsgruppen BOV (Byggenskap och Vårdhygien): Stockholm, Sweden, 2003.
62. Gedda, M. Evidence-based medicine, evidence-based practice and evidence index 1.0 (EVID-i). *Kinésithér. Rev.* **2017**. [CrossRef]
63. Polit, D.F.; Beck, C.T. *Nursing Research: Generating and Assessing Evidence for Nursing Practice*; Lippincott Williams & Wilkins: Philadelphia, PA, USA, 2008.
64. Huovila, P.; Hyvärinen, J.; Granath, K.; Johansson, P.; Bruun, C.; Annerstedt, J. *Value Driven Procurement in Building and Real Estate-Final Report*; VTT Technical Research Centre of Finland: Espoo, Finland, 2012.

Article

BIM-Based Research Framework for Sustainable Building Projects: A Strategy for Mitigating BIM Implementation Barriers

Bilal Manzoor [1], Idris Othman [1], Syed Shujaa Safdar Gardezi [2], Haşim Altan [3,*] and Salem Buhashima Abdalla [4]

[1] Department of Civil & Environmental Engineering, University Technology PETRONAS (UTP), Seri Iskandar 32610, Perak, Malaysia; bilal_18003504@utp.edu.my (B.M.); idris_othman@utp.edu.my (I.O.)
[2] Department of Civil Engineering, Capital University of Science and Technology (CUST), Islamabad 44000, Pakistan; dr.shujaasafdar@cust.edu.pk
[3] Department of Architecture, Faculty of Design, Arkin University of Creative Art and Design (ARUCAD), Kyrenia 99300, Cyprus
[4] Department of Architectural Engineering, College of Engineering, University of Sharjah (UOS), Sharjah 27272, United Arab Emirates; sabdalla@sharjah.ac.ae
* Correspondence: hasim.altan@arucad.edu.tr

Abstract: Although Building Information Modeling (BIM) can enhance efficiency of sustainable building projects, its adoption is still plagued with barriers. In order to incorporate BIM more efficiently, it is important to consider and mitigate these barriers. The aim of this study is to explore and develop strategies to alleviate barriers in developing countries, such as Malaysia, to broaden implementation of BIM with the aid of quantitative and qualitative approaches. To achieve this aim, a comprehensive literature review was carried out to identify the barriers, and a questionnaire survey was conducted with construction projects' stakeholders. The ranking analysis results revealed the top five critical barriers to be "unavailability of standards and guidelines", "lack of BIM training", "lack of expertise", "high cost", and "lack of research and BIM implementation". Comparative study findings showed that "lack of research and BIM implementation" is the least important barrier in other countries like China, United Kingdom, Nigeria, and Pakistan. Furthermore, qualitative analysis revealed the strategies to mitigate the BIM implementation barriers to enhance sustainable goals. The final outcome of this study is the establishment of a framework incorporated with BIM implementation barriers and strategies namely, the "BIM-based research framework", which can assist project managers and policymakers towards effective sustainable construction.

Keywords: building information modeling; sustainable building; construction projects; BIM implementation; stakeholders; barriers

Citation: Manzoor, B.; Othman, I.; Gardezi, S.S.S.; Altan, H.; Abdalla, S.B. BIM-Based Research Framework for Sustainable Building Projects: A Strategy for Mitigating BIM Implementation Barriers. *Appl. Sci.* **2021**, *11*, 5397. https://doi.org/10.3390/app11125397

Academic Editors: Lavinia Chiara Tagliabue and Ibrahim Yitmen

Received: 19 May 2021
Accepted: 8 June 2021
Published: 10 June 2021

Publisher's Note: MDPI stays neutral with regard to jurisdictional claims in published maps and institutional affiliations.

1. Introduction

With the rapid growth of the construction industry, an increasing number of sustainable building projects are underway all over the world and have far-reaching consequences for national and regional economic development [1]. All other industries rely to some extent on the construction industry because of their high productivity flow through the economy [2]. Similarly, Malaysia's construction industry, like other developing countries, varies from cost, time constraints, and efficiency, resulting in delays [3]. Moreover, due to budget and schedule overrun, most sustainable building projects struggle to achieve their goals. It is therefore important to use advanced digital technologies such as BIM in order to achieve the desired results effectively [4]. Using BIM in sustainable building projects has various advantages, such as reducing errors, rework, and waste [5–7]. In addition, the use of BIM in sustainable building projects facilitates multidisciplinary cooperation between different project teams, achieves the project objective, and increases the productivity of construction activities [8–10]. BIM can be used in any phase of sustainable building projects, such as visualization, cash detection, code checking, communication, collaboration, monitoring, time and cost management [11,12]. BIM also has the potential as a significant technical advancement in traditional CAD, offering more information and interoperability capabilities [13]. BIM has the power to transform the construction industry and is therefore considered to be the future of the construction industry [14].

Since buildings are of high economic value and have a major effect on the environment and quality of life, the construction industry can be considered one of the key elements for society's long-term growth [15]. Buildings are deemed sustainable if their environmental, economic, and social effects on the community have been properly addressed and have contributed to society's long-term growth [16]. In the past, it was anticipated that sustainable buildings would initially cost around 15% more than conventional ones [17].

In addition, it has been found that few studies have been conducted on barriers to the implementation of BIM in sustainable building projects. Memon et al. [4] used a survey of sample size questionnaires (n = 95) to report the barriers and found that BIM adoption in Malaysian construction was very low. However, the current analysis expands the scale of the study by increasing the number of survey respondents (n = 185). In addition, Zahrizan et al. [18] used the quantitative analysis approach through questionnaire surveys to identify barriers and to report that the lack of BIM awareness is a crucial barrier to the implementation of BIM. Subsequently, Hamid et al. [19] performed a report on barriers but limited to establishing only nine barriers. Wong et al. [20] conducted a recent study that emphasized the importance of transitioning from outdated approaches to sophisticated methodologies such as BIM in order to merge design and construction workflows with the goal of enhancing productivity. Therefore, strategies are needed to make the project successful and help construction stakeholders to perform effectively in order to achieve the sustainable goals.

Hence, this research aims to provide strategies for alleviating barriers with the approach of qualitative analysis and a BIM-based research framework for sustainable building projects. To fulfill the aim of the study, there were four research objectives as follows: (a) to identify barriers from the literature, (b) to rank the barriers with the aid of quantitative approach, (c) to provide strategies in order to mitigate the barriers with the aid of qualitative approach, and (d) to establish a BIM-based research framework integrating barriers and strategies for sustainable building projects. The intent of the research was to explore and uncover new knowledge gaps and practical needs of a BIM-based research framework for sustainable building projects. It would also serve as a theoretical foundation for adopting BIM in sustainable building projects.

The remainder of the paper is arranged in the following way: Section 2 addresses the theoretical background and research gaps. Section 3 explains the related works. Section 4 sets out the research methodology. Subsequently, Section 5 elaborates on the results and discussion. Section 6 explains the comparison of outcomes with other countries. Section 7 introduces the BIM-based research framework and is followed by Section 8, the conclusion, limitations, and future directions.

2. Theoretical Background and Research Gaps

BIM is defined as "a model of building information that provides full and necessary information to support all life-cycle processes and that can be directly interpreted by computer applications. It includes information on the building itself and its components, and involves information on the properties of the building, such as its structure, shape, material and life-cycle processes" [21]. The term "BIM" has several contradictory and misleading interpretations. Definitions may differ for different people in distinct organizations, depending on their point of views, work types, and functions. From a design perspective, for example, BIM is described as the digital representation of a project's physical and functional qualities, which relates to the methodology and technologies required to generate a model [22]. In the construction industry, BIM is defined as the creation and application of a computer software model to simulate the construction and operation of a facility [23].

The genesis, creation, and expansion of BIM ultimately represented the growth profile of computerization. In the late 1950s, Itek Corporation, a U.S. defense contractor, developed a computer graphics technology suitable for engineering design. This helped design visual representation technologies that had been integrated into commercial engineering design and design products. Subsequently, the idea was transformed into an Electronic Drafting Machine (EDM) [24]. By the mid-1960s, EDMs had been marketed for use by other organizations. During the 1970s and early 1980s, Applicon (founded in 1969 as Analytics, Inc. in Burlington, Massachusetts by a group of MIT Lincoln Laboratory programmers) provided 2D products for electrical design tasks. This included the concept of the printed circuit board and a 3D product named BRAVO! Computer Art and Design [24]. In the early 1980s, the BRAVO! The product has been considerably advanced. Since the early 1980s, Autodesk had been a significant competitor of Applicon and other CAD suppliers. Autodesk continues to be a pioneer in this area [25]. Distinctive characteristics such as structural analysis, monitoring and analysis of energy buildings, construction management, and performance tracking and even worker safety have recently been provided on AEC computer platforms [26]. The word BIM attracted TM/®/©, who began to promote it with their products [27]. In order to increase infrastructure development, reduce costs, and provide general management assistance during any step of construction, the BIM idea was introduced to the construction industry [28].

In the construction industry, barriers to BIM implementation are a challenge for stakeholders to improve sustainable goals. Almost the bulk of research in literature is related to the identification of barriers. For instance, BIM implementation barriers have been highlighted in Hong Kong but are not capable of providing strategies [29]. Similarly, a study was conducted in China and Australia to define, classify, and prioritize these barriers but not to establish strategies [30]. Researchers have recently performed barrier-related research for promoting sustainable construction. For example, the barriers to strengthening team coordination in BIM-based construction networks and the construction of a conceptual model. This conceptual model offers an intermediate theory, that is, a theoretical basis for guiding further attempts at knowledge formation on the subject [8]. In addition, the advantages and barriers to the implementation of BIM have been applied with a quantitative approach but are not capable of having a strategy to overcome [31]. Therefore, to fill the aforementioned research gap, this study focuses on providing strategies in reducing the barriers for sustainable building projects with the aid of a BIM-based research framework.

3. Related Work

In this section, related works in the domain of barriers in the global and Malaysian context are discussed.

3.1. BIM Implementation Barriers in Global Context

Globally, fast growth demands and the overwhelming majority of construction companies have allocated BIM to improve the sustainability goals [32]. In sustainable building projects around the globe, the use of BIM has been advocated by government and professional bodies to provide more collaboration and cooperation between stakeholders in the construction sector and to ensure the quality of the projects [33]. Moreover, countries like the United Kingdom, the United States, Australia have adopted BIM in-depth research and other field areas like project management, facility management, and safety management [34,35]. In Australia and New Zealand, the implementation of BIM is only at level 2, with the main focus being on 2D and 3D collaboration [36]. The researchers revealed that "lack of faith in the integrity of BIM" and "lack of client demand" among other barriers were one of the factors behind the lack of BIM implementation in Australia [37]. Many researchers in Germany, the United Kingdom, Canada, the United States, Denmark, France, China, Brazil, South Korea, and the Middle East [38–46] have also found numerous BIM implementation barriers. It includes lack of BIM experience, investment costs, lack of awareness, lack of specified standards, market and cultural changes, interoperability problems, lack of specific guidelines for BIM implementation, and habitual resistance to change [31,44,47]. However, it has been found that the construction industry in the United States has engaged BIM in its ventures relative to other industries across the globe [48]. The government's BIM initiative became compulsory in projects in the public sector, beginning in 2016 in the UK [43]. It is therefore noticed that implementation of the BIM in different countries is growing for effective and productive construction [49]. However, there is a need to explore BIM in sustainable building projects for the sake of an eco-friendly environment and prompt sustainable goals.

3.2. BIM Implementation Barriers in Malaysian Context

In Malaysia, BIM implementation is still in the developing stage. However, the construction industry in Malaysia has taken major steps to encourage and boost construction efficiency at the national level with the introduction of BIM [50,51]. In addition, BIM is mandatory for public projects with a budget of RM 100 million or more from 2018 [52]. According to Datuk Seri Dr Roslan Md Taha (PWD Director-General), a total of 18 construction projects have been initiated by BIM in various construction phases up to 2017 such as SMK Meru Raya Ipoh Perak, Health Clinic Maran Pahang, MACC Selangor Shah Alam, UTHM Batu Pahat Johor and Parit Buntar Hospital [53]. With the continuous implementation of BIM against sustainable building projects in Malaysia, the implementation of BIM still faces numerous barriers and BIM implementation in sustainable building projects has many concerns and challenges. Hence, there is a need to better understand barriers in order to make the construction smoother and more effective for boosting sustainability goals. Various researchers have also highlighted the barriers to BIM implementation. The most critical barrier which was identified by many researchers is high cost [54–57]. Likewise, in Hong Kong and China, high cost is also considered as the BIM implementation barrier [47,58]. Moreover, it was also described as the topmost critical barrier in the Middle East [59]. As a result, high costs are included as one of the barriers since they are commonly available in the literature. In the last year, another study was conducted to explore the level of adoption of BIM in Malaysia. This study found that only 13% of government and private participants use BIM in their organization, which is negative evidence that Malaysia remains a long way from the role it should play in implementing BIM [60]. Therefore, it is recommended to implement BIM by eliminating barriers and providing the strategies for sustainable building projects.

4. Research Methodology

In this section, a brief explanation of the research methodology has been discussed. The research methodology section comprises data collection and data analysis. In addition, the research design flowchart, as shown in Figure 1, comprises four phases to fulfill the aim and objectives of this study. In the first step, the research aim and objectives were formulated by evaluating the relevant literature and identifying research gaps in previous research performed by researchers. The second phase of the study consists of collecting data from the literature review by defining the implementation barriers of the BIM and developing a questionnaire survey to be distributed among stakeholders (contractors, clients, and consultants). In the third step, the triangulation approach was introduced to achieve the study goal and the study objectives. The method of triangulation consists of two forms of analysis, i.e., quantitative analysis and qualitative analysis. According to Altricher et al. [61], triangulation "gives the more detailed and fair image of the circumstances". In quantitative analysis, the SPSS software package was used to assess the feedback of the respondents. The SPSS software was used to analyze the values of Cronbach's alpha coefficient, Kendall's coefficient of concordance (Kendall's W), Chi-square analysis, and mean score analysis, while NVivo software was used in qualitative analysis to analyze the data collected from interviews. In the final and fourth phase of the study, the recommendation and conclusion were drawn from the analysis results.

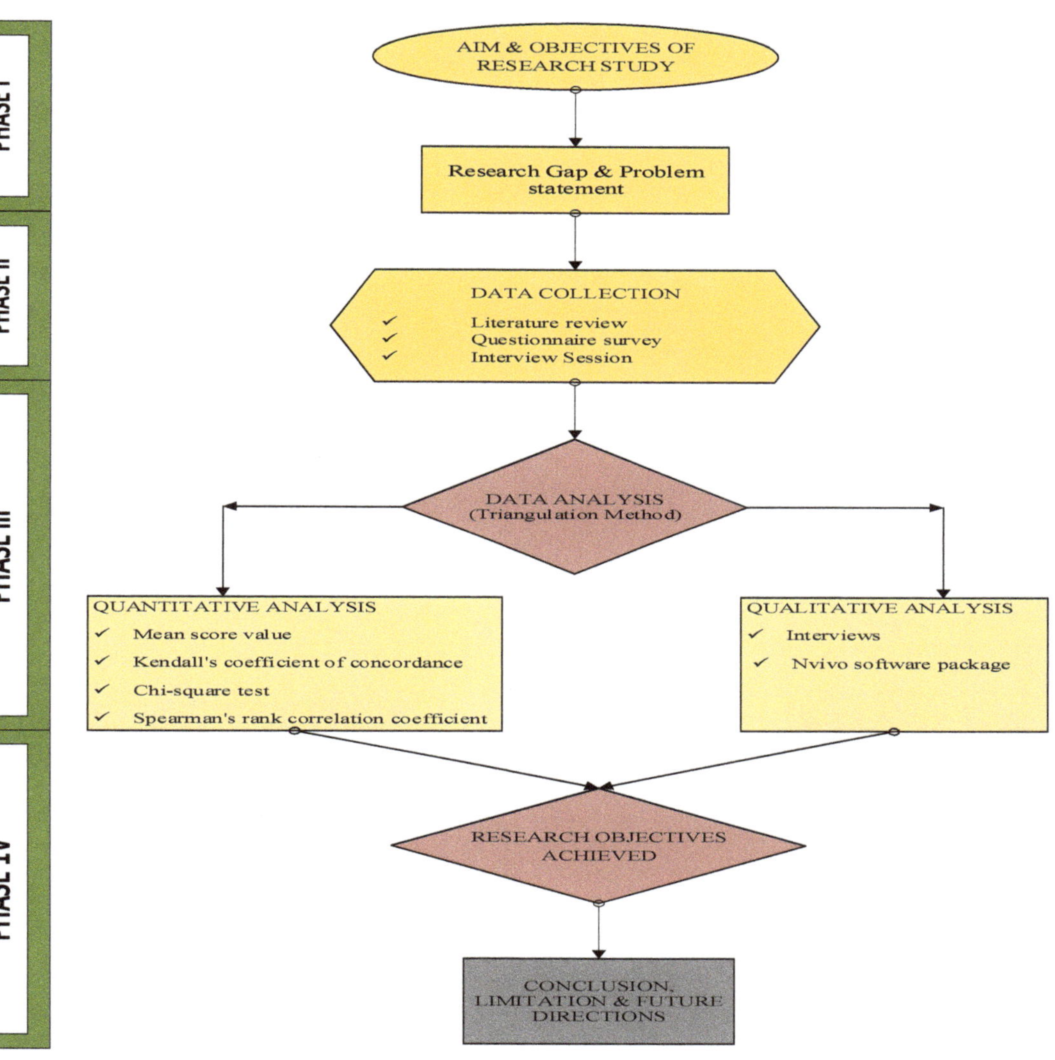

Figure 1. Research design flowchart.

4.1. Data Collection

After establishing the aim and objectives of the study in the introduction section, phase 2 of the study was conducted, which consisted of literature review, questionnaire survey, and interview session.

4.1.1. Literature Search Parameters

The purpose of the literature review was to identify barriers with various databases, such as Scopus, Web of Science, Google Scholar, and other relevant publications such as Elsevier, American Society of Civil Engineers (ASCE), Emerald and Taylor and Francis. Keywords such as "building information modeling", "construction industry", "construction projects", and "barriers" were used. Subsequently, the listed publications were screened, primarily by concentrating on the title, abstract, and conclusions, as well as the figures and tables. Table 1 elaborates the BIM implementation barriers that were used in this research study based on the existing literature.

Table 1. Building information modeling (BIM) implementation barriers.

BIM Implementation Barriers	References
High cost	[54–57]
Lack of expertise	[58,62–64]
Inadequate government policies	[47,65,66]
Lack of clients demand	[62,67]
Poor collaboration among stakeholders	[56,68,69]
Lack of vision of benefits	[4,38,70]
Unavailability of standards and guidelines Lack of BIM training	[18,58,71–73]
Lack of promotion	[44,74,75]
Lack of initiative and hesitance	[76–78]
Incompatibility and interoperability problems	[79–81]
Lack of industry standards	[71,82,83]
Competing initiatives	[84–86]
Lack of well-developed practical strategies	[77,87,88]
Licensing issues	[30,82,89–91]
Security issues	[82,92–94]
Insufficient external motivation (insufficient customer and market demands)	[95–98]
Misunderstanding of BIM	[99–102]
Lack of research and BIM implementation	[60,103,104]
Cultural barrier (resistance to change)	[63,105–107]

After rigorous review of current literature relating to BIM implementation barriers, various factors were identified. Table 1 elaborates a list of 20 factors that are well documented and hence more applicable. Prior to identification of BIM implementation barriers, it is kept in mind that the selection of well-known factors is more reliable and also easy for the respondents to understand the theme and feedback easily.

Previous studies were used to identify the aforementioned BIM implementation barriers. Following a thorough evaluation of these studies, this study identified 20 potential barriers to BIM implementation, which are presented in Table 1. For example, cost, lack of expertise, and lack of promotion are commonly acknowledged in the literature as crucial barriers to BIM implementation.

4.1.2. Questionnaire Survey

In this section, questionnaire design and responses were examined. The questionnaire survey method was used to collect the data through a quantitative approach. Therefore, the questionnaire was designed after careful consideration of existing literature in the domain of BIM implementation barriers. Prior to the final questionnaire survey, a semi-structured interview was arranged to ensure the potential and appropriateness of the questionnaire in the context of BIM implementation barriers. In the pilot study, the feedback from five participants was used to further improve the questions prior to the actual questionnaire survey. The pilot survey consisted of two professors, two assistant professors, and one postgraduate researcher. The questionnaire was finalized based on feedback from the pilot study. The final questionnaire consists of three parts: (a) research aim and objectives, (b) general information of respondents, (c) rank of the barriers using a Likert scale (1—Strongly Disagree, 2—Disagree, 3—Moderately Agree, 4—Agree, 5—Strongly Agree). A total of 300 questionnaires were sent via email, and 185 feedbacks were received, yielding a response rate of 61%. In addition, the sample size greater than 30 is reliable for further statistical analysis based on central limit theorem [108,109]. Figure 2 elaborates the breakdown of 185 returned questionnaires consisting of positions, working experience, education, and company type.

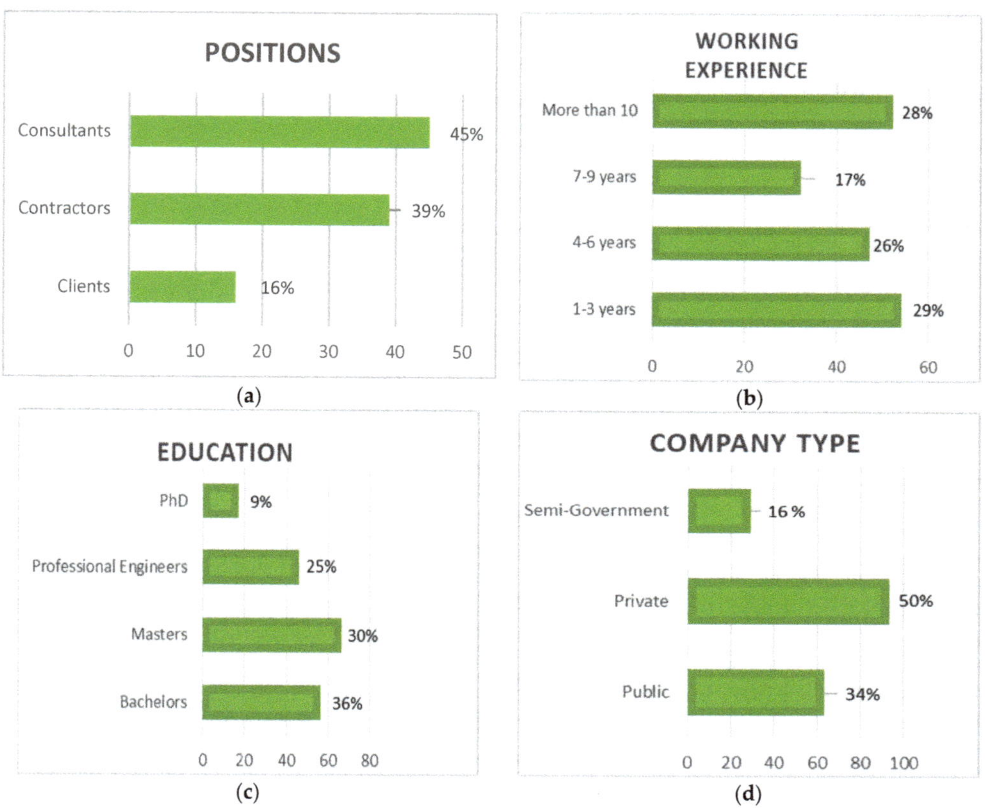

Figure 2. Breakdown of 185 returned questionnaires by Positions, Working Experience, Education, and Company Type (**a**–**d**).

Figure 2 depicts the positions of respondents, which include consultants, contractors, and clients. Consultants accounted for 45% of the total, contractors accounted for 39%, and clients accounted for 16%. Furthermore, 28% had more than ten years of work experience. Similarly, 36% of respondents were Bachelor's degree holders, 30% of respondents were Master's degree holders, 25% of respondents were professional engineers, and 9% of respondents were PhD holders. Furthermore, 16% of respondents worked for a semi-government organization, 50% of respondents worked for a private organization and 34% of respondents worked for a public organization.

4.1.3. Interview Session

In this section, interview questions design and responses were examined. The method of interviewing was used to gather data using a qualitative approach.

Due to the covid-19 pandemic, the online interviews were conducted using ZOOM software. Prior to conducting an online interview with ZOOM software, a semi-structured interview was created to ensure that all information related to barriers in construction projects could be obtained. In addition, previous studies have shown that the sample size greater than fifteen is effective and accurate for qualitative analysis [110]. The sample size of twenty for this qualitative analysis was therefore deemed to be appropriate. Twenty BIM experts with experience working in academia, industry, and construction projects were selected as participants in this qualitative analysis. Before the interview started, participants received a short introduction to the study and were informed of the anonymity and confidentiality of the information collected. Before the interviews started, demographic information such as gender, educational qualifications, and job experience was collected from interviewees. Interviewees were interviewed on the basis of the intended interview questions and after each answer, additional follow-up questions were asked as needed. The criteria for selection of participants depends upon: (a) possess working experience more than five years, (b) having excellent educational qualifications, (c) possess strong knowledge of BIM in construction projects. The demographic detail of participants is shown in Table 2.

Table 2. Participant demographics ($n = 20$).

Item	Description	Number of Participants	Percentage (%)
Gender	Male	20	100
	Female	-	-
Educational qualification	Bachelors	3	15
	Masters	5	25
	PhD	6	30
	Professional Engineers	6	30
Working experience	5–7 years	4	20
	8–9 years	9	45
	More than 10 years	7	35

It was seen that 100% of the participants were male, while none were female. Similarly, 15% of the participants were Bachelor's degree holders, while 25%, 30%, and 30% of the participants were Master's degree holders, PhD holders, and professional engineers, respectively. In addition, 35% of the participants had career experience of more than 10 years. The demographic analysis indicates that they had broad working experience, qualifications, and skills. Thus, they were considered as suitable interviewees for this research study.

4.2. Data Analysis

The third stage of study was data analysis. For this purpose, two types of data analysis were elaborated: (a) qualitative analysis via questionnaire survey and (b) qualitative analysis via interview.

4.2.1. Quantitative Analysis

Cronbach's alpha coefficient is used to calculate the internal consistency of the different variables in order to determine the strength of the five-point scales. The value for Cronbach's alpha coefficient in this study is 0.876, which is greater than the threshold of 0.7, which suggests that the data are reliable for further statistical analysis [111]. Furthermore, the mean value technique is used to determine the relative importance of individual barriers. The mean values of individual barriers are computed, ranked, and compared between the three groups (contractors, consultants, clients). Mean value analysis is a technique used to effectively identify key factors among various factors [112]. In addition, Kendall's coefficient of concordance (Kendall's W) is also calculated to measure the agreement of responses in particular groups [113]. The range of the value of Kendall's coefficient of concordance (W) is from 0 to 1. The higher value of W indicates the high level of consensus among the respondents within the group [114,115]. In addition, chi-square analysis should be carried out, if the number of items is greater than seven [116]. In addition, the Spearman's rank correlation coefficient was introduced to calculate the strength of a relationship between two groups [117]. The range of the Spearman's rank correlation coefficient (r_s) is from -1 to +1. The higher the positive/negative value of r_s, the stronger positive/negative linear correlation [90].

4.2.2. Qualitative Analysis

For analysis of the qualitative study, NVivo 11 was used as one of the available software packages [118]. Using software packages such as NVivo improves the degree of deeper understanding and makes qualitative data analysis quicker and more flexible [119]. In qualitative analysis, coding is considered to be an integrated part of the analysis. The theme of the interview is further summarized through coding for a better analysis of the idea put forward by the participants. The ideas of the interviewees were further listed in 200 codes for qualitative analysis. For instance, the "government must establish a BIM cell unit and allocate a defined budget for BIM implementation." This theme and concept were broken down into "government policies". The details of the coding theme are shown in Table 3.

Table 3. Coding theme and appropriate definition.

Categories (n = 200)	Subcategories	Definition
Development BIM guidelines (49)	• Positive attitude • Appropriate learning environment	Favor in developing BIM guidelines.
Enhance BIM seminars/course/workshops (27)	• Compulsion • Appreciation • Professional bodies	Made compulsion to adopt BIM seminars.
Hiring BIM experts (26)	• Awareness • Understanding	To develop the know-how of BIM implementation, hire BIM experts.
Budget allocation (44)	• Government policies	Government should enhance the budget allocation to BIM organizations and establish BIM cell.
BIM and academic curriculum (54)	• Collaboration	To overcome the lack of research, adopt more collaboration and exchange idea with different countries.

It was shown that interviewees emphasized the enhancement of BIM integrated with academic curricula to overcome the lack of BIM research and implementation. Likewise, a positive attitude and learning environment together with the creation of BIM guidelines are essential for grooming BIM implementation. Furthermore, interviewees suggested that the professional bodies should conduct BIM workshops, courses, and seminars to convey the benefits of using BIM in construction projects. Stakeholders should take part in workshops, seminars, and courses to grow the concept of incorporating BIM in their construction projects. Moreover, interviewees facilitated the development of know-how of BIM implementation, the dissemination of BIM information, and the development of awareness and understanding. Interviewees indicated that the government should assign the budget to BIM and set up a separate BIM-based allocation cell to track the proper use of the allocation budget. However, cooperation with other universities and countries is necessary in order to address the lack of study in the implementation of BIM and in organizations.

5. Results and Discussion

The follow-up section explores in detail the results of SPSS statistical package, which consists of mean value analysis, Kendall's coefficient of concordance (Kendall's W), chi-square test, and Spearman's rank correlation coefficient, and are summarized in the form of ranking.

5.1. Ranking of BIM Implementation Barriers

The ranking of BIM implementation barriers is categorized into four parts with the data analysis results using mean value: (a) overall ranking of BIM implementation barriers, (b) ranking according to contractors' perspectives, (c) ranking according to consultants' perspectives, and (d) ranking according to clients' perspectives. Table 4 illustrates the detailed picture of mean value analysis.

Table 4. Ranking of BIM implementation barriers.

BIM Implementation Barriers	Overall Respondents		Contractors' Perspectives		Consultants' Perspectives		Clients' Perspectives	
	Mean Value	Rank	Mean Value	Rank	Mean Value	Rank	Mean Value	Rank
Unavailability of standards and guidelines	4.89	1	4.02	4	3.85	5	4.27	5
Lack of BIM training	4.78	2	3.89	5	4.00	4	4.60	3
Lack of expertise	4.52	3	4.65	1	4.32	1	4.71	2
High cost	4.45	4	4.54	2	4.10	3	4.07	6
Lack of research and BIM implementation	4.03	5	4.21	3	3.41	7	4.83	1
Cultural barrier (resistance to change)	3.89	6	3.61	7	4.22	2	4.42	4
Lack of promotion	3.62	7	3.61	7	2.88	10	3.43	10
Incompatibility and interoperability problems	3.43	8	3.02	11	2.88	10	3.23	11
Insufficient external motivation (insufficient customer and market demands)	3.22	9	3.44	9	2.65	12	3.23	11
Lack of well-developed practical strategies	3.09	10	2.83	13	3.01	9	3.01	13
Security issues	2.98	11	2.61	14	3.70	6	2.98	14
Lack of clients demand	2.82	12	2.90	11	1.95	16	2.87	15
Lack of vision of benefits	2.67	13	3.23	10	3.20	9	2.87	15
Poor collaboration among stakeholders	2.44	14	3.73	6	2.21	15	3.88	7
Lack of initiative and hesitance	2.32	15	2.01	17	2.42	13	3.88	7
Misunderstanding of BIM	2.27	16	1.92	18	2.42	13	3.88	7
Inadequate government policies	2.18	17	1.88	19	1.65	19	1.90	20
Licensing issues	2.13	18	1.88	19	1.72	17	2.01	19
Lack of industry standards	1.98	19	2.32	16	1.72	17	2.45	18
Competing initiatives	1.92	20	2.43	15	1.54	20	2.61	17

Table 4 indicates that "unavailability of standards and guidelines" had the highest mean value of M = 4.89, while "competing initiatives" had the lowest mean value of M = 1.92. Likewise, the top five barriers were "unavailability of standards and guidelines", with M = 4.89, "lack of BIM training", with M = 4.78, "lack of expertise" with M = 4.52, "high cost", with M = 4.45, and "lack of research and BIM implementation", with M = 4.03.

5.2. *Kendall's Coefficient of Concordance (Kendall's W)*

The value of Kendall's coefficient of concordance (Kendall's W) of overall respondents, contractors' perspectives, consultants' perspectives, and clients' perspectives were 0.183, 0.175, 0.169, and 0.145, respectively. The null hypothesis was rejected because all significance levels were 0.000, which was less than the threshold level of 5%. As a consequence, a substantial level was found among the respondents.

5.3. *Chi-Square Test*

The chi-square test was conducted as there were 20 factors in the analysis. The chi-square test should be performed if there are more than seven factors in the study, i.e., the chi-square test values for contactors' perspectives, consultants' perspectives, and clients' perspectives were 68,432, 92,569, and 123,324, respectively. According to the analytical findings, the groups of respondents had significant levels dependent on each other in each group.

5.4. *Spearman's Rank Correlation Coefficient*

Spearman's rank correlation coefficient (r_s) is calculated by the following equation:

$$r_s = 1 - \left[6 \sum_{n=1}^{5} \frac{\left(d^2\right)}{n(n^2 - 1)} \right]$$

where r_s = Spearman's rank correlation coefficient; d = the difference between ranks assigned to items; n = the number of respondents.

Using Spearman's rank correlation coefficient, the relationship between the perspective of clients, consultants, and contractors with the implementation barriers was shown. Spearman's rank correlation coefficients between all parties were established. The coefficient value between the consultant and contractor was 0.932. The correlation between the client and consultant was 0.892, whereas that between the client and contractor was 0.821. In addition, there was also a significant relationship among the three parties, namely 0.000, 0.000, and 0.005, respectively, which was smaller than the permissible amount of significance (5%). In other words, it can be said that there was a strong correlation between consultant and contractor, client and consultant, and client and contractor.

5.5. Results of Factor Analysis

Factor analysis was carried out to further explore the barriers of BIM implementation in this study. The KMO value of this study was 0.512, which is acceptable as it satisfies the threshold of 0.50. Values below 0.50 should lead the researcher "to either collect more data or rethink which variables to include" [120]. These results are illustrated in Table 5. The detailed picture of factor analysis of barriers to BIM implementation in sustainable building projects was composed of five groups. These groups were Group 1—Government-related barriers, Group 2—Market-related barriers, Group 3—Personal-related barriers, Group 4—Construction environment-related barriers, and Group 5—Cost–risk barriers.

Table 5. Results of factor analysis (FA) on barriers to BIM implementation in sustainable building projects.

Variables (Barriers)	1	2	3	4	5	Grouping
Inadequate government policies	0.987	-	-	-	-	Government-related barriers
Unavailability of standards and guidelines	0.820	-	-	-	-	
Lack of industry standards	0.712	-	-	-	-	
Lack of well-developed practical strategies	0.620	-	-	-	-	
Licensing issues	0.567	-	-	-	-	
Lack of expertise	-	0.834	-	-	-	Market-related barriers
Lack of clients demand	-	0.621	-	-	-	
Poor collaboration among stakeholders	-	0.601	-	-	-	
Lack of promotion	-	0.542	-	-	-	
Insufficient external motivation (insufficient customer and market demands)	-	0.521	-	-	-	
Lack of vision of benefits	-	-	0.821	-	-	Personal-related barriers
Lack of BIM training	-	-	0.743	-	-	
Misunderstanding of BIM	-	-	0.612	-	-	
Cultural barrier (resistance to change)	-	-	0.590	-	-	
Lack of initiative and hesitance	-	-	-	0.798	-	Construction Environment-related barriers
Incompatibility and interoperability problems	-	-	-	0.656	-	
Competing initiatives	-	-	-	0.532	-	
Lack of research and BIM implementation	-	-	-	0.432	-	
High cost	-	-	-	-	0.712	Cost-risk barriers
Security issues	-	-	-	-	0.677	

6. Results Comparison with Other Countries

The findings of the current study were contrasted with other countries like China, the United Kingdom, Nigeria, and Pakistan. In addition, Table 6 elaborates the comparison of BIM implementation barriers with other countries.

By contrasting Malaysia's current study with China, "unavailability of standards and guidelines" was ranked first, while in Nigeria, "unavailability of standards and guidelines" was ranked fourth. However, "unavailability of standards and guidelines" could not gain the attention of researchers in the United Kingdom and in Pakistan and thus was marked as not identified. The "lack of BIM training" was not ranked in the top five in the United Kingdom, Nigeria, or Pakistan. However, "lack of BIM training" was labeled as not identified in China. Furthermore, "lack of expertise" was ranked third in the current study in Malaysia as well as in the United Kingdom. The "lack of expertise" in Nigeria was perceived to be the most crucial barrier. Moreover, "high cost" was ranked fourth in the current study in both Malaysia and China, while in Nigeria, "high cost" was ranked third. In addition, "lack of research and BIM implementation" was the least consideration barrier in the United Kingdom, Nigeria, and Pakistan.

These findings provide evidence that more attention needs to be paid to BIM research and implementation in Pakistan and that "lack of research and implementation of BIM" is a crucial barrier that is largely unidentified by researchers. It is also understandable that variations in ranks are due to different cultures and environmental factors in different countries. From this analysis, the conclusion can be drawn that barriers are similar among different countries but rank distinctively.

Table 6. Comparison of BIM implementation barriers.

Top Five BIM Implementation Barriers	Malaysia (Current Study)	China [47]	United Kingdom [107]	Nigeria [74]	Pakistan [121]
Unavailability of standards And guidelines	(Rank 1)	(Rank 1)	Not Identified	(Rank 4)	Not Identified
Lack of BIM training	(Rank 2)	Not Identified	(Rank 7)	(Rank 15)	(Rank 12)
Lack of expertise	(Rank 3)	(Rank 5)	(Rank 3)	(Rank 1)	(Rank 7)
High cost	(Rank 4)	(Rank 4)	(Rank 3)	(Rank 11)	(Rank 9)
Lack of research and BIM implementation	(Rank 5)	(Rank 7)	Not Identified	Not Identified	Not Identified

7. BIM-Based Research Framework

In this section, strategies to mitigate the BIM implementation barriers, namely "unavailability of standards and guidelines", "lack of BIM training", "lack of expertise", "high cost", and "lack of research and BIM implementation" were developed with the aid of the BIM-based research framework. This framework consists of two layers, namely the outer layer and the inner layer. The inner layer consists of barriers, while the outer layer, called the "mitigation layer", consists of strategies to overcome barriers.

The strategy to address the "unavailability of standards and guidelines" focused on the proper development of the BIM guidance plan and a positive attitude towards the learning environment. In developing countries, there is a lack of sound guidelines for sustainable building projects, which leads to cost overruns, delays, and waste. There is a need to establish a BIM guidance plan that can encourage and enhance the efficiency of sustainable construction, such as that in developing countries. It is important for owners to draw up a BIM execution plan during the pre-operation process. In addition, the strategy to reduce the "lack of BIM training" barrier is to strengthen BIM courses (linking with academia), seminars, and workshops. It is the liability of the professional bodies to control and track the essential eyes of construction stakeholders and organizations. Training should begin by educating workers on the relevance and usefulness of technology. Employee information on BIM technological innovation can be measured by a written analysis, such as an exit survey. These conclusions are based on responses such as "I think that recruiting skilled workers can be the best option", "I think that professional bodies should track construction organizations", and "I think that BIM courses, seminars and workshops should be made compulsory for stakeholders". In addition, the approach to reduce the "lack of expertise" barrier is to attract BIM experts and generate possible ideas for the appreciation of the use of BIM technology. Hiring BIM experts from developed countries will help to mitigate the barrier. In order to ensure that adequate technical support and expertise is available from the technology partner, construction companies can obtain all relevant information from the manufacturer before promising to incorporate and use the BIM technology. These conclusions are based on responses such as "I think the appropriate approach is to recruit BIM experts from developing countries", "I think the know-how to use BIM needs to be disseminated". Furthermore, a strategy to reduce the barrier of "high cost" is to assign a specific budget for a BIM implementation cell. The government should set up a BIM cell to increase the trust and enthusiasm of building stakeholders. It is the utmost duty of the BIM cell unit to adequately track and monitor the allocation budget and prepare the report by the end of each month. In order to enable construction stakeholders to use BIM technology, vendors should consider modifying their business models to reduce the initial costs of using these technologies. A subscription-based model or monthly payment plan over a fixed period may be a preferred model, as construction stakeholders may adjust periodic payments to match their usual billing cycle—shifting most upstream costs. However, daily payments should be reasonable. This is supported by responses such as "I think the government should allocate a budget for the implementation of BIM", "I think the government should adjust its policies", and "I think there should be a BIM cell unit". Furthermore, BIM-related topics have been included in the curriculum of civil and architectural engineering. Collaboration between universities and the promotion of a research culture will improve the implementation of the BIM. It is the government's duty to expand opportunities for students and faculty members through the development of scholarship programs, as indicated by responses such as "I think by cooperation", "I think fostering understanding of BIM technology", and "I think student exchange programs between universities in developing countries". In addition, a detailed picture of the BIM-based research framework is elaborated in Figure 3.

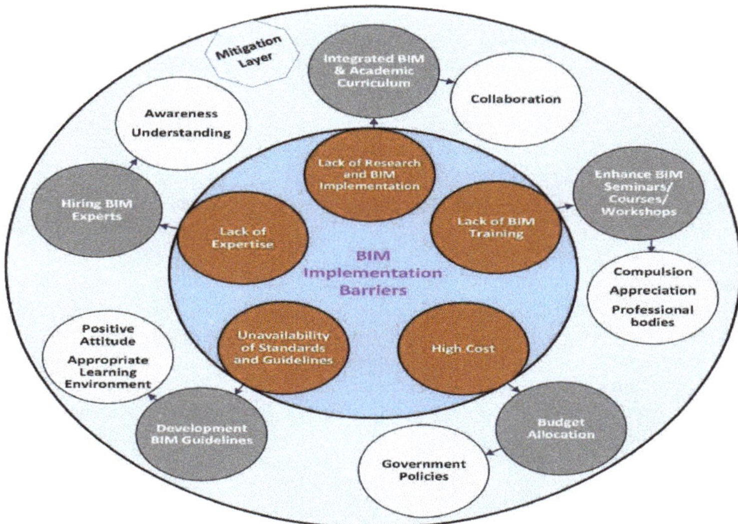

Figure 3. BIM-based research framework incorporating barriers and strategies for sustainable building projects.

8. Conclusions, Limitation, and Future Directions

BIM has the ability to enhance and ease construction, but there are various barriers that hinder the effectiveness of sustainable building projects. These barriers need to be tackled in order to boost successful and efficient construction. In addition, a comprehensive literature review was undertaken to illustrate the barriers in order to achieve the objectives of this study. Twenty barriers were established with a thorough analysis and proceeded to the questionnaire survey. In this study, the triangulation method consisting of quantitative analysis and qualitative analysis was adopted. In quantitative analysis, the top most barriers were "unavailability of standards and guidelines", with M = 4.89, "lack of BIM training", with M = 4.78, "lack of expertise", with M = 4.52, "high cost", with M = 4.45, and "lack of research and BIM implementation", with M = 4.03, whereas, "competing initiatives" had the lowest mean value of M = 1.92. In qualitative analysis, strategies to mitigate barriers were explored with the help of interviews of BIM experts in academia and in sustainable building projects. It was revealed that BIM workshops, courses, and seminars could enhance and promote the culture of BIM implementation. In addition, collaboration and budget allocation could increase the incentive and trust level of research to address the lack of research and implementation of BIM.

The theoretical contribution of this study is not only to fill the research gap but also to provide a valuable reference for helping stakeholders to mitigate the barriers. As far as the authors' knowledge, the novelty of this study is that there is no comprehensive study conducted in Malaysia to explore strategies in mitigating BIM implementation barriers for sustainable building projects. For practical implications, this study suggests that the BIM-based research framework be executed with ongoing real projects to enable stakeholders to complete sustainable construction effectively. This study also recommends that researchers create slightly different frameworks on the basis of the same collection of quantitative and qualitative data. While the aim and objectives of this research were accomplished, this study still has some limitations that are worth noting. First, this study was conducted in Malaysia, and therefore the findings in this study might not be applicable in other countries because of cultural differences. Second, in the qualitative study, the sample size could be expanded in order to gain further insight into the reduction of BIM implementation barriers. Finally, these limitations offer a route for potential researchers to verify this research study through case studies of successful construction projects.

Author Contributions: B.M.: Conceptualization, investigation, data curation, writing—original draft. I.O.; supervision, writing—review and editing. S.S.S.G., H.A. and S.B.A.; resources, review and editing. All authors have read and agreed to the published version of the manuscript.

Funding: This research received no external funding.

Institutional Review Board Statement: Not applicable.

Informed Consent Statement: Not applicable.

Data Availability Statement: Data sharing not applicable.

Conflicts of Interest: The authors declare no conflict of interest.

References

1. Mok, K.Y.; Shen, G.Q.; Yang, J. Stakeholder management studies in mega construction projects: A review and future directions. *Int. J. Proj. Manag.* **2015**, *33*, 446–457. [CrossRef]
2. Chia, F.C.; Skitmore, M.; Runeson, G.; Bridge, A. Economic development and construction productivity in Malaysia. *Constr. Manag. Econ.* **2014**, *32*, 874–887. [CrossRef]
3. Rahman, I.A.; Memon, A.H.; Nagapan, S.; Latif, Q.B.A.I.; Azis, A.A.A. Time and cost performance of construction projects in southern and central regions of Peninsular Malaysia. In *2012 IEEE Colloquium on Humanities, Science and Engineering (CHUSER)*; IEEE: Peninsular, Malaysia, 2012; pp. 52–57.
4. Memon, A.H.; Rahman, I.A.; Memon, I.; Azman, N.I.A. BIM in Malaysian construction industry: Status, advantages, barriers and strategies to enhance the implementation level. *Res. J. Appl. Sci. Eng. Technol.* **2014**, *8*, 606–614. [CrossRef]
5. Cheng, J.C.P.; Won, J.; Das, M. Construction and demolition waste management using BIM technology. In Proceedings of the 23rd Annual Conference of the International Group for Lean Construction, Perth, Australia, 29–31 July 2015; pp. 381–390.
6. Hwang, B.-G.; Zhao, X.; Yang, K.W. Effect of BIM on rework in construction projects in Singapore: Status quo, magnitude, impact, and strategies. *J. Constr. Eng. Manag.* **2019**, *145*, 4018125. [CrossRef]
7. Ham, N.; Moon, S.; Kim, J.-H.; Kim, J.-J. Economic analysis of design errors in BIM-based high-rise construction projects: Case study of Haeundae L project. *J. Constr. Eng. Manag.* **2018**, *144*, 5018006. [CrossRef]
8. Oraee, M.; Hosseini, M.R.; Edwards, D.J.; Li, H.; Papadonikolaki, E.; Cao, D. Collaboration barriers in BIM-based construction networks: A conceptual model. *Int. J. Proj. Manag.* **2019**, *37*, 839–854. [CrossRef]
9. Oraee, M.; Hosseini, M.R.; Papadonikolaki, E.; Palliyaguru, R.; Arashpour, M. Collaboration in BIM-based construction networks: A bibliometric-qualitative literature review. *Int. J. Proj. Manag.* **2017**, *35*, 1288–1301. [CrossRef]
10. Doumbouya, L.; Gao, G.; Guan, C. Adoption of the Building Information Modeling (BIM) for construction project effectiveness: The review of BIM benefits. *Am. J. Civ. Eng. Archit.* **2016**, *4*, 74–79.
11. Latiffi, A.A.; Mohd, S.; Kasim, N.; Fathi, M.S. Building Information Modeling (BIM) Application in Malaysian Construction Industry. *Int. J. Constr. Eng. Manag.* **2013**, *2*, 1–6. [CrossRef]
12. Latiffi, A.A.; Brahim, J.; Fathi, M.S. Transformation of Malaysian construction industry with building information modelling (BIM). In *MATEC Web of Conferences*; EDP Sciences: Kuala Lumpur, Malaysia, 2016; p. 22.
13. Lee, G.; Sacks, R.; Eastman, C.M. Specifying parametric building object behavior (BOB) for a building information modeling system. *Autom. Constr.* **2006**, *15*, 758–776. [CrossRef]
14. Qureshi, A.H.; Alaloul, W.S.; Manzoor, B.; Musarat, M.A.; Saad, S.; Ammad, S. Implications of Machine Learning Integrated Technologies for Construction Progress Detection Under Industry 4.0 (IR 4.0). In Proceedings of the 2020 Second International Sustainability and Resilience Conference: Technology and Innovation in Building Designs, Sakheer, Bahrain, 11–12 November 2020.
15. Ortiz, O.; Castells, F.; Sonnemann, G. Sustainability in the construction industry: A review of recent developments based on LCA. *Constr. Build. Mater.* **2009**, *23*, 28–39. [CrossRef]
16. Lützkendorf, T.; Lorenz, D. Sustainable property investment: Valuing sustainable buildings through property performance assessment. *Build. Res. Inf.* **2005**, *33*, 212–234. [CrossRef]

17. Hong, J.; Shen, G.Q.; Li, Z.; Zhang, B.; Zhang, W. Barriers to promoting prefabricated construction in China: A cost–Benefit analysis. *J. Clean. Prod.* **2018**, *172*, 649–660. [CrossRef]
18. Zahrizan, Z.; Ali, N.M.; Haron, A.T.; Marshall-Ponting, A.; Abd, Z. Exploring the adoption of Building Information Modelling (BIM) in the Malaysian construction industry: A qualitative approach. *Int. J. Res. Eng. Technol.* **2013**, *2*, 384–395.
19. Hamid, A.B.A.; Taib, M.Z.M.; Razak, A.H.N.A.; Embi, M.R. Building information modelling: Challenges and barriers in implement of BIM for interior design industry in Malaysia. In *IOP Conference Series: Earth and Environmental Science*; IOP Publishing: Langkawi Malaysia, 2018; p. 12002.
20. Wong, J.H.; Rashidi, A.; Arashpour, M. Evaluating the Impact of Building Information Modeling on the Labor Productivity of Construction Projects in Malaysia. *Buildings* **2020**, *10*, 66. [CrossRef]
21. van Nederveen, S.; Beheshti, R.; Gielingh, W. Modelling concepts for BIM. In *Handbook of Research on Building Information Modeling and Construction Informatics: Concepts and Technologies*; IGI Global: Delft, The Netherlands, 2010. [CrossRef]
22. Latiffi, A.A.; Brahim, J.; Fathi, M.S. The development of building information modeling (BIM) definition. In *Applied Mechanics and Materials*; Trans Tech Publications Ltd.: Nilai, Malaysia, 2014; pp. 625–630.
23. Heermann, D.W. Computer-simulation methods. In *Computer Simulation Methods in Theoretical Physics*; Springer: Berlin/Heidelberg, Germany, 1990; pp. 8–12. [CrossRef]
24. Weisberg, D.E. The engineering design revolution: The people, companies and computer systems that changed forever the practice of engineering. *Cyon Res. Corp.* **2008**, *12*, 1–26.
25. Becerik-Gerber, B.; Jazizadeh, F.; Li, N.; Calis, G. Application areas and data requirements for BIM-enabled facilities management. *J. Constr. Eng. Manag.* **2012**, *138*, 431–442. [CrossRef]
26. EGarcia, G.; Zhu, Z. Interoperability from building design to building energy modeling. *J. Build. Eng.* **2015**, *1*, 33–41.
27. Hannus, M. FIATECH Capital Projects Technology Roadmap, ECTP FA PICT. 2007. Available online: http//www.fiatech.org/Tech-Roadmap (accessed on 25 May 2012).
28. Succar, B. Building information modelling framework: A research and delivery foundation for industry stakeholders. *Autom. Constr.* **2009**, *18*, 357–375. [CrossRef]
29. Chan, C.T.W. Barriers of implementing BIM in construction industry from the designers' perspective: A Hong Kong experience. *J. Syst. Manag. Sci.* **2014**, *4*, 24–40.
30. Liu, S.; Xie, B.; Tivendal, L.; Liu, C. Critical barriers to BIM implementation in the AEC industry. *Int. J. Mark. Stud.* **2015**, *7*, 162. [CrossRef]
31. Chan, D.W.M.; Olawumi, T.O.; Ho, A.M.L. Perceived benefits of and barriers to Building Information Modelling (BIM) implementation in construction: The case of Hong Kong. *J. Build. Eng.* **2019**, *25*, 100764. [CrossRef]
32. Olawumi, T.O.; Chan, D.W.M. Identifying and prioritizing the benefits of integrating BIM and sustainability practices in construction projects: A Delphi survey of international experts. *Sustain. Cities Soc.* **2018**, *40*, 16–27. [CrossRef]
33. Sepasgozar, S.M.E.; Hui, F.K.P.; Shirowzhan, S.; Foroozanfar, M.; Yang, L.; Aye, L. Lean Practices Using Building Information Modeling (BIM) and Digital Twinning for Sustainable Construction. *Sustainability* **2021**, *13*, 161. [CrossRef]
34. Olawumi, T.O.; Chan, D.W.M.; Wong, J.K.W. Evolution in the intellectual structure of BIM research: A bibliometric analysis. *J. Civ. Eng. Manag.* **2017**, *23*, 1060–1081. [CrossRef]
35. Wong, J.K.W.; Zhou, J. Enhancing environmental sustainability over building life cycles through green BIM: A review. *Autom. Constr.* **2015**, *57*, 156–165. [CrossRef]
36. von Both, P. Potentials and barriers for implementing BIM in the German AEC market: Results of a current market analysis. In Proceedings of the 30th eCAADe Conference, Prague, Czech Republic, 12–14 September 2012; pp. 151–158.
37. Aibinu, A.; Venkatesh, S. Status of BIM adoption and the BIM experience of cost consultants in Australia. *J. Prof. Issues Eng. Educ. Pract.* **2014**, *140*, 4013021. [CrossRef]
38. Arayici, Y.; Khosrowshahi, F.; Ponting, A.M.; Mihindu, S.A. Towards implementation of building information modelling in the construction industry. In Proceedings of the Fifth International Conference on Construction in the 21st Century: Collaboration and Integration in Engineering, Management and Technology, Istanbul, Turkey, 20–22 May 2009; pp. 1342–1351.
39. Porwal, A.; Hewage, K.N. Building Information Modeling (BIM) partnering framework for public construction projects. *Autom. Constr.* **2013**, *31*, 204–214. [CrossRef]
40. Cao, Y.; Zhang, L.H.; McCabe, B.; Shahi, A. The Benefits of and Barriers to BIM Adoption in Canada. In Proceedings of the 36th International Association for Automation and Robotics in Construction (ISARC), Banff, AB, Canada, 21–24 May 2019; pp. 152–158.
41. Ku, K.; Taiebat, M. BIM experiences and expectations: The constructors' perspective. *Int. J. Constr. Educ. Res.* **2011**, *7*, 175–197. [CrossRef]
42. Jensen, P.A.; Jóhannesson, E.I. Building information modelling in Denmark and Iceland. *Eng. Constr. Archit. Manag.* **2013**, *20*, 99–110. [CrossRef]
43. Davies, R.; Crespin-Mazet, F.; Linne, A.; Pardo, C.; Havenvid, M.I.; Harty, C.; Ivory, C.; Salle, R. BIM in Europe: Innovation Networks in the Construction Sectors of Sweden France and the UK. In Proceedings of the 31st Annual ARCOM Conference, Lincoln, UK, 7–9 September 2015; pp. 1135–1144.
44. Tan, T.; Chen, K.; Xue, F.; Lu, W. Barriers to Building Information Modeling (BIM) implementation in China's prefabricated construction: An interpretive structural modeling (ISM) approach. *J. Clean. Prod.* **2019**, *219*, 949–959. [CrossRef]

45. Obi, L.; Awuzie, B.; Obi, C.; Omotayo, T.S.; Oke, A.; Osobajo, O. BIM for Deconstruction: An Interpretive Structural Model of Factors Influencing Implementation. *Buildings* **2021**, *11*, 227. [CrossRef]
46. Al-Yami, A.; Sanni-Anibire, M.O. BIM in the Saudi Arabian construction industry: State of the art, benefit and barriers. *Int. J. Build. Pathol. Adapt.* **2019**, *39*, 33–47. [CrossRef]
47. Zhou, Y.; Yang, Y.; Yang, J.-B. Barriers to BIM implementation strategies in China. *Eng. Constr. Archit. Manag.* **2019**, *26*, 554–574. [CrossRef]
48. Yan, H.; Demian, P. Benefits and Barriers of Building Information Modelling. In Proceedings of the 12th International Conference on Computing in Civil Building Engineering (ICCCBE XII), Beijing, China, 16–18 Octoner 2008.
49. Volk, R.; Stengel, J.; Schultmann, F. Building Information Modeling (BIM) for existing buildings—Literature review and future needs. *Autom. Constr.* **2014**, *38*, 109–127. [CrossRef]
50. Musa, S.; Marshall-Ponting, A.; Nifa, F.A.A.; Shahron, S.A. Building information modeling (BIM) in Malaysian construction industry: Benefits and future challenges. In *AIP Conference Proceedings*; AIP Publishing LLC: Melville, NY, USA, 2018; p. 20105. [CrossRef]
51. Latiffi, A.A.; Mohd, S.; Brahim, J. Application of building information modeling (BIM) in the Malaysian construction industry: A story of the first government project. In *Applied Mechanics and Materials*; Trans Tech Publications Ltd.: Bäch, Switzerland, 2015; pp. 996–1001. [CrossRef]
52. Hasni, M.I.A.K.; Ismail, Z.; Hashim, N.; Hassan, A.A. A Review of Building Information Modelling (BIM) Documents in the Malaysian Construction Industry: Public Works Department (PWD) and Construction Industry Development Board (CIDB). *Adv. Sci. Lett.* **2018**, *24*, 8913–8916. [CrossRef]
53. CIDB Malaysia. *Construction Industry Transformation Programme 2016–2020*; CIDB: Kuala Lumpur, Malaysia, 2015.
54. Wong, A.K.D.; Wong, F.K.W.; Nadeem, A. Attributes of building information modelling and its development in Hong Kong. *HKIE Trans.* **2009**, *16*, 38–45. [CrossRef]
55. Hong, Y.; Hammad, A.; Zhong, X.; Wang, B.; Akbarnezhad, A. Comparative modeling approach to capture the differences in BIM adoption decision-making process in Australia and China. *J. Constr. Eng. Manag.* **2020**, *146*, 4019099. [CrossRef]
56. Won, J.; Lee, G. Identifying the consideration factors for successful BIM projects. In Proceedings of the 17th International Workshop on Intelligent Computing in Engineering (EG-ICE 2010), Nottingham, UK, 12–13 July 2019.
57. Czmoch, I.; Pękala, A. Traditional design versus BIM based design. *Procedia Eng.* **2014**, *91*, 210–215. [CrossRef]
58. Ma, X.; Darko, A.; Chan, A.P.C.; Wang, R.; Zhang, B. An empirical analysis of barriers to building information modelling (BIM) implementation in construction projects: evidence from the Chinese context. *Int. J. Constr. Manag.* **2020**. [CrossRef]
59. Mehran, D. Exploring the Adoption of BIM in the UAE Construction Industry for AEC Firms. *Procedia Eng.* **2016**, *145*, 1110–1118. [CrossRef]
60. Othman, I.; Al-Ashmori, Y.Y.; Rahmawati, Y.; Amran, Y.H.M.; Al-Bared, M.A.M. The level of Building Information Modelling (BIM) Implementation in Malaysia. *Ain Shams Eng. J.* **2020**, *12*, 455–463. [CrossRef]
61. Altricher, H.; Feldman, A.; Posch, P.; Somekh, B. *Teachers Investigate Their Work: An Introduction to Action Research across the Professions*; Routledge: New York, NY, USA, 2005.
62. Wong, S.Y.; Gray, J. Barriers to implementing Building Information Modelling (BIM) in the Malaysian construction industry. In *IOP Conference Series: Materials Science and Engineering*; IOP Publishing: Bristol, UK, 2019.
63. Becerik-Gerber, B.; Kensek, K. Building information modeling in architecture, engineering, and construction: Emerging research directions and trends. *J. Prof. Issues Eng. Educ. Pract.* **2010**, *136*, 139–147. [CrossRef]
64. Kassem, M.; Brogden, T.; Dawood, N. BIM and 4D planning: A holistic study of the barriers and drivers to widespread adoption. *J. Constr. Eng. Proj. Manag.* **2012**, *2*, 1–10. [CrossRef]
65. Olanrewaju, O.I.; Chileshe, N.; Babarinde, S.A.; Sandanayake, M. Investigating the barriers to building information modeling (BIM) implementation within the Nigerian construction industry. *Eng. Constr. Archit. Manag.* **2020**, *27*, 2931–2958. [CrossRef]
66. Shafiq, M.T.; Afzal, M. Improving Construction Job Site Safety with Building Information Models: Opportunities and Barriers. In *International Conference on Computing in Civil and Building Engineering*; Springer: Cham, Switzerland, 2020; pp. 1014–1036.
67. Zahrizan, Z.; Ali, N.M.; Haron, A.T.; Marshall-Ponting, A.J.; Hamid, Z.A. Exploring the barriers and driving factors in implementing building information modelling (BIM) in the Malaysian construction industry: A preliminary study. *J. Inst. Eng. Malays.* **2014**, *75*, 1–10.
68. Harding, J.; Suresh, S.; Renukappa, S.; Mushatat, S. Do building information modelling applications benefit design teams in achieving BREEAM accreditation? *J. Constr. Eng.* **2014**, *2014*, 1–8. [CrossRef]
69. Hope, A.; Alwan, Z. Building the future: Integrating building information management and environmental assessment methodologies. In Proceedings of the First UK Academic Conference on BIM, Newcastle, UK, 5–7 September 2012.
70. Lee, S.-K.; Kim, K.-R.; Yu, J.-H. BIM and ontology-based approach for building cost estimation. *Autom. Constr.* **2014**, *41*, 96–105. [CrossRef]
71. Rogers, J.; Chong, H.-Y.; Preece, C. Adoption of building information modelling technology (BIM). *Eng. Constr. Archit. Manag.* **2015**, *22*, 424–445. [CrossRef]
72. Gerges, M.; Austin, S.; Mayouf, M.; Ahiakwo, O.; Jaeger, M.; Saad, A.; El Gohary, T. An investigation into the implementation of Building Information Modeling in the Middle East. *J. Inf. Technol. Constr.* **2017**, *22*, 1–15.

73. Gu, N.; London, K. Understanding and facilitating BIM adoption in the AEC industry. *Autom. Constr.* **2010**, *19*, 988–999. [CrossRef]
74. Babatunde, S.O.; Udeaja, C.; Adekunle, A.O. Barriers to BIM implementation and ways forward to improve its adoption in the Nigerian AEC firms. *Int. J. Build. Pathol. Adapt.* **2020**, *39*, 48–71. [CrossRef]
75. Li, H.; Wang, Y.; Yan, H.; Deng, Y. Barriers of BIM application in China—Preliminary research. In *ICCREM 2016 BIM Application and Off-Site Construction*; American Society of Civil Engineers: Reston, VA, USA, 2017; pp. 37–41.
76. Ding, Z.; Zuo, J.; Wu, J.; Wang, J.Y. Key factors for the BIM adoption by architects: A China study. *Eng. Constr. Archit. Manag.* **2015**, *22*, 732–748. [CrossRef]
77. Jin, R.; Hancock, C.M.; Tang, L.; Wanatowski, D. BIM investment, returns, and risks in China's AEC industries. *J. Constr. Eng. Manag.* **2017**, *143*, 4017089. [CrossRef]
78. Matthews, J.; Love, P.E.D.; Mewburn, J.; Stobaus, C.; Ramanayaka, C. Building information modelling in construction: Insights from collaboration and change management perspectives. *Prod. Plan. Control* **2018**, *29*, 202–216. [CrossRef]
79. Bradley, A.; Li, H.; Lark, R.; Dunn, S. BIM for infrastructure: An overall review and constructor perspective. *Autom. Constr.* **2016**, *71*, 139–152. [CrossRef]
80. Bui, N.; Merschbrock, C.; Munkvold, B.E. A review of Building Information Modelling for construction in developing countries. *Procedia Eng.* **2016**, *164*, 487–494. [CrossRef]
81. Arayici, Y.; Coates, P.; Koskela, L.; Kagioglou, M.; Usher, C.; O'Reilly, K. Technology adoption in the BIM implementation for lean architectural practice. *Autom. Constr.* **2011**, *20*, 189–195. [CrossRef]
82. Azhar, S.; Khalfan, M.; Maqsood, T. Building information modelling (BIM): Now and beyond. *Constr. Econ. Build.* **2012**, *12*, 15–28. [CrossRef]
83. Matarneh, R.; Hamed, S. Barriers to the adoption of building information modeling in the Jordanian building industry. *Open J. Civ. Eng.* **2017**, *7*, 325–335. [CrossRef]
84. Abubakar, M.; Ibrahim, Y.M.; Kado, D.; Bala, K. Contractors' perception of the factors affecting Building Information Modelling (BIM) adoption in the Nigerian Construction Industry. In Proceedings of the Computing in Civil and Building Engineering, Orlando, FL, USA, 23–25 June 2014; pp. 167–178.
85. Akerele, A.O.; Etiene, M. Assessment of the level of awareness and limitations on the use of building information modeling in Lagos State. *Int. J. Sci. Res. Publ.* **2016**, *6*, 229–233.
86. SBabatunde, O.; Perera, S.; Ekundayo, D.; Adeleke, D.S. An investigation into BIM uptake among contracting firms: An empirical study in Nigeria. *J. Financ. Manag. Prop. Constr.* **2020**, *26*, 23–48. [CrossRef]
87. Bernstein, P.G.; Pittman, J.H. Barriers to the adoption of building information modeling in the building industry. *Autodesk Build. Solut.* **2004**, *32*, 1–14.
88. Gardezi, S.S.S.; Shafiq, N.; Nuruddin, M.F.; Farhan, S.A.; Umar, U.A. Challenges for implementation of building information modeling (BIM) in Malaysian construction industry. In *Applied Mechanics and Materials*; Trans Tech Publications Ltd.: Nilai, Malaysia, 2014; pp. 559–564.
89. Thompson, D.B.; Miner, R.G. Building Information Modeling-BIM: Contractual Risks Are Changing with Technology. 2006. Available online: http//www.aepronet.org/ge/no35.html (accessed on 20 April 2020).
90. Olawumi, T.O.; Chan, D.W.M.; Wong, J.K.W.; Chan, A.P.C. Barriers to the integration of BIM and sustainability practices in construction projects: A Delphi survey of international experts. *J. Build. Eng.* **2018**, *20*, 60–71. [CrossRef]
91. Ahmed, S. Barriers to implementation of building information modeling (BIM) to the construction industry: A review. *J. Civ. Eng. Constr.* **2018**, *7*, 107–113. [CrossRef]
92. Ahn, K.-U.; Kim, Y.-J.; Park, C.-S.; Kim, I.; Lee, K. BIM interface for full vs. semi-automated building energy simulation. *Energy Build.* **2014**, *68*, 671–678. [CrossRef]
93. Sardroud, J.M.; Mehdizadehtavasani, M.; Khorramabadi, A.; Ranjbardar, A. Barriers Analysis to Effective Implementation of BIM in the Construction Industry. In *ISARC, Proceedings of the International Symposium on Automation and Robotics in Construction*; IAARC Publications: Berlin, Germany, 2018; pp. 1–8.
94. Hamada, H.M.; Haron, A.; Zakiria, Z.; Humada, A.M. Benefits and barriers of BIM adoption in the Iraqi construction firms. *Int. J. Innov. Res. Adv. Eng.* **2016**, *3*, 76–84.
95. Singh, V.; Gu, N.; Wang, X. A theoretical framework of a BIM-based multi-disciplinary collaboration platform. *Autom. Constr.* **2011**, *20*, 134–144. [CrossRef]
96. Liu, Y.; van Nederveen, S.; Hertogh, M. Understanding effects of BIM on collaborative design and construction: An empirical study in China. *Int. J. Proj. Manag.* **2017**, *35*, 686–698. [CrossRef]
97. Edirisinghe, R.; London, K.A.; Kalutara, P.; Aranda-Mena, G. Building information modelling for facility management: Are we there yet? *Eng. Constr. Archit. Manag.* **2017**, *24*, 1119–1154. [CrossRef]
98. Xu, Y.Q.; Kong, Y.Y. Analysis of the influence factors of application and promotion of BIM in China. *J. Eng. Manag.* **2016**, *30*, 28–32.
99. Khosrowshahi, F.; Arayici, Y. Roadmap for implementation of BIM in the UK construction industry. *Eng. Constr. Archit. Manag.* **2012**, *19*, 610–635. [CrossRef]
100. Chien, K.-F.; Wu, Z.-H.; Huang, S.-C. Identifying and assessing critical risk factors for BIM projects: Empirical study. *Autom. Constr.* **2014**, *45*, 1–15. [CrossRef]

101. Xu, H.; Feng, J.; Li, S. Users-orientated evaluation of building information model in the Chinese construction industry. *Autom. Constr.* **2014**, *39*, 32–46. [CrossRef]
102. Zhang, X.; Azhar, S.; Nadeem, A.; Khalfan, M. Using Building Information Modelling to achieve Lean principles by improving efficiency of work teams. *Int. J. Constr. Manag.* **2018**, *18*, 293–300. [CrossRef]
103. Nawi, M.N.M.; Haron, A.T.; Hamid, Z.A.; Kamar, K.A.M.; Baharuddin, Y. Improving integrated practice through building information modeling-integrated project delivery (BIM-IPD) for Malaysian industrialised building system (IBS) Construction Projects. *Malays. Constr. Res. J.* **2014**, *15*, 1–15.
104. Yaakob, M.; Wan, W.N.A.; Radzuan, K. Critical success factors to implementing building information modeling in Malaysia construction industry. *Int. Rev. Manag. Mark.* **2016**, *6*, 252–256.
105. Aranda-Mena, G.; Crawford, J.; Chevez, A.; Froese, T. Building information modelling demystified: Does it make business sense to adopt BIM? *Int. J. Manag. Proj. Bus.* **2009**, *2*, 419–434. [CrossRef]
106. Enshassi, A.; AbuHamra, L.; Mohamed, S. Barriers to implementation of building information modelling (BIM) in the palestinian construction industry. *Int. J. Constr. Proj. Manag.* **2016**, *8*, 103.
107. Eadie, R.; Odeyinka, H.; Browne, M.; McKeown, C.; Yohanis, M. Building information modelling adoption: An analysis of the barriers to implementation. *J. Eng. Archit.* **2014**, *2*, 77–101.
108. Ott, R.L.; Longnecker, M.T. *An Introduction to Statistical Methods and Data Analysis*; Cengage Learning: Boston, MA, USA, 2015.
109. Ling, F.Y.Y.; Low, S.P.; Wang, S.Q.; Lim, H.H. Key project management practices affecting Singaporean firms' project performance in China. *Int. J. Proj. Manag.* **2009**, *27*, 59–71. [CrossRef]
110. Flick, U. *The Sage Qualitative Research Kit: Collection*; SAGE Publications Limited: London, UK, 2009.
111. Pallant, J.; Manual, S.S. A step by step guide to data analysis using SPSS for windows. In *SPSS Survival Manual*; Open University Press: New York, NY, USA, 2007.
112. Lam, P.T.I.; Chan, E.H.W.; Ann, T.W.; Cam, W.C.N.; Jack, S.Y. Applicability of clean development mechanism to the Hong Kong building sector. *J. Clean. Prod.* **2015**, *109*, 271–283. [CrossRef]
113. Field, A.P. K endall's Coefficient of Concordance. *Wiley StatsRef Stat. Ref. Online* **2014**. [CrossRef]
114. Chan, D.W.M.; Choi, T.N.Y. Difficulties in executing the Mandatory Building Inspection Scheme (MBIS) for existing private buildings in Hong Kong. *Habitat Int.* **2015**, *48*, 97–105. [CrossRef]
115. Daniel, W.M.; Hung, H.T.W. An empirical survey of the perceived benefits of implementing the Mandatory Building Inspection Scheme (MBIS) in Hong Kong. *Facilities* **2015**, *33*, 337–366.
116. Hartmann, A.; Love, P.E.D.; Hon, C.K.H.; Chan, A.P.C.; Chan, D.W.M. Strategies for improving safety performance of repair, maintenance, minor alteration and addition (RMAA) works. *Facilities* **2011**, *29*, 591–610.
117. Chen, P.Y.; Smithson, M.; Popovich, P.M. *Correlation: Parametric and Nonparametric Measures*; SAGE: London, UK, 2002.
118. Lewin, A.; Silver, C. *Using Software in Qualitative Research*; SAGE: London, UK, 2014.
119. Bazeley, P.; Jackson, K. *Qualitative Data Analysis with NVivo*; SAGE Publications Limited: London, UK, 2013.
120. Field, A. *Discovering Statistics Using IBM SPSS Statistics*; SAGE: London, UK, 2013.
121. Farooq, U.; Rehman, S.K.U.; Javed, M.F.; Jameel, M.; Aslam, F.; Alyousef, R. Investigating BIM Implementation Barriers and Issues in Pakistan Using ISM Approach. *Appl. Sci.* **2020**, *10*, 7250. [CrossRef]

Article

IoT Open-Source Architecture for the Maintenance of Building Facilities

Valentina Villa [1,*], Berardo Naticchia [2], Giulia Bruno [3], Khurshid Aliev [1], Paolo Piantanida [1] and Dario Antonelli [3]

[1] Department of Structural, Geotechnical and Building Engineering, Polytechnic of Turin, 10129 Turin, Italy; Aliev.khurshid@polito.it (K.A.); paolo.piantanida@polito.it (P.P.)
[2] Department of Construction, Civil Engineering and Architecture, Marche Polytechnic University, 60121 Ancona, Italy; b.naticchia@univpm.it
[3] Department of Management and Production Engineering, Polytechnic of Turin, 10129 Turin, Italy; giulia.bruno@polito.it (G.B.); dario.antonelli@polito.it (D.A.)
* Correspondence: valentina.villa@polito.it

Abstract: The introduction of the Internet of Things (IoT) in the construction industry is evolving facility maintenance (FM) towards predictive maintenance development. Predictive maintenance of building facilities requires continuously updated data on construction components to be acquired through integrated sensors. The main challenges in developing predictive maintenance tools for building facilities is IoT integration, IoT data visualization on the building 3D model and implementation of maintenance management system on the IoT and building information modeling (BIM). The current 3D building models do not fully interact with IoT building facilities data. Data integration in BIM is challenging. The research aims to integrate IoT alert systems with BIM models to monitor building facilities during the operational phase and to visualize building facilities' conditions virtually. To provide efficient maintenance services for building facilities this research proposes an integration of a digital framework based on IoT and BIM platforms. Sensors applied in the building systems and IoT technology on a cloud platform with opensource tools and standards enable monitoring of real-time operation and detecting of different kinds of faults in case of malfunction or failure, therefore sending alerts to facility managers and operators. Proposed preventive maintenance methodology applied on a proof-of-concept heating, ventilation and air conditioning (HVAC) plant adopts open source IoT sensor networks. The results show that the integrated IoT and BIM dashboard framework and implemented building structures preventive maintenance methodology are applicable and promising. The automated system architecture of building facilities is intended to provide a reliable and practical tool for real-time data acquisition. Analysis and 3D visualization to support intelligent monitoring of the indoor condition in buildings will enable the facility managers to make faster and better decisions and to improve building facilities' real time monitoring with fallouts on the maintenance timeliness.

Keywords: digital twin; facility management; building information modeling; HVAC; fan coil; Internet of Things; predictive maintenance; fault detection; smart building

Citation: Villa, V.; Naticchia, B.; Bruno, G.; Aliev, K.; Piantanida, P.; Antonelli, D. IoT Open-Source Architecture for the Maintenance of Building Facilities. *Appl. Sci.* **2021**, *11*, 5374. https://doi.org/10.3390/app11125374

Academic Editor: Lavinia Chiara Tagliabue

Received: 30 April 2021
Accepted: 30 May 2021
Published: 9 June 2021

1. Introduction

Correctly managing the maintenance phase is the pivot to lower costs and energy waste [1] and to preserve the asset value over time and, above all, to maintain the level of performance required by users. Real Estate property managers are now realizing that investing in this activity brings tangible benefits in terms of investment preservation and a strong perception by customers/users. One of the early frameworks used to visualize the collected data was the building management system (BMS), developed in the 1980s [2]. The BMS is a system for controlling and monitoring a building's facilities, such as heating, lighting, electrical and mechanical services, safety and security [3]. Since the introduction of BIM [4], the industry's awareness of the importance of adopting digital models for FM is increasing [5]. Thus, there is a need for faster, more efficient, and possibly error-free real-time visualization and analysis of collected data. In fact, building control together with maintenance operators need to analyze both stored and current data measured by sensors to track and monitor the overall performance of the building [6].

Maintenance is generally classified as corrective (run-to-failure), scheduled, preventive, predictive and proactive [7]. Corrective maintenance operates only in case of failures and in the event of interruption of a service; scheduled maintenance uses the estimated life span of each component to provide for its replacement according to predefined schedules [8]. Preventive maintenance involves system inspection and control at fixed intervals to lessen the likelihood of it failing unexpectedly [9]. Predictive maintenance, through sensors and machine learning algorithms, allows detection of abnormal behaviors well before accidents happen [10]. Presently, this is mainly applied in manufacturing industries [11]: out-of-range data [12] are identified, analyzed and intervention procedures are implemented to restore the system's correct behavior. Proactive maintenance focuses on the root causes of failure, not fault symptoms, in order to improve the system operation.

The breakthrough for improving the efficiency and effectiveness of FM is the integration with IoT technologies, i.e., building environmental data [13] and the use of wireless sensors to collect data from building systems [14].

Several technologies are introduced: for example, data visualization using a BIM platform that helps operators to efficiently navigate through buildings [15], or the use of BIM combined with wireless sensors [16] to monitor temperatures in the subway stations [17], or CO_2 levels [18] or occupancy levels [19] in the rooms.

Therefore, the technologies now developed should allow facility managers real-time analysis, optimization, and visualization of large data sets to better manage energy consumption, operation costs, and user comfort. In general, there is still limited research on the application of these new technologies during the building's service life [20], particularly concerning the FM sector. In most cases, even if the use of automated sensors/devices and databases is present, the information collected is not fully exploited [21,22].

Similar research has already been introduced [23–25], where some authors [23] represent automated IoT and BIM-based alert systems for comfort monitoring in buildings. They used humidity and temperature sensors to detect discomfort in rooms, whereas other authors [26] integrated IoT sensor data (temperature, illuminance and power consumption) to monitor indoor conditions of the buildings. The weak point of all the above works is that they do not provide building facility management services. This unavoidably results in laborious and inefficient processes [20].

In addition, FM operators still rely on paper maintenance and control sheets, and this again greatly increases both the time required for compilation and processing of the information [21]. Therefore, the proposal of a stream-lined and technology-integrated workflow methodology is essential for the improvement of FM processes and the effective improvement of the efficiency of plant systems. In the present case study, the proposed methodology will be tested on a HVAC device, specifically on a FC.

The study goal is to create an IoT network for FM in order to monitor components of the HVAC system and detect their failures. This work presents part of a larger research focused on developing a proof-of-concept implementation of a digital framework based on the connection between existing technologies, to support the building data digital transaction. The specific purpose of this paper is to define a data management methodology integrated with fault detection, tailored for the FM. A cloud-based user interface (UI), an integrated IoT and BIM model provides the building 3D model with information on room temperature, humidity and luminosity, and fan coil service condition. On the edge of the system, this integrated fault detection methodology notifies and sends alerts to the facility managers when any anomaly is detected. Moreover, maintenance operators can easily find the location of the faulty component on the building 3D model by room ID or room name, both of them integrated into the application. Using a cloud-based service, building supervisors and facility managers can remotely monitor the thermal condition of fan coils and take necessary actions, e.g., if the operating temperature exceeds the pre-defined thresholds. As a result, this paper is organized as follows: a review of the literature related to the current research topic is reported in Section 2, the design architecture of the framework is presented in Section 3 together with the FM fault detection method. IoT and BIM application with data visualization within the case study is presented in Section 4, and finally the results of the IoT and BIM application for the considered case study are discussed.

2. Literature Review

2.1. Operation and Maintenance in Facility Management

Considering the whole life cycle, operation and maintenance (O&M) is the longest and most costly phase compared to others [27]. Therefore, the efficiency of control and monitoring systems is becoming increasingly important. An overall improvement in operational efficiency will be achieved due to improved technologies for data collection and inter-device communication, besides the decreasing cost of sensors with good data collection capabilities [28]. As the complexity and number of devices in HVAC systems increases, errors due to incorrect system configuration may increase; it is estimated that 40% of buildings have improperly configured devices [29] and there is a potential for 40% energy savings from correcting building errors [30]. This not only results in energy waste, but also in system malfunctions leading to discomfort indoors and also reduces air quality levels [31]. For this reason, fault detection and diagnosis (FDD) is becoming more and more important.

As FM is worth up to 85% of the building's entire life cycle cost [32], efficiency gains become crucial for cost containment and building quality preservation. Furthermore, facility management is a multidisciplinary topic involving many stakeholders and requires the collaboration and coordination of several different teams [33]. ISO 41011:2017 defines FM as an "organizational function which integrates people, place and process within the built environment with the purpose of improving the quality of life of people and the productivity of the core business" [34]. Most buildings currently do not operate efficiently due to standard procedures, fittings not customized for specific use, that generate a lack of information for operators [35]. In addition, there is habitually no historical data available to compare situations and to properly search for the cause for malfunctions or failures; the tools that come along with digital management are often still manually updated spreadsheets, making performance tracking challenging. Moreover, this mixed process is prone to error and confusion. Sometimes some important information is included in the digital models but they cannot describe the building systems' operation accurately. On the other hand, understanding and using real-time data is crucial nowadays, especially in high-performance buildings equipped with complex and multiple systems that require dynamic management to optimize their energy performance and provide adequate indoor environmental conditions for a large number of occupants [6].

HVAC systems are essential for efficient building operation and indoor air quality (IAQ) is critical to the habitability and comfort of spaces; many research efforts are now focusing on air quality monitoring for post-COVID-19 recovery [36]. HVAC systems also consume about 30% of primary energy in Europe [37], 14% of primary energy in the United States [38], and about 32% of the total amount of electricity generated in the United States [39]. Calculated energy waste as a result of defects in building HVAC systems is due to inappropriate operating procedures, equipment malfunctions, or design issues [40], while controlled and efficient management of HVAC systems is expected to save an average of 5–15% of total energy consumption if fully adopted in existing buildings [41,42].

2.2. BIM and IoT as Digital Twins

With the introduction of sensing and IoT in the construction industry, a new vision of building management tools has been emerging. In particular, a significant solution taken from the manufacturing sector has arisen: cyber-physical systems (CPS) [33]. CPS, better known as digital twins (DT), are a revolutionary digital vision, where the most innovative technologies come together to create a virtual model that performs exactly like its physical "twin". This continuous information exchange between the data collected in the real building and those processed by the virtual model through self-learning algorithms, makes it possible to mirror the life of the real twin and the corresponding building, to predict the building components' health, their useful life, failures [43] and, in general, the building performance [44]. In the Industry 4.0 era, physical and virtual worlds are really growing together [45].

Asghari, Rahmani, and Javadi [46] define IoT as "an ecosystem that contains smart objects equipped with sensors, networking and processing technologies integrating and working together to provide an environment in which smart services are taken to the end-users".

IoT is an unprecedented disruptive technology that has led to interconnection between people and objects on a scale and pace unimaginable even a decade ago [47]. It enables new strategies to improve the quality of life [48], allows autonomous decisions to connected devices through algorithms and machine learning, and can properly inform users to make the best decisions [49], for example, in case of emergencies or major failures. IoT and data networks have great potential in optimizing FM activities, including document management, historical data cataloging, logistics and material tracking, building component lifecycle monitoring, and building energy controls [20]. Several studies have been conducted on the use of data from IoT devices (e.g., [50]), although many of them do not include any BIM integration, and all smart buildings and homes [51] are examples of IoT technology integration.

Actually, the DT can be simply configured as a dashboard with critical performance indicators (KPIs), e.g., data on temperature, pressure, tilt, power, voltage, etc., representing inputs from sensors located on systems' components or in specific environments. This allows the DT to actively participate from the design to the functional phase of any product or process. According to [52], the first appearance of the DT was in NASA's Apollo program. In that case, DT was defined as "an integrated multi-physics, multi-scale, probabilistic simulation of a vehicle or system that uses the best available physical models, sensor updates, fleet history, etc., to mirror the life of its flying twin" [53].

In this situation, NASA needed to reproduce the useful life of its assets in a precise manner through the statistical analysis of the data collected by the sensors [54]. Thus, a DT consists of an entity composed of a physical space, a digital space, and a connection layer.

The digitization permeates across all sectors, including construction, and aims to build systems as well as approaches for helping operators not only in the conceptualization, prototyping, testing, and design optimization phase, but also during the operational phase. The importance of numerical simulation tools in the first phase is unquestionable, however, the potential of real-time data availability in the operational phase is opening new possibilities to monitor and improve operations throughout the life cycle of a product. Grieves [55], in his white-paper, called the presence of this virtual representation a "Digital Twin".

The digital twin arose from the integration of sensor networks and the digitization of machinery and manufacturing systems in the manufacturing industry [56]. The main difference between a design-phase simulation and a digital twin is that the latter requires a physical asset and sensor network, whereas simulation lives in a completely virtual environment [43]. Accordingly, the study in [56] presented an extended definition: "digital twins will facilitate the means to monitor, understand, and optimize the functions of all physical entities, living and non-living, by enabling seamless transmission of data between the physical and virtual worlds." Research in [57] described the simulation aspect of digital twins as the collection of relevant digital artifacts involving engineering and operational data, as well as the description of behavior using various simulation models. Digital twins use these specific simulation models based on their ability to solve problems, derive solutions relevant to real-life systems, and describe behavior. In general, the study in [57] defined the digital twin view as "a complete physical and functional description of a component, product, or system along with all available operational data."

Hicks [58] differentiates a digital twin from a virtual prototype and redefines its concept as an appropriately synchronized object of useful information (structure, function, and behavior) of a physical entity in virtual space, with information flows enabling convergence between the physical and virtual states. According to the authors in [44], digital twins represent real objects or subjects with their data, functions, and communication capabilities in the digital world.

Researchers in [59] defined the digital twin of a building as "the interaction between the real-world indoor environment of the building and a digital but realistic virtual representation model of the building environment, which provides the opportunity for real-time monitoring and data acquisition." In their definition, an indoor environment indicates information about air temperature, airflow, relative humidity, and lighting conditions, while a digital virtual environment indicates computational fluid dynamics and luminance levels. Furthermore, based on the study presented in [59], some of the notable advantages of creating a building digital twin are as follows: (1) collection, generation, and visualization of the building environment; (2) analysis of data irregularities; and (3) optimization of building services. The digital twin is continuously updated with sensor data in near real-time, and the data can be reprocessed with algorithms that make the data concise and more usable, even for customers or users. In addition, the large amount of collected data makes decision-making more informed, giving the ability to make predictions about how the object will behave in the future [60].

The DT can be applied to an asset for a technical simulation, with the purpose of integrating different model components to simulate almost every aspect of a complex system [45]. In a manufacturing plant in the industrial sector, for example, a DT offers the ability to simulate and improve the production system, considering the logistical aspects and optimization of the production process.

According to an Oracle report [61], the digital twin has many application areas: real-time remote monitoring and control; increased efficiency and safety at work; risk assessment; synergy and collaboration among team members; informed decision support; customization of products and services; better documentation, data collection and communication; and, most importantly for us, predictive maintenance and scheduling. The IoT paradigm is paving the way for smart cities and smart buildings. The definition of a common model of web-based protocols for data exchange allows transferring this data between objects, making the various network parts interactive. This is where the digital twin concept was born, which is a candidate to revolutionize the building management model through predictive analysis and dynamic simulations based on real-time data [62].

A full digital twin will ensure sensors are collecting data and the onset of failures can be detected well in advance through intelligent analysis of that data. This will allow for better maintenance scheduling.

Digital twin is a concept that can be exported from manufacturing to many fields and technologies [63]. Moreover, digital twin is one of the top ten strategic technology trends, and, according to future research predictions, the digital twin market will reach USD 15 billion by 2023 [57,64].

Once DT has been introduced, a look at how it can be linked to BIM methodology is provided. The U.S. National Institute of Building Sciences (NIBS) defines BIM as "The digital representation of the physical and functional characteristics of a structure. As such, it serves as a shared knowledge resource for information about a structure, forming a re-liable basis for decisions throughout its life cycle from inception onward" [65].

BIM, up from DT is widely used in the construction industry. BIM models allow for integrated information management throughout the building life cycle, including the FM phase [66]. BIM provides a collaborative platform to manage not only the project and its components, but also the relationships between stakeholders [67]. The BIM model acts as a collector of all building information, thus for facility managers it is of great support, working on information that is always up-to-date and shared by all team members, overcoming the uncertainty of paper data and information fragmentation.

However, BIM is much more widely used in the design phase, while in the FM phase it is still underutilized.

The most significant causes hindering this integration are: (1) the use of BIM only as a three-dimensional model, which has no added value in the maintenance management phase [68]; (2) FM workers are not involved in the creation of the model so the information contained within the BIM system is not useful [69]; (3) the need for interoperability between BIM and FM technologies and the lack of open systems [70]; (4) the lack of clear roles, responsibilities, contract and accountability framework [70]; and (5) the information contained in BIM models is static and not dynamic, as FM requires. Data is provided during the design phase but is not updated during the building life cycle [71].

Currently, as reported above, the BIM process is mainly used for visualization, construction, coordination, quantity calculation, planning, and project cost evaluation [72]. BIM is basically a repository for project information, with the major limitation of not transferring all this information to the facility management phase [73] and not being able to handle real-time data.

However, the great potential of a three-dimensional BIM model is the integration with all the virtual reality (VR), augmented reality (AR) and mixed reality (MR) systems. Research can be found discussing the development of collaborative BIM approaches based on AR/MR/VR [74–76]. One example is the creation of a framework based on building information modeling, mixed reality, and a cloud platform to support information flow in facility management [77].

3. Materials and Methods

3.1. IoT and BIM System Architecture

Automated data acquisition (DAQ) technology has seen significant advancements in hardware and software in recent years, even if the majority of available technologies are still expensive and not open-source, namely, a sort of "black box" where users have no access to alter and modify the implemented algorithms. Furthermore, a free access to the data previously stored is frequently impossible without buying a specific software. To address the aforementioned issues and overcome the limits of commercial technologies, studies were recently carried out to investigate and develop customized design of automated DAQ systems. This chapter proposes an automated data acquisition system within the IoT and BIM integration methodology based on open-source technologies.

All the sensors for building facilities and rooms are hosted by Arduino based microcontrollers and a Raspberry Pi single board computer with integrated Wi-Fi module to acquire the planned data in real time and store them on the server. The collected data is then processed to monitor and detect anomalies in building facilities operation.

A digital twin is next implemented by linking the stored data with a dashboard for trend visualization; a BIM model is used to visualize the component position and to have an overall view of the building.

The general architecture of the IoT integrated into the BIM is shown in Figure 1. The architecture exhibits IoT building facilities' sensors continuously sending data to the sensor nodes. In the hardware and networking section, raw data coming from sensors are processed by the sensor nodes' microcontroller and raw data becomes human-readable data. The following sensor nodes send processed data to the server that is connected to the gateway. Users can gain access to the local and cloud sensor node dashboards through this gateway providing data visualization and fault detection features. In the last section of the architecture, IoT data is integrated into BIM and combined IoT-BIM data is visualized on the BIM dashboard.

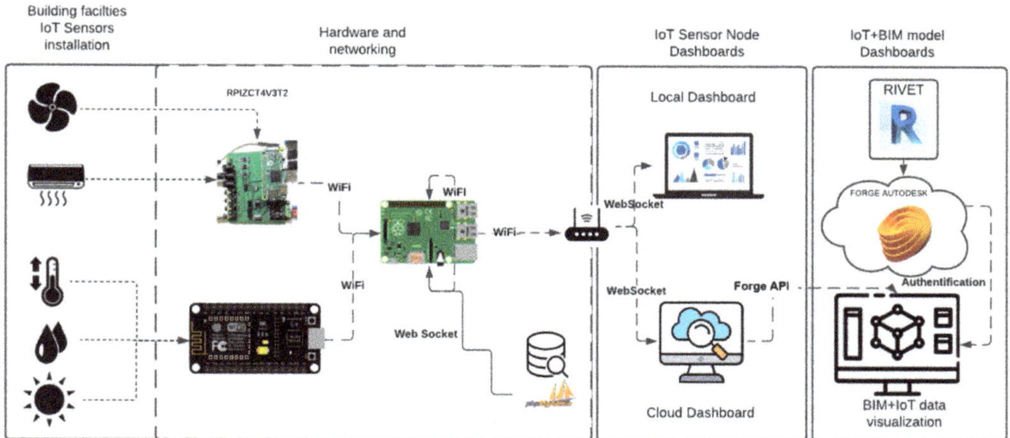

Figure 1. IoT and BIM integration system architecture.

3.2. IoT System Components

The proposed IoT system is based on open-source tools and market components. This system is flexible to set up, to add extra sensors and sensor nodes in real case-building facilities, as well as to add IoT sensors data into the 3D digital model of the building, according to any requirement of end-users.

Coming to the specific application, physical parameters must be acquired to monitor the room's internal conditions and to control the FC operation. For that reason, sensors shown in Table 1 are placed in the room and connected to their sensor boards.

Table 1. Applied sensors with technical specifications.

Sensor Board	Sensor Name	Accuracy	Measuring Ranges	Units
RPIZCT4V3T2	Current (SCT-013-000)	±3	0–100 A	Ampere (A)
	Voltage (EU: 77DE-06-09)	±5	0–230 (50 Hz)	Volt (V)
	Temperature (DS18B20)	±0.5	0–90 °C	Celsius (C)
	Temperature (PT100)	±0.05	−200 to 550 °C	Celsius (C)
ESP8266	Ambient temperature (DHT22)	±0.5	−40 to 80 °C	Celsius (C)
	Humidity (DHT22)	±2	0–100% RH	Relative Humidity (%RH)
	Photoresistor (LDR)	Resistant dependent	0–1 MΩ	lx

To oversee the temperature, relative humidity, and illuminance of the room, the DHT22 sensor and a light-dependent resistor (LDR), are connected to the ESP8266 module. FC monitoring sensors are connected to the RPIZCT4V3T2 board. Furthermore, both sensor boards are connected to the server board Raspberry Pi 3B (Rpi3B) to store and display incoming data locally and remotely using wireless sensor nodes (WSN).

The networking and hardware connectivity block diagram of the system is displayed in Figure 2.

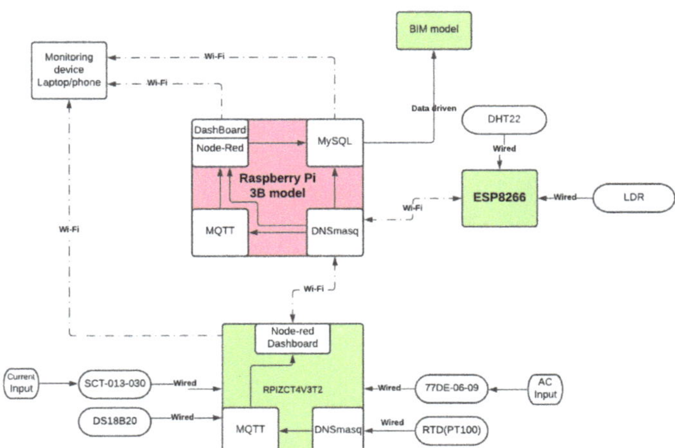

Figure 2. Hardware and connectivity block diagram.

The ESP8266 module is powered by a simple universal serial bus (USB) cable and automatically connects to the Rpi3B by means of installed credentials. In case of wireless network failure or unavailability, the board tries to connect again every five seconds until a successful connection is established. After that, the ESP8266 starts to collect data from the deployed sensors through the defined pins and the acquired data are sent through a serial monitor to the server with a sampling frequency of 1 Hz.

An RPIZCT4V3T2 sensor board is used to monitor the FC of the room; Figure 3 shows a simplified block diagram of the sensor board.

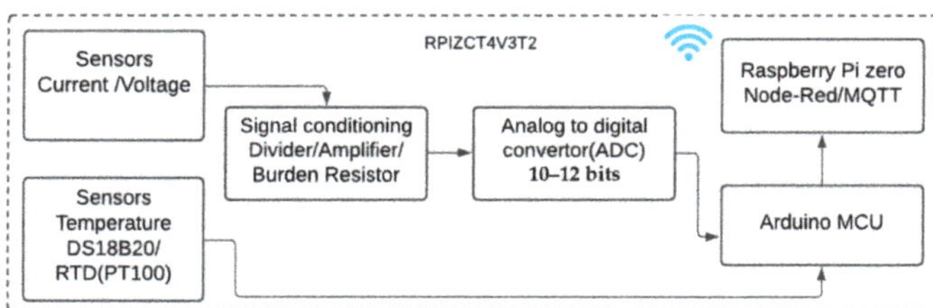

Figure 3. Simplified block diagram of the RPIZCT4V3T2 sensor board.

The RPIZCT4V3T2 board hosts an Arduino microcontroller (MCU) that is connected to two types of temperature sensors and current/voltage sensors that are connected to the MCU through an amplifier and analog to digital converter (ADC). Additionally, the RPIZCT4V3T2 board is connected to the Raspberry Pi (Rpi) Zero W. The MCU collects all the raw data, computes necessary values, sending the final computation to the Rpi Zero W using the universal asynchronous receiver-transmitter (UART) serial port; Rpi Zero W supports Wi-Fi and the board connects to the server Rpi3B through IP address. The sampling frequency of each sensor is specific and explained later in the next section.

The main component of the IoT and BIM system is the Rpi3B, which is a small single-board computer; it operates as server, networking router, middle communicator, hosts a dashboard, and a database. The Rpi3B system connects with other sensor nodes through the Wi-Fi network and logs the received data from the sensors to a database. On the Rpi3B system, Node-Red [6] is installed providing the ability to access all the sensor variables through serial protocols and displaying them on its own local dashboard. For each sensor node, a topic is assigned that is responsible for sending (publishing) a message to the main server (Rpi3B) which will act as a receiver (subscriber) using Message Queue Telemetry Transport (MQTT) protocol. MQTT protocol is an OASIS standard messaging protocol for IoT; it is designed as an extremely lightweight publish/subscribe messaging transport that presents as being ideal for connecting remote devices with a small code footprint and minimal network bandwidth [27].

Moreover, DNSmasq free software is installed to use the Rpi3B as a router and to provide a communication bridge between internal (sensor nodes) and external (internet network) components using Internet Protocol (IP) addresses. By setting the SSID, password, and an IP address on the Rpi3B using DNSmasq, the system becomes visible on the network to the publisher and subscriber.

To store sensor data locally on the Rpi3B, MySQL database, PHP interpreter, and Apache web server are utilized. By powering the Rpi3B, MySQL gets the IP address with a configured port number and waits for Node-Red to send the data to be collected; then, MySQL allocates the received data to the linked tables.

Later MQTT, MySQL, and Node-Red utilize the same credentials to background run and connect to the network, getting the IP address from DSNmasq. Communication between sensor nodes being established, the MQTT protocol on the configured port of the server receives all the subscribed topic's data from publishers through TCP protocol and provides publishing devices access to the port. Simultaneously, Node-Red starts on port 1880 using the same IP address to manage and monitor the data flow to the server and also sends data to the BIM dashboard by using forge nodes in JSON format or using a data driven approach, in CSV format.

Thus, acquired data from sensors will be displayed on the dashboard of the Rpi3B and using MQTT on the internet, and MySQL can be monitored and viewed through any device connected to the same network by opening the IP address followed by the port number.

3.3. FC Fault Detection Methodology

In HVAC systems, FCs are used as heating/cooling elements of rooms, which are a very common systems, especially in office buildings, hospitals, and schools. In order to have precise control of the FC operation, sensors must be inserted in the various components of the FC. The choice of which elements to monitor was made in relation to the identification of anomalies through collected data. Figure 4 shows the methodology for detecting faults in the FC and subsequent maintenance planning according to the condition of the FC. The image also shows the possible failures of the FC linked to the collection of anomaly data of the FC parts. The most common anomalies of the FC system are blocked motor, insufficient air flow, dirty filters, capacitor failure, insufficient water flow, etc. Each of these anomalies require an appropriate action. Particularly, the maintenance actions to be executed by operators are to clean the filter and battery, to change bearings, to replace capacitors, and to check valve adjustment and presence of pipe sediments. Thus, based on the anomaly types and data collected by sensors, a FC condition monitoring system is implemented. In addition, centralized collected data can be used to inform FM about the condition of any FC and provide preventive maintenance services, if needed.

Figure 5 demonstrates the anomaly detection algorithm and alarm system implemented on the RPIZCT4V3T2 sensor board. The maintenance alarming system is composed of three main sections: installed sensors on the FC, management and monitoring system of the FC components, and alarming FMs or end-users when an anomaly has occurred. In the sensors section, three voltage sensors, type EU: 77DE-06-09 are responsible for acquiring data from the three motor speeds (v1, v2, v3); three current sensors SCT-013-000 acquire data (i1, i2, i3) from the three speeds; and temperature sensors DS18B20 are responsible for monitoring T1 (delivery water temperature), T2 (return water temperature), and T4 (outlet air temperature) in the range 0–90 °C and T3 (inlet air temperature) in the range 0–50 °C. T5 (motor case temperature) is monitored with an PT100 temperature sensor in the range of 0–200 °C. On the controller of the RPIZCT4V3T2 sensor board, the management and monitoring algorithm of each component of the FC is implemented. The algorithm makes decisions depending on the sensors' signal values compared to the installation conditions.

FAN COIL	Edge machine	Maintenance		Centralized data collection			
Fault kinds	Autodecision (local)	Extemporaneous request (under condition)	Programming / Planning	For component reliability rating	To verify an anomaly in the environment	To check "zone" anomalies	Notes
False Voltage		■ (only controller, no fan coil)		■ (only controller, no fan coil)		■ (only controller, no fan coil)	
Motor blocked		■		■		■	
Operating hours / exhaustion filters			■ cumulative (filter cleaning, bearing lubrication if necessary)		■	■	the environmental anomaly is signaled when an environment is different from a "similar" activation period of more than "X" time the average(deleted sample extreme values). The "zone" anomaly is signaled when a large part of the area of the fan presents a high number of hours in relation to the predictable(as a function of ambient data and user programming)
Unbalancing of rotating masses				■			
Excessive resistant torque (motor overheating)	■ limit value exceeded	■ exceeded recommended value		■		■	the anomaly is signaled if most of the fan coils in the area show overheating
Insufficient water circulation		■				■	the "zone" anomaly is signaled when most of the fan coils in the zone have insufficient water flow
Delivery temperature anomaly (water)		■			■	■	the room anomaly is signaled when a single fan coil does not receive water supply(delivery temp. = return temp.) during the service activation period. The "zone" anomaly is signaled when a large part of the fan coils of the area is supplied with water temperature that does not correspond to that provided by the management system
Insufficient air circulation		■		■			

Figure 4. Kinds of faults and data collection for preventive maintenance of the FC.

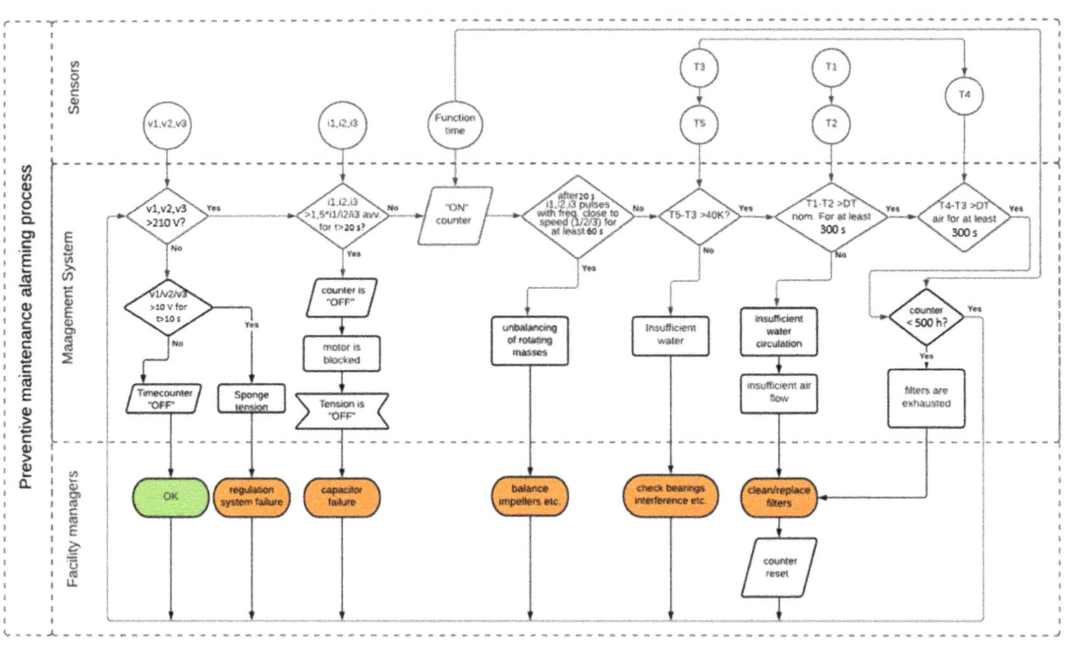

Figure 5. Integrated anomaly detection flow chart.

Likewise, Table 2 shows the sampling frequency for each FC component depending on the power condition. For example, on the server side, delivery pipe, return pipe, air inlet, and air outlet components temperatures are monitored every 30 min by sending average values when the voltage is at least 200 V, while the motor temperature must be monitored every 10 s and send average values of 30 min to the server. The motor voltage must be monitored more frequently and the sensor board must send average values to the server every 3 min if the voltage is in the range between 0 and 200 V. On the other side, electric power (current) shall be monitored in the first 10 s, when the voltage is above 200 V. More detailed sampling frequencies on the server-side of the system integrated on the microcontroller of the sensor board and sensor locations in the FC is given in Table 2.

The given measure ranges are implemented on the RPIZCT4V3T2 board using Node-Red flows and displayed on the dashboard too. The system also informs the end-user if any anomaly is detected on the FC.

Table 2. Sampling frequency and FC sensor allocation.

Sensors	Frequency	Sensor Allocation	Note
T1	180″	delivery pipe	every 1800″ send to the server the average detected values in the interval in which one of the values is at least 200 volts
T2	180″	return pipe	every 1800″ send to the server the average detected values in the interval in which one of the values is at least 200 volts
T3	180″	air inlet	every 1800″ send to the server the average detected values in the interval in which one of the values is at least 200 volts
T4	180″	air outlet	every 1800″ send to the server the average detected values in the interval in which one of the values is at least 200 volts
T5	10″	motor case	every 1800″ send to the server the average detected values in the interval in which one of the values is at least 200 volts
v1	0.1″	motor voltage	send voltage value to the server only if for at least 180″ v1 is greater than 0 and less than 200 volts
v2	0.1″	motor voltage	send voltage value to the server only if for at least 180″ v1 is greater than 0 and less than 200 volts
v3	0.1″	motor voltage	send voltage value to the server only if for at least 180″ v1 is greater than 0 and less than 200 volts
i1	0.05″/3″	motor current	0.05″ for the first 10′ from v1 equal to at least 200 volts (anti-unbalance)/3″ in normal operation/no fields-on if v1 < 200 volts/send to the server the integral of i1 ×dt on the range where v1 is at least 200 volts (power consumption control)/send to alarm server if i1 > 1.5 × i1 rated for more than 6″
i2	0.05″/3″	motor current	0.05″ for the first 10′ from v2 equal to at least 200 volts (anti-unbalance)/3″ in normal operation/no fields-on if v2 < 200 volts/send to the server the integral of i1 × dt on the range where v2 is at least 200 volts (power consumption control)/send to alarm server if i2 > 1.5 × i2 rated for more than 6″
i3	0.05″/3″	motor current	0.05″ for the first 10′ from v3 equal to at least 200 volts (anti-unbalance)/3″ in normal operation/no fields-on if v3 < 200 volts/send to the server the integral of i3 × dt on the range where v3 is at least 200 volts (power consumption control)/send to alarm server if i3 > 1.5 × i1 rated for more than 6″

3.4. Data Transmission and Visualization 3D Model and IoT Data

The 2D and 3D models of the fan coil unit and the building model used for this research are both built in Autodesk Revit authoring tool. To connect IoT sensor data with BIM, the Autodesk Forge Platform is utilized: it is a cloud-based platform and provides application programming interface (API) services. A free trial of Forge APIs lasts 90 days and gives access to 100 cloud credits, 5 GB storage, and services such as authentication (two-legged authentication), data management, Model Derivative, Model Viewer, all of them necessary and sufficient to create the customized application. By registering on the Forge platform, the end-user receives a client ID, client password, and all the services necessary for application development. BIM–IoT integration and the visualization process using the Forge platform is described in Figure 6: the right column records the services used for the custom application on the Forge platform, while the left column shows the applications layer.

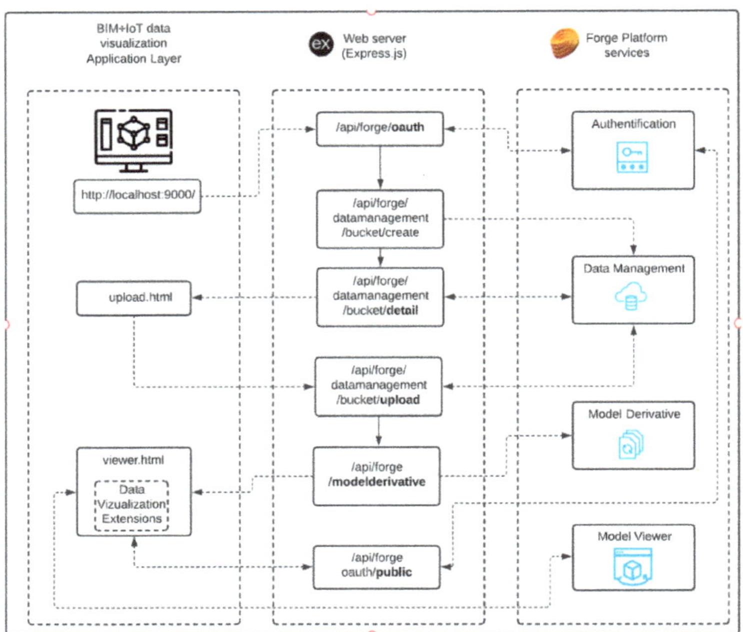

Figure 6. BIM–IoT data integration and visualization approach diagram.

The 3D building model can be uploaded onto the Forge platform by inserting credentials provided by the authentication API and using the upload.html file. Afterwards, the Model Derivative API translates the uploaded 3D model using the viewer.html file, so the uploaded model can be accessed by typing a static address (e.g., localhost:9000) on any browser.

The middle column of Figure 6 displays the browser visualization on the Forge Viewer: it can be customized by inserting different extensions, buttons, or plots using the JavaScript programming language that supports the Open-source VS Code editor. Furthermore, IoT sensor data in JSON or in CSV formats can be added through plugins to the uploaded 3D building model. The model properties will appear in the viewer as soon the model itself is uploaded on the Forge Viewer.

A custom web application of Forge Viewer can be created by accessing the object's properties in the model and relying on the database identifier (DbId) for the model objects.

4. Case Study

This section explains the case study setup, the allocation of sensors, the algorithms to monitor and detect anomalies in building facilities, and summarize the achieved results.

The case study has been operated at the Politecnico di Torino in the DISEG Laboratory. The laboratory room is located below ground in the department. Figure 7 shows the case study building: (a) geolocation and (b) BIM model Revit software representation. In all the rooms, fan coils are located under the windows.

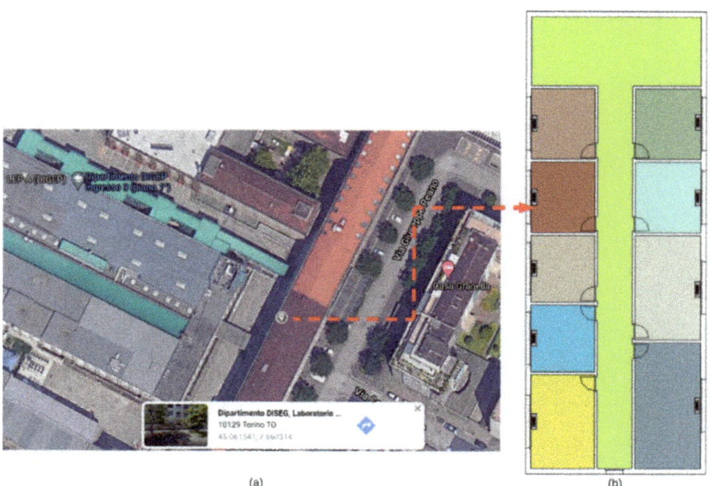

Figure 7. Case study building: (a) geolocation on the maps.google.it and (b) 2D view of the case study room with the location of the FC.

The case study fan-coil motor is made by EMI (EuroMotors Italia), type FC83M-2014/1 with a 4 possible speeds. The FC motor rotates in an anti-clockwise direction at 1100 revolutions per minute (RPM) at maximum speed. Moreover, FC has a cooling or heating battery and a filter that must often be monitored; to provide an automatic preventive maintenance system, the FC is equipped with sensors to acquire real-time data.

More technical details of the FC are provided in Table 3.

Table 3. Technical specifications of FC Motor FC 83M-2014/1.

Model Number	V/HZ	Ampere	Num Speeds	Power	RPM
FC 83M-2014/1	230–240V/50	0.23 A	4	14/53 W	1100

The ESP8266 sensor node is placed on the FC together with sensors. All the collected data from the room facilities are sent to the BIM using the Forge API; Autodesk Forge API provides the client ID and client password to access the uploaded customized BIM on the cloud; using the callback URL everyone can access the IoT-integrated BIM dashboard. A general schematic diagram of sensors' locations of the FCs, sensor boards, and data acquisition system of the BIM model is showed in Figure 8.

The system is powered with simple USB cables and starts collecting data from the installed sensor nodes in the room. The Rpi3B mainboard then connects to the ESP8266 and RPIZCT4V3T2 sensor nodes and starts acquiring data from sensors and at the same time registers these data to the MySQL database. Using the developed extensions on the Visual Studio Code for Autodesk Forge Viewer, collected IoT data and the 3D building model of Revit can be visualized together on the Forge Viewer through a static URL and port provided by Forge API. The functional flowchart of the whole system is shown in Figure 9.

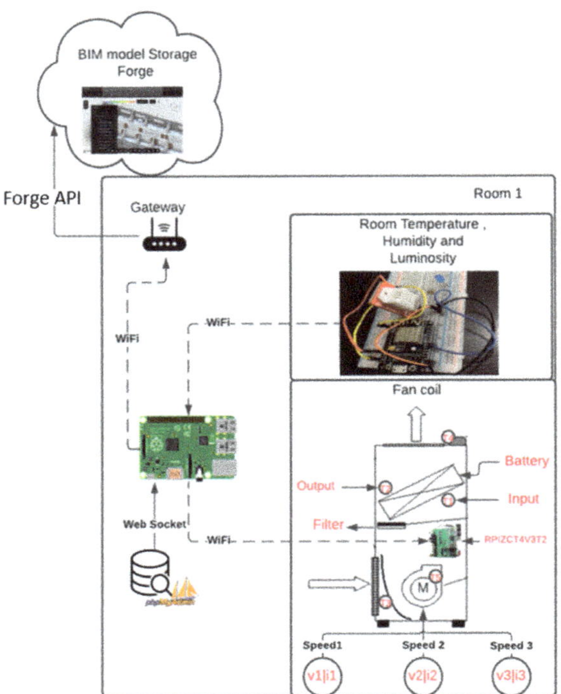

Figure 8. Sensors' location in the FC and building facilities.

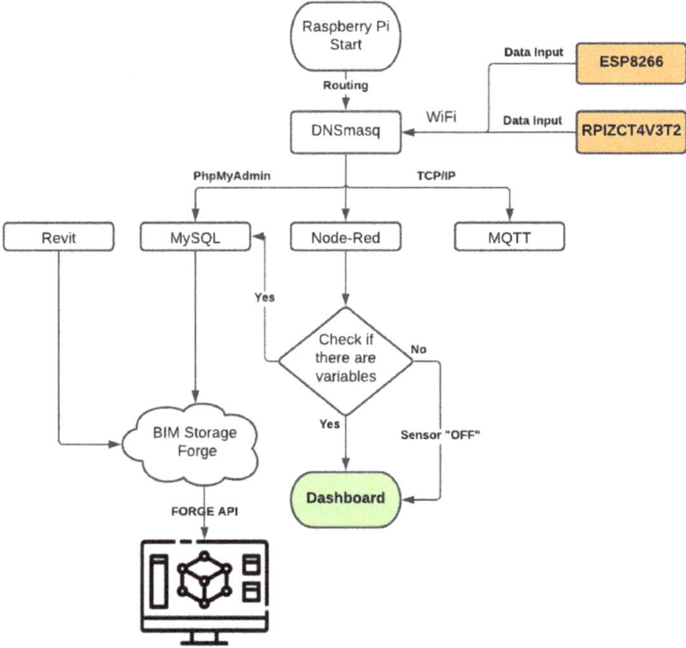

Figure 9. Functional flowchart of the system.

FC with wireless sensor nodes (b), FC battery with installed temperature sensors (a), the motor with a temperature sensor (c), and room sensors with the sensor node (d) are shown in Figure 10.

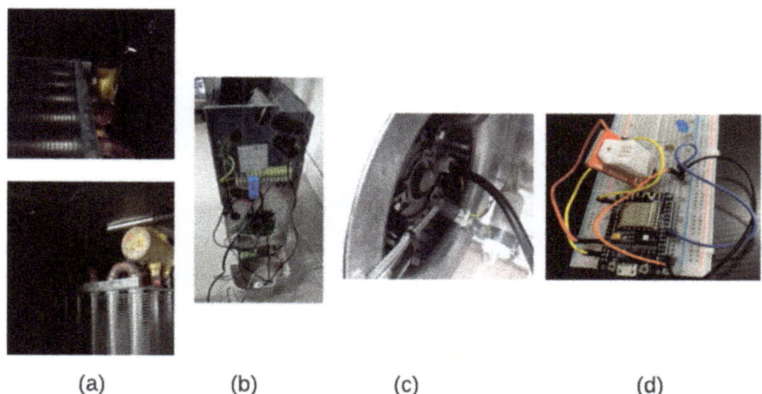

Figure 10. Case study sensor nodes placement to the room and FC: (**a**) FC battery and temperature sensor, (**b**) FC with RPIZCT4V3T2 board, (**c**) motor with temperature sensor, and (**d**) room equipped with ESP8266 sensor node.

Data are measured by fan coil and room sensors and then displayed on the dashboard of the Rpi3B server using Node-Red flows in Figure 11.

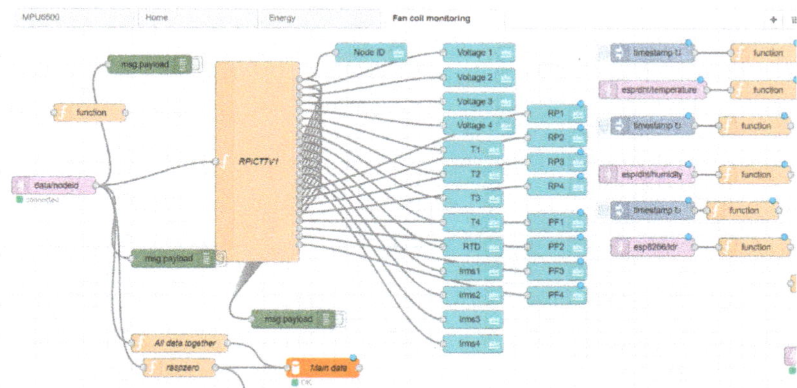

Figure 11. Node-Red flows of the FC and room sensors data acquisition system.

Measured and integrated variables with their descriptions on the Node-Red flow besides the dashboard are provided in Table 4.

To verify the IoT and BIM integrated application dashboard, one voltage (V1) and one current (I1) sensors are connected to the first port of the sensor's board and all temperature sensors including RTD are allocated to the FC as shown in Figure 8.

Table 4. FC Measured physical parameters.

Measured Parameters	Formulas with Measuring Units
Real Power (RP)	RP (W)
Apparent Power (AP)	AP = Irms × Vrms (W)
Vrms	Rms Voltage (V)
Irms	Rms current (mA)
Estimated Power (EP)	EP = Irms × Vest (W)
Power Factor (PF)	PF = RP/AP (no units)
Temperature	Temperature (°C)
Frequency	Frequency (Hz)
RTD Temperature	Temperature (°C)

5. Results

The proposed framework was tested on the fan coil in the room with connected sensors. The FC sensors were linked to the set-up speed and each time the speed was adjusted. The final result of the dashboard with the acquired data at speed 1 is shown in Figure 12a—the motor power consumption values are as follows: V1 = 244.1, I1 = 0.1 A, RP1 = 32.3 and PF = 0.899. Fan coil data at speed 2 on the dashboard are shown in Figure 12b—the motor power consumption values are V1 = 241.5, I1 = 0.2 A, RP1 = 38.1 and PF1 = 0.924. Finally speed 3 on the dashboard is shown in Figure 12c—the motor power was measured as V1 = 244, I1 = 0.2 A, RP1 = 47.8 and PF1 = 0.944. The average temperature of the FC motor was RTD = 23.43 °C and average outlet temperature was indicating as T2 = 29.45 °C because the fan coil was heating the room. Rooms' dashboards with physical parameters acquired from sensors are visualized in Figure 12d.

The Rpi3B server board memory was not sufficient for the big data storage To avoid overloading the local server memory, only daily maximum and minimum values were sent from the sensor board to the cloud and BIM server. The fault detection methodology described in the previous section was applied to the system. If the values supplied by sensors exceed the predictable ranges, the system sends alarm signals or real-time notifications to the facility managers and operators using Telegram or SMS.

On the Forge platform, IoT data are integrated into the BIM using the Forge reference application (https://github.com/Autodesk-Forge/forge-dataviz-iot-reference-app, accessed on 6 April 2021) and two NPM modules (React UI components and Client-Server Data-Module-Components): NPM is the packet manager for Node.js and it is an open-source project helping to support JavaScript developers.

Installation and running of the Forge reference application started by cloning the application from GitHub repository (git clone https://github.com/Autodesk-Forge/forge-dataviz-iot-reference-app.git, accessed on 6 April 2021) to the VScode using lines in the console of the editor.

The project folder is composed of client-side codes, a guide on how to upload the Revit model, router configuration, and client-server configuration files that speed up work for developers on their custom applications. The structure of the folders and files of the reference application is shown in Figure 13.

In the reference application an .env file is created and added to Forge CLIENT_ID, CLIENT_SECRET, and Forge_BUCKET; in this way the custom 3D building model is added to the project (Figure 14). By running the project, the custom 3D model is then uploaded in a static browser (http://localhost:9000/upload, accessed on 6 April 2021).

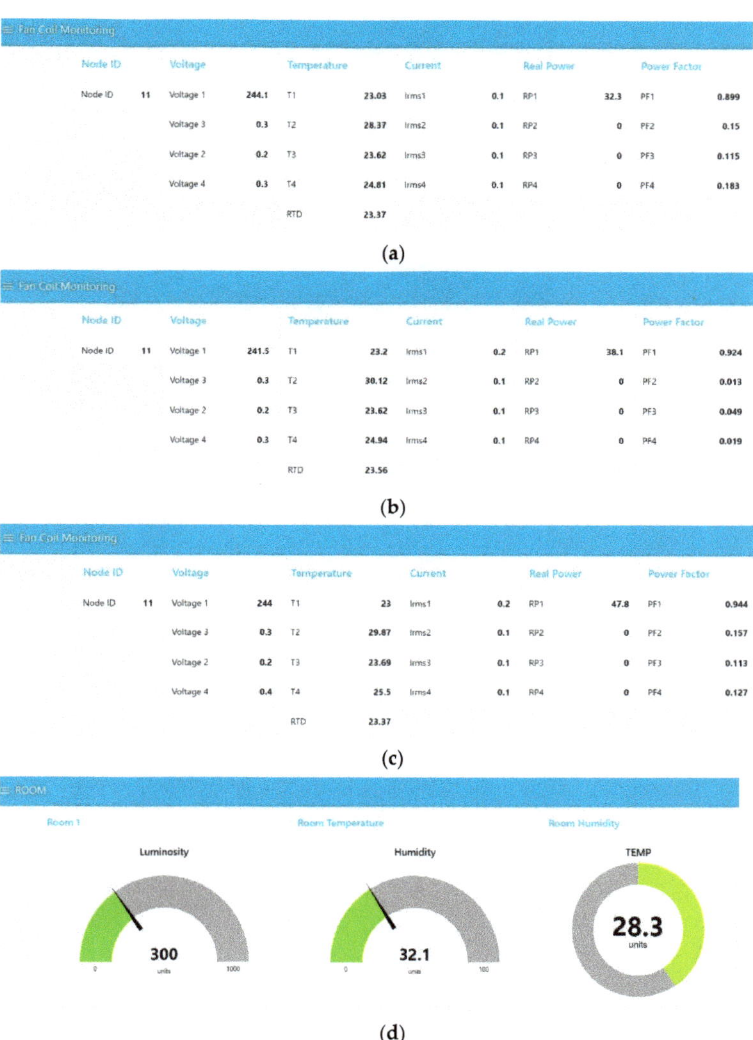

Figure 12. Node-Red dashboard results: (**a**) FC at speed 1; (**b**) FC at speed 2; (**c**) FC at speed 3; and (**d**) room dashboard values coming from the ESP8266 sensor node.

```
.
├── assets              # Static .svg and .png files
├── client              # Client-side code + configuration
├── docs                # Additional guide on how to upload a Revit model
├── scss                # SCSS files
├── server              # Server-side, router configuration, CSV data etc.
├── shared              # Config files shared between client and server
├── tools               # Useful tools for own webpack
├── package.json
├── webpack.config.js
├── LICENSE
└── README.md
```

Figure 13. Reference application directory structure.

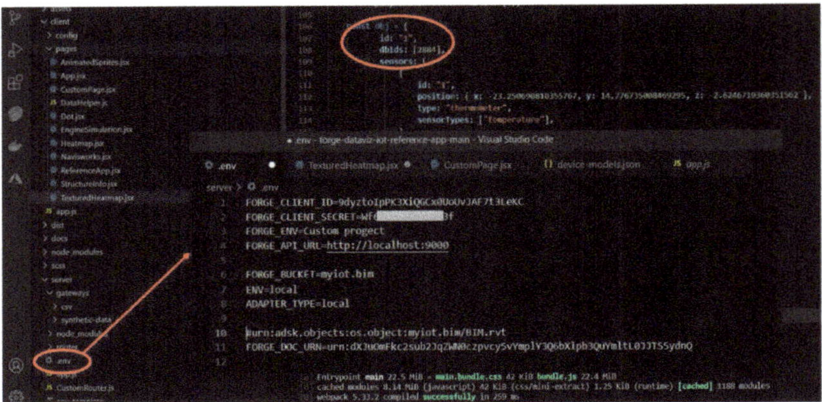

Figure 14. The .env file with authentication credentials and the added sensor values in the model DbId objects in JavaScript.

To interact with uploaded model objects on the Forge Viewer, DbIds are required. This is because most API methods to manipulate entities require the argument DbId (or array), such as isolate, hide, highlight, etc.; knowing the DbId array, a map with the model hierarchy node, a unique ID can be built and then custom functions can be written to connect IoT sensor variables to the BIM model.

The objects; dbIds can be acquired using functions getSelection() that returns a list of DbIds, and getProperties() expecting DbId as input, etc. The extraction process of DbIds from model objects is shown Figure 15.

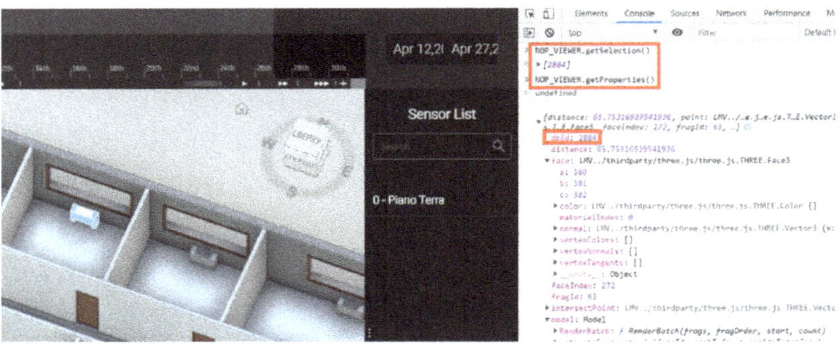

Figure 15. Uploaded custom model DbId's extraction.

To identify each fan coil DbId's was added as a list in customized JavaScript code. It included sensor parameters such as position, type, etc. The added custom IoT sensor data were implemented through additional functions as shown in Figure 15.

Finally, Figure 16 shows IoT and BIM applications developed on the Forge platform. In the application, the fan coil color changes according to data coming from the sensor nodes. The color is "green" if the fan coil is in a good condition, "red" if overheated, and "blue" if the FC is cooling.

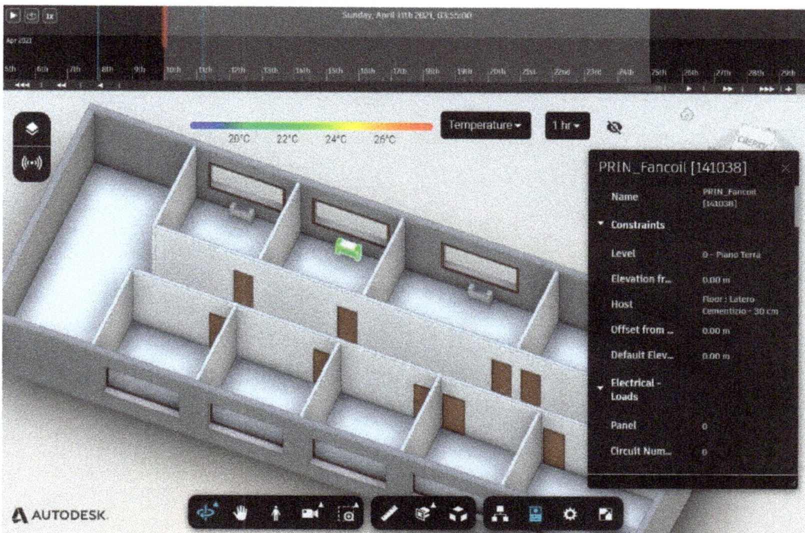

Figure 16. Final result of implemented heatmap function to the fan coil data and BIM.

6. Discussion and Conclusions

Cloud services for data analytics, including platforms for data visualization, are now available from providers like Microsoft Azure, Amazon Web Services and Google Cloud. These services can gather data and generate alerts, notifications, and graphs, but they are not flexible and cannot interact with the BIM model or sensor data.

This paper introduced a fully automated integrated framework for the maintenance of building facilities using open source IoT technologies. The IoT part of the proposed framework is composed of sensors to be installed in building facilities and wireless sensor network nodes that continuously send received data through gateways to the local and cloud servers according to configured sampling frequencies. The proposed fault detection methodology was integrated into the server and sends alarms to the end user or managers when any anomaly occurs for it to be fixed effectively; BIM was utilized to view the monitored HVAC FC component's condition and the room's physical parameters such as temperature, humidity, and luminosity using wireless-connected remote devices. A case study was used to test the new framework implementation. The integration of an FC monitoring system into the BIM model would improve the building's maintenance plan by helping the facility managers to inspect the monitored building environments inside the 3D model. Thus, facility managers can profit from the proposed framework to resolve maintenance problems in the following aspects: (a) using building facility anomaly or failure signals, facility managers can plan and schedule maintenance work in advance; (b) conditional and real-time data coming from sensors allow for more accurate maintenance; (c) data visualization and real-time data on the dashboard create the possibility of avoiding the risk of disastrous breakdowns and reduce unplanned forced outages of building components; (d) IoT sensor data for building components within the BIM model makes maintenance work more convenient, e.g., it would be easy to find the location of the failure component in the real-time BIM model; and finally, (e) collected indoor and facility sensor data is the initial tool to perform predictive maintenance actions.

The study conducted in this paper aims to fill the gaps of the following researches: limited number of sensors [23], an absence of automation [24], and lack of data acquisition system [25]. The final results of the fully automated framework composed of IoT sensors, dashboard, IoT and BIM integrated application, and implemented preventive maintenance methodology of the building facilities on the server prove the proposed framework's applicability.

Nevertheless, this work has several limitations, which are as follows:

- In terms of building monitored facilities, the study is fairly restricted. Additional research is needed to perform a large-scale inquiry by connecting more building facilities and evaluating the system's integration with facility managers and clients to assess its reliability, repeatability, robustness, and simplicity of use.
- A small number of sensors were used to test the proposed framework in this study. Other sensors, such as indoor air quality sensors (e.g., an oxide gas sensor, a particle dust sensor, etc.) and facility management sensors can be added to the system (e.g., motion sensor, occupancy sensor, etc.). The developed system should also be tested with a larger number and range of sensors.

Future research work will attempt to use collected data from the proposed system for predictive maintenance management of building facilities. Decision support systems for facility predictive maintenance management systems, artificial intelligence (AI), specifically machine learning (ML) tools, and algorithms should be used.

Time series sensor data, either historical failures, anomalies, or both, data can be used as ML inputs. For example, building facility anomalies or failures can be predicted using classification analyses. The use of regression analyses can utilize time series sensor data to forecast physical parameters of the building facilities' components. Thus, either predicted values, failures, anomalies, or combination, of building facilities offer the FM managers the ability to provide even more effective maintenance services before failures.

Author Contributions: Conceptualization, V.V., B.N., and G.B.; methodology, V.V. and P.P..; software, K.A.; validation, P.P. and D.A.; formal analysis, P.P.; investigation, G.B., B.N. and V.V.; resources, P.P. and K.A.; data curation, D.A. and P.P.; writing—original draft preparation, V.V. and K.A.; writing—review and editing, B.N., D.A. and P.P.; visualization, V.V.; supervision, V.V.; project administration, B.N. and V.V.; funding acquisition, B.N. and V.V. All authors have read and agreed to the published version of the manuscript.

Funding: This research was funded by the Italian government, through the PRIN2017 project of the Ministero dell'Istruzione, dell'Università e della Ricerca (MIUR). The project entitled "Distributed Digital Collaboration Framework for Small and Medium-Sized Engineering and Construction Enterprises" is coordinated at the national level by Berardo Naticchia.

Institutional Review Board Statement: Not applicable.

Informed Consent Statement: Not applicable.

Data Availability Statement: Not applicable.

Conflicts of Interest: The authors declare no conflict of interest.

References

1. Lauro, F.; Moretti, F.; Capozzoli, A.; Khan, I.; Pizzuti, S.; Macas, M.; Panzieri, S. Building fan coil electric consumption analysis with fuzzy approaches for fault detection and diagnosis. In Proceedings of the 2014 6th International Conference on Sustainability in Energy and Buildings, SEB-14, Energy Procedia 62, Cardiff, UK, 25–27 June 2014; pp. 411–420.
2. Allen, P.A. *An Integrated Power and Building Services Management System, INTELEC '86, International Telecommunications Energy Conference*; IEEE: Piscataway, NJ, USA, 1986; pp. 525–530. [CrossRef]
3. Malatras, A.; Asgari, A.H.; Baugé, T. Web Enabled Wireless Sensor Networks for Facilities Management. *IEEE Syst. J.* **2008**, *2*, 500–512. [CrossRef]
4. Pishdad-Bozorgi, P.; Gao, X.; Eastman, C.; Self, A.P. *Planning and Developing Facility Management-Enabled Building Information Model (FM-enabled BIM), Automation in Construction*; Elsevier: Amsterdam, The Netherlands, 2018; Volume 87, pp. 22–38. ISSN 0926-5805. [CrossRef]
5. Becerik-Gerber, B.; Jazizadeh, F.; Li, N.; Calis, G. Application Areas and Data Requirements for BIM-Enabled Facilities Management. *J. Constr. Eng. Manag.* **2012**. [CrossRef]
6. Kazado, D.; Kavgic, M.; Eskicioglu, R. Integrating Building Information Modeling (BIM) and Sensor Technology for Facility Management in 2019. *J. Inf. Technol. Constr.* **2019**, *24*, 440–458. Available online: https://www.itcon.org/2019/23 (accessed on 6 June 2021).

7. Zhou, X.; Lifeng, X.; Lee, J. Reliability-centered predictive maintenance scheduling for a continuously monitored system subject to degradation. *Reliab. Eng. Syst. Saf.* **2007**, *92*, 530–534. [CrossRef]
8. Shalabi, F.; Turkan, Y. IFC BIM-based facility management approach to optimize data collection for corrective maintenance. *J. Perform. Constr. Facil.* **2017**, *31*, 04016081. [CrossRef]
9. Bousdekis, A.; Magoutas, B.; Apostolou, D.; Mentzas, G. A proactive decision making framework for condition-based maintenance. *Ind. Manag. Data Syst.* **2015**, *115*, 1225–1250. [CrossRef]
10. Bouabdallaoui, Y.; Lafhaj, Z.; Yim, P.; Ducoulombier, L.; Bennadji, B. Predictive Maintenance in Building Facilities: A Machine Learning-Based Approach. *Sensors* **2021**, *21*, 1044. [CrossRef]
11. Aliev, K.; Antonelli, D. Proposal of a Monitoring System for Collaborative Robots to Predict Outages and to Assess Reliability Factors Exploiting Machine Learning. *Appl. Sci.* **2021**, *11*, 1621. [CrossRef]
12. Martí, L.; Sanchez-Pi, N.; Molina, J.M.; Garcia, A.C.B. Anomaly Detection Based on Sensor Data in Petroleum Industry Applications. *Sensors* **2015**, *15*, 2774–2797. [CrossRef] [PubMed]
13. Dave, B.; Buda, A.; Nurminen, A.; Främling, K. A framework for integrating BIM and IoT through open standards. In *Automation in Construction*; Elsevier: Amsterdam, The Netherlands, 2018; pp. 35–45. [CrossRef]
14. Krishnamurthy, S.; Anson, O.; Sapir, L.; Glezer, C.; Rois, M.; Shub, I.; Schloeder, K. Automation of facility management processes using machine-to-machine technologies. In *Lecture Notes in Computer Science (Including Subseries Lecture Notes in Artificial Intelligence and Lecture Notes in Bioinformatics), 4952 LNCS*; Springer: Berlin/Heidelberg, Germany, 2008; pp. 68–86. [CrossRef]
15. Araszkiewicz, K. Digital technologies in Facility Management–the state of practice and research challenges. *Procedia Eng.* **2017**, *196*, 1034–1042. [CrossRef]
16. Cahill, B.; Menzel, K.; Flynn, D. BIM as a Centre Piece for Optimised Building Operation. In *eWork and eBusiness in Architecture, Engineering and Construction: ECPPM 2014*; Taylor & Francis Group: London, UK, 2012; pp. 549–555. [CrossRef]
17. Marzouk, M.; Abdelaty, A. BIM-based framework for managing performance of subway stations. In *Automation in Construction*; Elsevier B.V.: Amsterdam, The Netherlands, 2014; pp. 70–77. [CrossRef]
18. Tagliabue, L.C.; Re Cecconi, F.; Rinaldi, S.; Ciribini, A.L.C. Data Driven Indoor Air Quality Prediction in Educational Facilities Based on IoT Network. *Energy Build.* **2021**, *236*, 110782. [CrossRef]
19. Rinaldi, S.; Flammini, A.; Tagliabue, L.C.; Ciribini, A. *An IoT Framework for the Assessment of Indoor Conditions and Estimation of Occupancy Rates: Results from a Real Case Study*; ACTA IMEKO: Benevento, Italy, 2019.
20. Wong, J.K.W.; Ge, J.; He, S.X. Digitisation in facilities management: A literature review and future research directions. *Autom. Constr.* **2018**, *92*, 312–326. [CrossRef]
21. Xiao, Y.-Q.; Li, S.-W.; Hu, Z.-Z. Automatically Generating a MEP Logic Chain from Building Information Models with Identification Rules. *Appl. Sci.* **2019**, *9*, 2204. [CrossRef]
22. Mannino, A.; Dejaco, M.C.; Re Cecconi, F. Building Information Modelling and Internet of Things Integration for Facility Management—Literature Review and Future Needs. *Appl. Sci.* **2021**, *11*, 3062. [CrossRef]
23. Valinejadshoubi, M.; Moselhi, O.; Bagchi, A.; Salem, A. Development of an IoT and BIM-based automated alert system for thermal comfort monitoring in buildings. *Sustain. Cities Soc.* **2021**, *66*, 102602. [CrossRef]
24. Wehbe, R.; Shahrour, I. Use of BIM and Smart Monitoring for buildings' Indoor Comfort Control. In *MATEC Web of Conferences*; EDP Science: Lille, France, 2019; Volume 295.
25. Natephra, W.; Motamedi, A. *Bim-Based Live Sensor Data Visualization Using Virtual Reality for Monitoring Indoor Conditions*; CumInCAD: Wellington, New Zealand, 2019.
26. Cahill, B.; Menzel, K.; Flynn, D. BIM and IoT Sensors Integration: A Framework for Consumption and Indoor Conditions Data Monitoring of Existing Buildings. *Sustainability* **2021**, *13*, 4496.
27. Guillen, A.; Crespo, A.; Gómez, J.; González-Prida, V.; Kobbacy, K.; Shariff, S. Building Information Modeling as Asset Management Tool. In *3rd IFAC Workshop on Advanced Maintenance Engineering, Services and Technology AMEST 2016*; IFAC-PapersOnLine; Elsevier B.V.: Amsterdam, The Netherlands, 2016; pp. 191–196. [CrossRef]
28. Piette, M.A.; Kinney, S.K.; Haves, P. Analysis of an information monitoring and diagnostic system to improve building operations. *Energy Build.* **2001**, *33*, 783–791. [CrossRef]
29. Piette, M.A.; Nordman, B.; Greenberg, S. Quantifying energy savings from commissioning: Preliminary results from the pacific northwest. In *Processings of the Second National Conference on Building Commissioning*; Lawrence Berkeley National Laboratory: Berkeley, CA, USA, 1994.
30. Schein, J.; Bushby, S.T.; Castro, N.S.; House, J.M. A rule-based fault detection method for air handling units. *Energy Build.* **2006**, *38*, 1485–1492. [CrossRef]
31. Mills, E. Building commissioning: A golden opportunity for reducing energy costs and greenhouse-gas emissions. In *Lawrence Berkeley National Laboratory*; Springer Nature: Cham, Switzerland, 2010.
32. Cheng, J.C.P.; Chen, K.; Chen, W.; Li, C.T. A Bim-Based Location Aware Ar Collaborative Framework for Facility Maintenance Management. *J. Inf. Technol. Constr.* **2019**, *24*, 30–39. Available online: http://www.itcon.org/2019/19 (accessed on 20 February 2021).
33. Chung, S.W.; Kwon, S.W.; Moon, D.Y.; Ko, T.K. Smart facility management systems utilising open BIM and augmented/virtual reality. In Proceedings of the 35th International Symposium on Automation and Robotics in Construction (ISARC2018), Berlin, Germany, 20–25 July 2018.

34. *International Standard Facility Management—Vocabulary*; ISO: Geneva, Switzerland, 2017.
35. Omar, N.S.; Najy, H.I.; Ameer, H.W. Developing of Building Maintenance Management by Using BIM. *Int. J. Civ. Eng. Technol.* **2018**, *9*, 1371–1383. Available online: https://www.iaeme.com/MasterAdmin/uploadfolder/IJCIET-09-11-132/IJCIET-09-11-132.pdf (accessed on 20 February 2021).
36. Arpaci, I.; Al-Emran, M.; Al-Sharafi, M.A.; Marques, G. Indoor Air Quality Monitoring Systems and COVID-19. *Emerg. Technol. COVID-19 Pandemic* **2021**, *348*, 133–147. [CrossRef]
37. Laustsen, J. Energy efficiency requirements in building codes, energy efficiency policies for new buildings. In *International Energy Agency (IEA)*; IEA Information Paper; IEA: Paris, France, 2008.
38. DoE, U.S. *Buildings Energy Data Book*; Energy Efficiency Renewable Energy Department, IEER: Washington, DC, USA, 2011.
39. Hyvarinen, J.; Karki, S. *Building Optimisation and Fault Diagnosis Source Book*; IEA Annex; International Energy Agency: Paris, France, 1996.
40. Haves, P.; Claridge, D.; Lui, M. *Report assessing the Limitations of EnergPlus and SEAP with Options for Overcoming Those Limitations*; HPCBS #E5P2.3T1; California Energy Commission Public Interest Energy Research Program: Sacramento, CA, USA, 2001.
41. *CIBSE Guide H: Building Control Systems*; Butterworth-Heinemann: Oxford, UK, 2000.
42. Friedman, H.; Piette, M.A. Comparative Guide to Emerging Diagnostic Tools for Large Commercial HVAC Systems. In *Ernest Orlando Lawrence Berkeley National Laboratory, LBNL 48629*; Ernest Orlando Lawrence Berkeley National Laboratory, LBNL: Berkeley, CA, USA, 2001.
43. Glaessgen, E.; Stargel, D. The Digital Twin Paradigm for Future NASA and US Air Force Vehicles. In Proceedings of the 53rd AIAA/ASME/ASCE/AHS/ASC Structures, Structural Dynamics and Materials Conference 20th AIAA/ASME/AHS Adaptive Structures Conference 14th AIAA, Honolulu, HI, USA, 23–26 April 2012.
44. Bonci, A.; Carbonari, A.; Cucchiarelli, A.; Messi, L.; Pirani, M.; Vaccarini, M. A cyber-physical system approach for building efficiency monitoring. *Autom. Constr.* **2019**, *102*, 68–85. [CrossRef]
45. Schluse, M.; Priggemeyer, M.; Atorf, L.; Rossmann, J. Experimentable digital twins. *IEEE Trans. Ind. Inform.* **2018**, *14*, 1722–1731. [CrossRef]
46. Asghari, P.; Rahmani, A.M.; Javadi, H.H.S. Internet of Things applications: A Systematic Review. *Comput. Netw.* **2019**, *148*, 241–261. [CrossRef]
47. Gubbi, J.; Buyya, R.; Marusic, S.; Palaniswami, M. Internet of Things (IoT): A vision, architectural elements, and future directions. *Futur. Gener. Comput. Syst.* **2013**, *29*, 1645–1660. [CrossRef]
48. Pradeep, S.; Kousalya, T.; Suresh, K.M.A.; Edwin, J. IoT and its connectivity challenges in smart home. *Int. Res. J. Eng. Technol.* **2016**, *3*, 1040–1043. Available online: http://www.indjst.org/index.php/indjst/article/view/109387 (accessed on 20 February 2021).
49. Rizal, R.; Hikmatyar, M. Investigation Internet of Things (IoT) Device using Integrated Digital Forensics Investigation Framework (IDFIF). *J. Phys. Conf. Ser.* **2019**, *1179*, 012140. [CrossRef]
50. Rafsanjani, H.N.; Ghahramani, A. Towards utilising internet of things (IoT) devices for understanding individual occupants' energy usage of personal and shared appliances in office buildings. *J. Build. Eng.* **2020**, *27*, 100948. [CrossRef]
51. Wang, M.; Zhang, G.; Zhang, C.; Zhang, J.; Li, C. An IoT-based appliance control system for smart homes. In Proceedings of the IEEE 4th International Conference on Intelligent Control and Information Processing (ICICIP), Beijing, China, 9–11 June 2013; pp. 744–747.
52. Schleich, B.; Anwer, N.; Mathieu, L.; Wartzack, S. Shaping the digital twin for design and production engineering. *CIRP Ann.* **2017**, *66*, 141–144. [CrossRef]
53. Negri, E.; Fumagalli, L.; Macchi, M. A review of the roles of digital twin in cps-based production systems. *Procedia Manuf.* **2017**, *11*, 939–948. [CrossRef]
54. Luo, W.; Hu, T.; Zhu, W.; Tao, F. Digital twin modeling method for cnc machine tool. In *2018 IEEE 15th Int Conf in Networking; Sensing and Control*; IEEE: Zhuhai, China, 2018; pp. 1–4.
55. Grieves, M. Digital Twin: Manufacturing Excellence Through Virtual Factory Replication. Available online: https://www.researchgate.net/publication/329311603_digital_twin_manufacturing_excellence_through_virtual_factory_replication (accessed on 6 June 2021).
56. Saddik, A.E. Digital twins: The convergence of multimedia technologies. *IEEE Multimed.* **2018**, *25*, 87–92. [CrossRef]
57. Boschert, S.; Rosen, R. Digital twin_The simulation aspect. In *Mechatronic Futures*; Springer: Cham, Switzerland, 2016; pp. 59–74.
58. Hicks, B. Industry 4.0 and Digital Twins: Key Lessons from NASA. Available online: https://www.thefuturefactory.com/blog/24 (accessed on 6 June 2021).
59. Nasaruddin, A.N.; Ito, T.; Tuan, T.B. Digital twin approach to building information management. In Proceedings of the Manufacturing Systems Division Conference, Stockholm, Sweden, 16–18 May 2018; p. 304.
60. Rasheed, A.; San, O.; Kvamsdal, T. Digital Twin: Values, Challenges and Enablers from a Modeling Perspective. *IEEE Access* **2020**, *8*, 21980–22012. [CrossRef]
61. Digital Twins for IoT Applications: A Comprehensive Approach to Implementing IoT Digital Twins. Available online: https://pdf4pro.com/view/digital-twins-for-iot-applications-oracle-5b4880.html (accessed on 6 June 2021).

62. Villa, V.; Chiaia, B. Digital Twin for Smart School Buildings: State of the Art, Challenges, and Opportunities. In *Handbook of Research on Developing Smart Cities Based on Digital Twins*; Del Giudice, M., Osello, A., Eds.; IGI Global: Hershey, PA, USA, 2021; pp. 320–340. [CrossRef]
63. Daily, J.; Peterson, J. Predictive maintenance: How big data analysis can improve maintenance. In *Supply Chain Integration Challenges in Commercial Aerospace*; Springer: Cham, Switzerland, 2017; pp. 267–278.
64. Market Research Future. Global Digital Twin Market Is Estimated to Grow at a Cagr of 37% from 2017 to 2023. Available online: https://www.marketresearchfuture.com/reports/digital-twin-market-4504 (accessed on 23 January 2019).
65. NIBS. National BIM Guide for Owners. Available online: https://www.nibs.org/files/pdfs/NIBS_BIMC_NationalBIMGuide.pdf (accessed on 20 February 2021).
66. Hilal, M.; Maqsood, T.; Abdekhodaee, A. A hybrid conceptual model for BIM in FM. *Constr. Innov.* **2019**, *19*, 531–549. [CrossRef]
67. Tagliabue, L.C.; Villa, V. Il Bim per le Scuole. In *Analisi del Patrimonio Scolastico e Strategie Di Intervento*; Il BIM per le Scuole: Hoepli, Italy, 2017.
68. Munir, M.; Kiviniemi, A.; Jones, S.W. Business value of integrated BIM-based asset management. *Eng. Constr. Arch. Manag.* **2019**, *26*, 1171–1191. [CrossRef]
69. Dixit, M.K.; Venkatraj, V.; Ostadalimakhmalbaf, M.; Pariafsai, F.; Lavy, S. Integration of facility management and building information modeling (BIM). *Facilities* **2019**, *37*, 455–483. [CrossRef]
70. Pärn, E.A.; Edwards, D.J.; Sing, M.C.P. The building information modelling trajectory in facilities management: A review. *Autom. Constr.* **2017**, *75*, 45–55. [CrossRef]
71. Patacas, J.; Dawood, N.; Vukovic, V.; Kassem, M. BIM for facilities management: Evaluating BIM standards in asset register creation and service life planning. *J. Inf. Technol. Constr.* **2015**, *20*, 313–331.
72. Gerrish, T.; Ruikar, K.; Cook, M.; Johnson, M.; Phillip, M. Using BIM capabilities to improve existing building energy modelling practices. *Eng. Constr. Archit. Manag.* **2017**, *24*, 190–208. [CrossRef]
73. Thabet, W.; Lucas, J.D. A 6-Step Systematic Process for Model-Based Facility Data Delivery. *J. Inf. Technol. Constr.* **2017**, *22*, 104–131, ITcon, New Zealand. Available online: http://www.itcon.org/2017/6 (accessed on 6 June 2021).
74. Wang, X.; Love, P.E.D.; Kim, M.J.; Park, C.S.; Sing, C.P.; Hou, L. A conceptual framework for integrating building information modeling with augmented reality. *Autom. Constr.* **2013**, *34*, 37–44. [CrossRef]
75. El Ammari, K.; Hammad, A. Remote interactive collaboration in facilities management using BIM-based mixed reality. *Autom. Constr.* **2019**, *107*, 102940. [CrossRef]
76. Saar, C.C.; Klufallah, M.; Kuppusamy, S.; Yusof, A.; Shien, L.C.; Han, W.S. Bim integration in augmented reality model. *Int. J. Technol.* **2019**, *10*, 1266–1275.
77. Naticchia, B.; Corneli, A.; Carbonari, A. Framework based on building information modeling, mixed reality, and a cloud platform to support information flow in facility management. *Front. Eng. Manag.* **2020**, *7*, 131–141. Available online: https://doi.org/10.1007/s42524-019-0071-y (accessed on 6 June 2021). [CrossRef]

Article

The Implementation of Visual Comfort Evaluation in the Evidence-Based Design Process Using Lighting Simulation

Anahita Davoodi *, Peter Johansson and Myriam Aries

Department of Construction Engineering and Lighting Science, Jönköping University, SE-551 11 Jönköping, Sweden; Peter.Johansson@ju.se (P.J.); myriam.aries@ju.se (M.A.)
* Correspondence: anahita.davoodi@ju.se; Tel.: +46-73-910-1581

Abstract: Validation of the EBD-SIM (evidence-based design-simulation) framework, a conceptual framework developed to integrate the use of lighting simulation in the EBD process, suggested that EBD's post-occupancy evaluation (POE) should be conducted more frequently. A follow-up field study was designed for subjective–objective results implementation in the EBD process using lighting simulation tools. In this real-time case study, the visual comfort of the occupants was evaluated. The visual comfort analysis data were collected via simulations and questionnaires for subjective visual comfort perceptions. The follow-up study, conducted in June, confirmed the results of the original study, conducted in October, but additionally found correlations with annual performance metrics. This study shows that, at least for the variables related to daylight, a POE needs to be conducted at different times of the year to obtain a more comprehensive insight into the users' perception of the lit environment.

Keywords: building performance simulation; lighting simulation; lighting quality; visual comfort; office field study; evidence-based design

Citation: Davoodi, A.; Johansson, P.; Aries, M. The Implementation of Visual Comfort Evaluation in the Evidence-Based Design Process Using Lighting Simulation. *Appl. Sci.* **2021**, *11*, 4982. https://doi.org/10.3390/app11114982

Academic Editors: Lavinia Chiara Tagliabue and Ibrahim Yitmen

Received: 29 April 2021
Accepted: 25 May 2021
Published: 28 May 2021

1. Introduction

In the field of architecture, evidence-based design (EBD) is defined as the process of basing design decisions about the built environment on credible research and learning from previous evidence to achieve the best possible outcomes [1,2]. One of the easiest, quickest, and inexpensive methods to provide evidence for predicting or evaluating values in a built environment is using computational modeling. In a framework for evaluating evidence in EBD developed by Pati [3], computer simulation is categorized under 'experiment level', which confirms the application of simulation for providing evidence. For example, in the study by Jakubiec and Reinhart [4], simulation tools were used for the prediction of occupants' visual comfort within daylit environments. The results illustrated that it is possible to use current simulation-based visual comfort predictions to predict occupants' long-term visual comfort assessments in a complex daylit space.

Investigation regarding the application of simulation tools in the EBD process is ongoing [5–9]. A conceptual framework was developed to integrate the use of lighting simulation within the EBD process in a systematic way: the EBD-SIM framework [6]. The study concluded that the translation between the user evaluation and the simulation-based evaluation was a critical step in the integration of lighting simulation with EBD. Therefore, an initial validation study [7] was dedicated to a first application of the EBD-SIM framework in a post-occupancy evaluation (POE) step.

POE is defined as 'the process of evaluating buildings in a systematic and rigorous manner after they have been built and occupied for some time' [10] (p.3). POE's purpose and methodology are varied, often depending on the type of building. For example, POEs of office buildings are, in most cases, interested in occupants' comfort and productivity, utilizing both subjective and objective evaluations of indoor environmental quality (IEQ) [11]. The recent literature review conducted by Dam-Krogh et al. [12] investigated methods applied in previously performed POEs in office buildings with special attention to IEQ and productivity to compare and evaluate successful practices of POEs in office buildings. In more specific and recent POE studies, lighting quality of office buildings was investigated by [13–17]. These studies were concerned with obtaining better insight on occupant satisfaction and acceptance about daylighting by conducting POEs while using photosensors, shading, and/or control systems. This is a key factor for proper lighting design by utilizing daylighting without neglecting human comfort and achieving responsive buildings. Having more information available about occupants' behaviors, needs, and preferences enables architects to design better responsive elements of a building for optimal rates, scales, and types of changes.

While POEs are popular tools for the evaluation of different aspects of building from the occupant's point of view, most POEs are one-off studies [11]. Additionally, using POEs in a framework of EBD is not studied well (yet). One attempt to strengthen the EBD knowledge base by developing standardized POE tools was conducted by [18]. In this study, based on a review of over 100 research publications, a conceptual framework and a set of standard tools were established to comprehensively evaluate building performance in terms of eight key design areas, including air quality, visual environment, thermal comfort, acoustic environment, hazardous materials, conservation of resources, overall climate response, and building envelope (façade).

A real-time case study in a fully operational office building was conducted to analyze visual comfort from a subjective ('the user') and an objective ('the simulation') point of view to provide a systematic performance evaluation using the EBD-SIM framework [7]. It covered both long-term and short-term evaluations of the light environment. The results showed that, although illuminance preference varies significantly among individuals, there was a positive correlation between the overall lighting quality perceived by the occupants and the amount of light on the task area. Regarding performance metrics, the results showed the highest correlation with point-in-time horizontal illuminance (E_h), especially on the task area ($E_{h\text{-task}}$) and human perception. The implementation of the study results in the EBD-SIM process model is schematically illustrated in Figure 1.

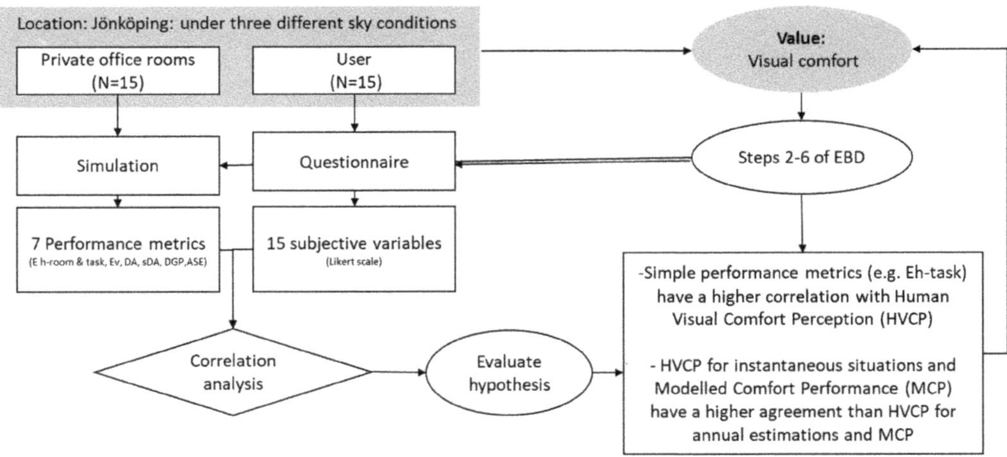

Figure 1. Study results [7] implementation in the EBD-SIM model addressing the workflow during the post-occupancy evaluation steps.

The initial validation study [7] suggested that a POE should be conducted more frequently to obtain a better insight into user perception of daylight and, subsequently, the use of the new evidence to improve the design of the EBD-SIM model further. Therefore, the field study was repeated to perform a POE for a second time, including the same participants but at a different time of year to investigate the possible improvement in the reliability of a POE study (within-subjects study [19]). In addition, a larger sample with a similar procedure was selected in which users experienced only one condition each (between-subject-study [19]) to investigate if previous findings are confirmed with a larger sample or that results are just a coincidence and more detailed simulation of the actual/intended use of a space is required to forecast the visual comfort.

Even though a long-term goal is to show how the EBD-SIM framework can be used to incrementally develop evidence through several projects, the research questions for this project were as follows:

1. How do the most frequently used visual comfort metrics correlate with perceived occupant visual comfort?
2. To what extent do instantaneous and annual human visual comfort perceptions correlate with simulated comfort assessment?
3. How much would the usefulness of a POE improve if its frequency were increased?

Note that the first two questions are repeated from the preceding study [7], though with a larger sample. The most frequently used visual comfort metrics were identified based on the state-of-the-art literature review conducted by [7].

The third question was added specifically for the follow-up POE study.

2. Materials and Methods

2.1. Research Design

A study in a fully operational office building focusing on visual comfort analysis was conducted to explain the application of the EBD-SIM framework in the POE step. Objective data were collected via computational modeling and subjective data by using an online questionnaire. Calculated performance metrics and questionnaire results were collected similarly to the previous study [7]. Previously collected data were included in the study as part of the repeated measures method.

2.2. Procedure

At first, the physical environment was modeled, and performance metrics were calculated. The simulation results were compared against a limited and randomized set of illuminance measurements in the real building to check if they were in the same range.

The results showed a difference between simulated and real values of $20 \pm 11\%$, which is within normal variability levels for simulated objects [20]. Secondly, using an online questionnaire, occupant characteristics were recorded, and their feedback regarding visual comfort was gathered. Finally, user feedback and simulation output were compared. This comparison was performed to investigate how metrics measured by simulation tools (to assess visual comfort) correlate with actual visual comfort perceived by the users, and to answer the research questions. The first sample of the population ($N = 15$) filled out the questionnaire in October, and the second sample of the population ($N = 46$) filled it out in June. For $N = 10$ of these people, all working in one wing of the survey location, the questionnaire was a repetition. In total, $N = 51$ unique people participated in the study, which means, in total, $N = 61$ complete survey results were gathered.

An overview of the three research questions and the results of the comparison between POE and the performance metrics (PM) for different groups of participants at different data collection moments is illustrated in Figure 2.

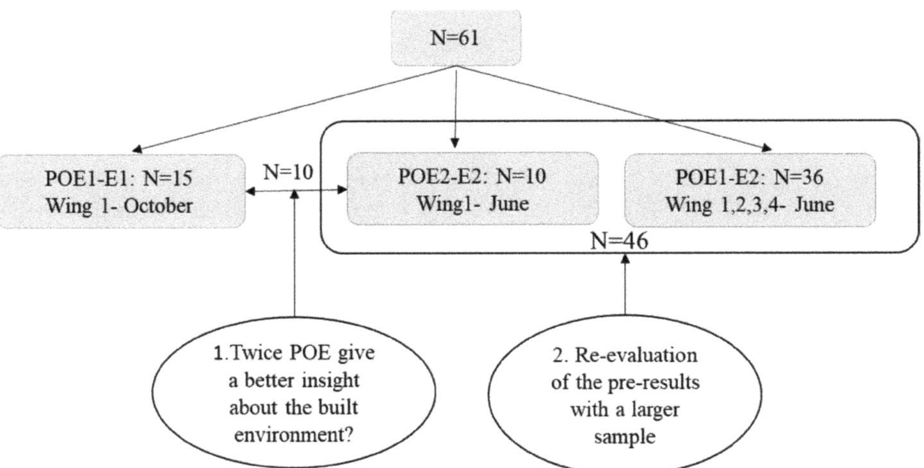

Figure 2. Three main research questions and the results of POE vs. PM comparison for different groups of participants during the first and second times (October and June).

2.3. Site

The study was carried out in a four-story academic building (with a basement) located in Jönköping, Sweden (lat. 57.778168° N, long. 14.163526° E). To the left and right of the building are buildings of comparable height. Physical data were collected from 51 private office rooms on the second, third, and fourth floors. Most of the rooms are approximately 15 ± 5 m^2 and are furnished mainly by a large desk, a chair, and one or more bookshelves. The rooms have windows in the north-east (21), south-west (19), north (9), and south-east orientations (7). Three rooms on the north-east and south-west side have a second window in the same direction. Five rooms on the south-west side have a second window on the south-east side. The number of rooms with two windows is highlighted in the parenthesis in Table 1. The size of the windows on the second floor on the north-east side is one third larger compared to that on the third and fourth floors. There is permanent solar shading at the building's south and south-west sides (see Figure 3). All rooms are equipped with conventional suspended luminaires and were used according to the users' needs during the survey, but they were not modeled nor further included in the study.

Table 1. Number of rooms based on the location of the windows; rooms with two windows are shown within parenthesis.

Window Location	Frequency	Percentage
North-East (NE)	21(3)	41.2
South-West (SW)	14(3)	37.3
North (N)	9	17.6
South-East (SE)	7(5)	3.9
Total	51	100

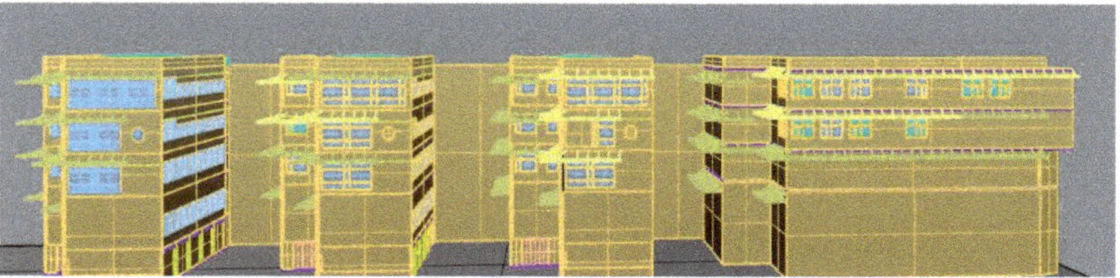

Figure 3. Photo of the real building (**top**) and a 3D render of the building model (**bottom**) showing the south and south-west façades including the permanent overhang. Note that on this side of the building, the basement level is above the ground.

2.4. Survey Participants

Occupants of four wings of the building were invited to participate in the survey. They were all academic employees whose work involves research and teaching/education, mainly using computers. The invitation email was sent to 150 people, of which 61 people (39 male/22 female, average age 45 ± 20 years) responded (response rate 40.7%). For $N = 36$ participants, this was the first time they filled out the survey; $N = 15$ people answered during the first study in early October [7], and $N = 10$ of this group participated for the second time in this study, see Table 2. (in parenthesis: the respondents who were inquired twice). Most participants in this study worked for more than one year in their current office room, and more than half of the participants ($N = 36$) reported wearing vision aids (near vision = 13, distance vision = 10, both = 11, trifocal = 1, other = 2).

Table 2. Number of respondents per floor and wing—within parenthesis: the respondents who were inquired twice.

	Wing 1	Wing 2	Wing 3	Wing 4	Total
Floor 2	12 (5)	4	3	4	23
Floor 3	16 (5)	4	2	8	30
Floor 4	1	1	2	4	8
Total	29	9	7	16	61

2.5. Performance Metrics for Visual Comfort

Seven currently used performance metrics for visual comfort were obtained, including five metrics related to illuminance values ($E_{v\text{-}eye}$, $E_{h\text{-}room}$, $E_{h\text{-}task}$, Daylight Autonomy DA, Spatial Daylight Autonomy sDA) and two glare indicators (Daylight Glare Probability DGP, Annual Sunlight Exposure ASE), see Table 3. Further details are explained in [7].

Table 3. Collected visual comfort metrics (daylighting).

Criteria	Metrics	Time
Illuminance (under CIE Standard Clear Sky, CIE Intermediate Sky, and CIE Standard Overcast Sky)	Mean/maximum/minimum illuminance value on the task area (Eh-Task)	Moment
	Mean/maximum/minimum illuminance value on room surface (Eh-Room)	Moment
	Vertical eye illuminance (Ev-eye)	Moment
	Daylight Autonomy (DA)	Annual
	Spatial Daylight Autonomy (sDA)	Annual
Glare (under CIE Standard Clear Sky, CIE Intermediate Sky, and CIE Standard Overcast Sky)	Daylight Glare Probability (DGP)	Moment
	Annual Sunlight Exposure (ASE)	Annual

2.6. Simulation Model

The building was modeled in Autodesk Revit. For detailed daylight analysis, the model was exported to Rhinoceros 3D to perform lighting analysis with the DIVA plugin. In addition to the simulated building wing, direct surroundings were also modeled to consider potential shading. The material properties and simulation parameters used for the calculations are described in Table 4.

Table 4. Material properties and simulation parameters.

Material	Reflectance/Visible Light Transmittance (Tvis) in %
Walls	White interior wall 70%
Floor	Generic floor 20%
Ceiling	Generic ceiling 70%
Desk	Generic furniture 50%
Glazing	Glazing double pan 80%
External Shading	Metal sheet
Surrounding Building	Outside faced 35%
Ground	Outside ground 20%
Simulation Parameters for Daylight Autonomy	
Weather Data	Goteborg—Landvetter Airport, Sweden
Occupancy Schedule	8–18 with Daylight Saving Time (DST) 60 min
Target Illuminance	300 lux
Radiance Parameters	-ab2 -ad 1000 -as 20 -ar 300 -aa
Sensor Density	Grade sensor—750 mm above 450 mm distance
Shading	No automated shading
Electric Lighting	No electric lighting

For each workplace area on the second, third, and fourth floors, a grid of 1.45 × 1.45 m at workplace height (h = 0.75 m) was used to calculate the mean, minimum, and maximum horizontal illuminance under CIE Standard Clear Sky, CIE Intermediate Sky, and CIE Standard Overcast Sky conditions [21]; for the rest of the paper, these are referred to as 'clear', 'overcast', and 'intermediate' sky conditions. The same setting was applied for the room area of each room for horizontal illuminance ($E_{h\text{-room}}$), ASE, and DA calculations. One point in the middle of each room with a viewing direction towards windows was selected for the vertical eye illuminance ($E_{v\text{-eye}}$) and DGP calculations under clear, intermediate, and overcast sky conditions. As the questionnaires were filled out in early October and early June, a day in early October (11) and early June (4) was chosen for the simulation to have a comparable day length and sun path. In total, 67.2% of the questionnaires were filled out in the morning; hence, for the simulation, a time in the morning was selected (10 a.m.).

2.7. Questionnaire

The questionnaire was a web-based form in English to collect feedback regarding the visual comfort perception of office occupants. It contained questions regarding office characteristics, satisfaction with the lit environment in general ('annual'), satisfaction with the lit environment at the time of response ('momentary'), user preferences, and personal information. For 15 satisfaction variables of the lit environment 'in general' (annual) and 'at the time of response' (momentary), a 7-point Likert satisfaction scale [22] was used (1 = very satisfactory to 7 = very unsatisfactory). All variables were included in the analysis, see Table 5.

Table 5. Description of the variables included in the analysis with their variable name. Satisfaction variables are all on a 7-point Likert scale (1 = very satisfactory to 7 = very unsatisfactory).

	Variable Description	Variable Name
Momentary	Satisfaction with light at the desk area	Light desk
	Satisfaction with natural light (daylight)	Natural
	Satisfaction with artificial (electric lighting)	Artificial
	Satisfaction with glare from sunlight	Glare sun
	Satisfaction with glare from artificial lighting	Glare artificial
	Satisfaction with lighting quality	Lighting quality
Annual	Satisfaction with light at the desk area	A-Light desk
	Satisfaction with natural light (daylight)	A-Natural
	Satisfaction with artificial (electric lighting)	A-Artificial
	Satisfaction with glare from sunlight	A-Glare sun
	Satisfaction with glare from artificial lighting	A-Glare artificial
	Satisfaction with lighting quality	A-Lighting quality
	Satisfaction with the view to outside	A-View
	Satisfaction with overall indoor environmental quality (i.e., thermal, acoustic)	A-IEQ
	Satisfied with job	A-Job

2.8. Analysis

The correlation of all 15 satisfaction variables as well as the correlation among these variables with seven groups of performance metrics was explored by calculating Pearson correlation coefficients using IBM SPSS Statistics 24. To determine whether there is statistical evidence that the mean difference between paired observations on a particular outcome is significantly different from zero, a paired sample t-test was run for ten participants who filled out the questionnaire two times during early October and June. Only significant correlations are reported.

3. Results

In this section, the objective, subjective, and correlation analysis of the follow-up POE study for ten participants who filled out the questionnaire two times during early October and June as well as the larger sample ($N = 46$) who filled out the questionnaire only once in June is reported.

3.1. Within-Subject Analysis (Follow-Up POE Investigation)

3.1.1. Objective Comparison

1. Horizontal illuminance on a room area ($E_{h\text{-room}}$)—The overview of the results for the mean value of horizontal illuminance $E_{h\text{-room}}$ of the calculated sensor points of the ten rooms' areas at the work plane height is shown in Figure 4a,b. Where the mean $E_{h\text{-room}}$ values in October were below 200 lux for all sky conditions, the amount of daylight increased, as expected, during the second POE in June. On average, the rooms received 200 lux or more horizontal illuminance during June. Since the participants mostly worked under the clear sky condition, more details of the results for the clear sky condition are illustrated in the right image.

2. Horizontal illuminance on a task area ($E_{h\text{-}task}$)—Based on the position of the desk area in each room, the mean horizontal illuminance received on task area $E_{h\text{-}task}$ was calculated. As presented in Figure 4c,d, on average, the amount of daylight received at the task area in June was approximately 100 lux higher compared to October. Under the clear sly condition, the areas received, on average, 300 lux with a variation between 100 and 800 lux (see Figure 4d).
3. Vertical eye illuminance ($E_{v\text{-}eye}$)—Vertical eye illuminance $E_{v\text{-}eye}$ was calculated at the center point of each room with a viewing direction towards the daylight opening. The results of $E_{v\text{-}eye}$ under three different sky conditions are illustrated in Figure 4e.
4. Daylight Autonomy (DA) and Spatial Daylight Autonomy (sDA)—The highest value for DA was found for the rooms on the second floor, with an average of 44 ± 8%. The rooms on the third floor at the north-east side had DA values around 29 ± 3%, and the south-west side had the lowest DA values around 2 ± 0.3%. sDA values of all rooms were lower than 50% of the space area.
5. Daylight Glare Probability (DGP)—The center point of each room with a direction towards windows was elected for DGP analysis. The results of this analysis show that, under the 'clear' sky condition, DGP values were in the range of intolerable glare (DGP > 0.45) for 70% of the rooms in October and for all the rooms in June. Under 'overcast' and 'intermediate' sky conditions, DGP values were below the 'imperceptible' level (DGP < 0.35) in October and at the 'perceptible' level (0.40 < DGP < 0.35) in June. The mean DGP values for the rooms under the three sky conditions are presented in Figure 5. The variations in DGP values under the clear sky condition are presented in the right image.
6. Annual Sunlight Exposure (ASE)—For all rooms, less than 10% of the areas had an illuminance value higher than 1000 lux/250 h per year, which means that, according to the LEED certification, the rooms are categorized as 'a room with not too much direct sunlight during the year' [23].

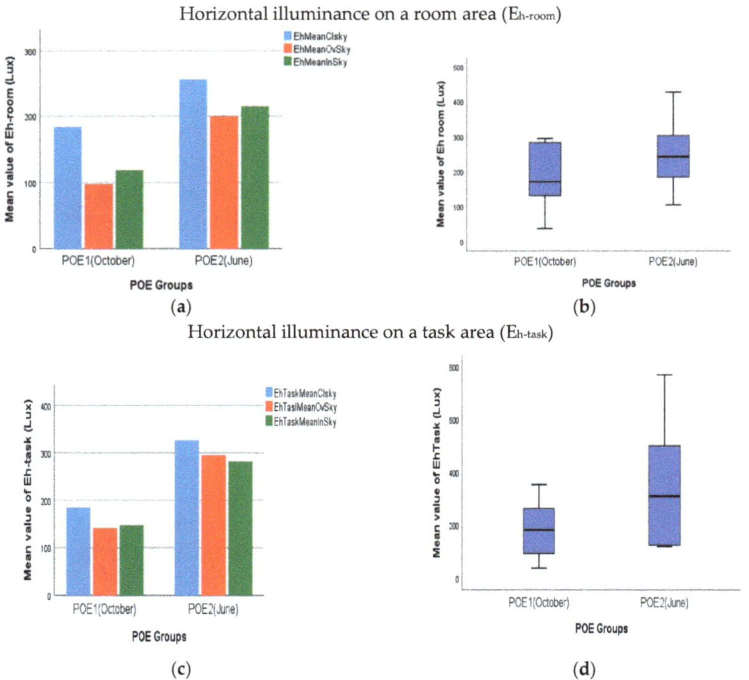

Figure 4. *Cont.*

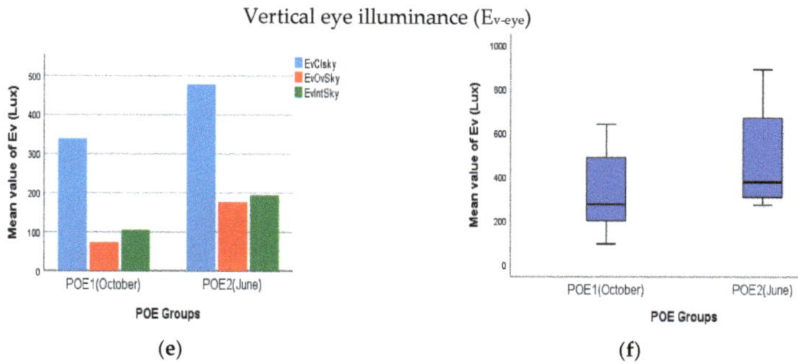

Figure 4. Overview of the results for the mean value of (**a**) horizontal illuminance of the calculated sensor points of the ten rooms' areas at the work plane height ($E_{h\text{-room}}$), (**c**) horizontal illuminance on the task area of the calculated sensor points of the ten rooms' areas ($E_{h\text{-task}}$), and (**e**) mean value of vertical illuminance in the center point of each room with viewing direction towards the daylight opening of the ten rooms (E_v) under three different sky conditions for 11 October and 4 June both for 10 a.m. (**a**,**c**,**e**). The details of the mean values under clear sky condition (**b**,**d**,**f**).

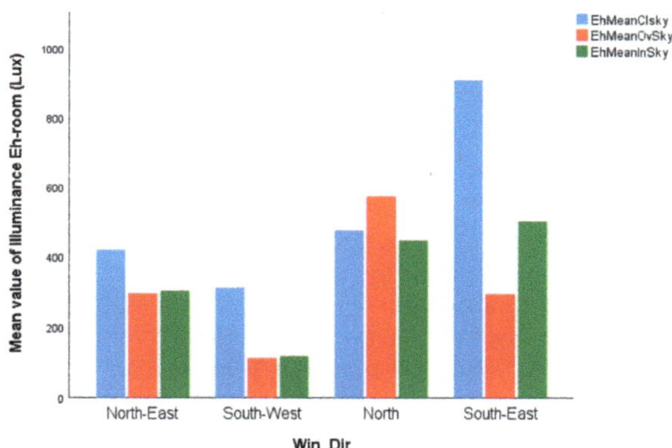

Figure 5. (**a**) DGP values under clear sky (blue), overcast sky (red), and intermediate sky (green) conditions: imperceptible level (DGP < 0.3), perceptible level (0.3 < DGP < 0.35), disturbing glare (0.35 < DGP < 0.4), intolerable glare (DGP > 0.45); (**b**) The details of the mean values under clear sky condition.

3.1.2. Subjective Comparison

Descriptive statistics—Descriptive statistics analysis for all 15 satisfaction variables shows that the mean values of all variables except 'Glare sun' (satisfaction with glare from the sun at the moment) and 'A-artificial' (annual satisfaction with artificial lighting) were a bit lower for the second POE in June compared to October. On average, for the satisfaction values, the change was 0.4 points. This means people were more satisfied with the built environment in general in June (1 = very satisfactory, 7 = very unsatisfactory). Similarly, in October, the highest satisfaction belonged to 'satisfaction with the job', which increased slightly in June. Then, 'satisfaction with daylight' (both 'annually' and 'at the moment') had the highest satisfaction rate in June ($M = 2.3$, $SD = 1.5$). Satisfaction with daylight 'at the moment' with a 1.8 point improvement from October shows the highest increase in satisfaction. The least satisfaction in June belonged to artificial light ($M = 3.5$, $SD = 1.58$). See Table 6 for more details on other variables.

Paired samples *t*-test—To determine whether there is statistical evidence that the mean difference between paired observations on a particular outcome is significantly different from zero, a paired sample *t*-test was run for ten participants who filled out the questionnaire two times during early October and June. The results show that, although the mean satisfaction values are increased from October to June for almost all variables, a significant difference was only found in scores for natural light from October to June (*sig* = 0.016, *M* = 1.80, *SD* = 1.93). In addition, a marginally significant difference was found for 'A-Glare sun' (*sig* = 0.052, *M* = 0.50 *SD* = 0.71) and 'Light desk' (*sig* = 0.053, *M* = 0.80 *SD* = 1.13).

Table 6. Descriptive statistics analysis of all 15 satisfaction variables (1 = very satisfactory, 7 = very unsatisfactory) of the first POE in October (*N* = 10) and the second POE in June (*N* = 10) of within-subjects study.

Variable	Mean ± Std. Deviation (October)	Mean ± Std. Deviation (June)
A-light desk	3.5 ± 0.97	3.2 ± 1.23
A-Natural	3.0 ± 1.33	2.3 ± 0.48
A-Artificial	3.4 ± 1.17	3.5 ± 1.72
A-Glare sun	3.4 ± 1.58	2.9 ± 1.52
A-Glare artificial	3.5 ± 1.65	3.0 ± 1.63
A-View	3.7 ± 2.58	3.4 ± 2.50
A-Lighting quality	3.5 ± 1.35	3.3 ± 0.95
A-Overall IEQ	3.3 ± 1.06	2.8 ± 1.23
A-Job	2.1 ± 2.08	1.9 ± 1.85
Light desk	3.9 ± 0.99	3.1 ± 1.45
Natural	4.1 ± 1.73	2.3 ± 1.50
Artificial	3.5 ± 0.97	3.5 ± 1.58
Glare sun	2.6 ± 1.35	2.9 ± 1.45
Glare artificial	3.3 ± 1.70	2.8 ± 1.62
Lighting quality	3.5 ± 1.18	3.0 ± 1.49
Valid N (listwise)	10	10

Correlation analysis—The results show that there were 'very strong' ($0.8 < r < 1$) to 'strong' ($0.6 < r < 0.79$) degrees of correlations for all the variables between annual and at the moment evaluations except for 'Natural light' in October. In June (*N* = 10), the highest correlation was found for 'Glare sunlight' (0.95, $p < 0.01$) and 'Glare artificial light' (0.92, $p < 0.01$). This means that participants either perceived glare as similar all year round or found it hard to recall the difference between glare at the moment and annually.

3.2. Between-Subjects Analysis

All participants who responded to the survey in June (*N* = 46) worked in individual office rooms. These rooms are located on the second, third, and fourth floors of the building's four wings. Based on the location of the window, these rooms are categorized into four groups which are presented in Table 7.

Table 7. Number of rooms based on the location of the windows—within parenthesis: the rooms with two windows.

Window Orientation	Frequency	Percentage
North-East (NE)	19(3)	41.3
South-West (SW)	11(1)	23.9
North (N)	9	19.6
South-East (SE)	7(5)	15.2
Total	46	100

There are three rooms at the north-east side and one room at the south-west side with two windows on the same side. Five rooms at the south-east side have a second window at the south-west side. These rooms are highlighted in parenthesis in Table 7.

3.2.1. Objective Evaluation Using Light Simulation

1. Horizontal illuminance in a room area ($E_{h\text{-room}}$)—The overview of the results for the mean value of horizontal illuminance $E_{h\text{-room}}$ of the calculated sensor points of the room areas at work plane height is illustrated in Figure 6. On average, the rooms received more than 300 lux light under the clear sky condition. Since most of the rooms on the south-east side have two windows, they receive the highest value of illuminance compared to the others. There is no surrounding building on the north side; therefore, this side received the second highest values with low variance for all three sky conditions.
2. Horizontal illuminance in a task area ($E_{h\text{-task}}$)—Based on the position of the desk area in each room, one sensor point was selected to calculate the horizontal illuminance incident on task areas. As illustrated in Figure 7, on average, the amount of daylight incident at the task areas is slightly higher compared to the mean illuminance value of the rooms under the clear sky condition.
3. Vertical eye illuminance (Ev-eye)—In the center point of each room with a viewing direction towards the daylight opening, the vertical eye illuminance was calculated under three different sky conditions, and the results are illustrated in Figure 8.
4. Daylight Autonomy (DA) and Spatial Daylight Autonomy (sDA)—The rooms on the north sides had DA values with a mean of (67.2 ± 3.5)%. The rooms on the south-east side had a mean DA value of (63 ± 13)%. At the north-east side, the mean value of DA was around (31.5 ± 9)%. The south-west side had the lowest mean DA value of (20 ± 14)%. The average of sDA values of all rooms was lower than 50% of the space area, but there are a few rooms on the south-east side with sDA values of around 70%.
5. Annual Sunlight Exposure (ASE)—None of the rooms at the north-east, north, and south-west sides exceed 1000 lux for more than 250 h per year for more than 10% of the space area of each room. The mean value of ASE for the rooms at the south-east side, however, received more than 1000 lux for 243 ± 156.5 h in a year, which means that some of the rooms on this side are categorized as 'rooms which receive too much direct sunlight during the year' (USG) [23].
6. Daylight Glare Probability (DGP)—The center point of each room with a viewing direction toward the windows was selected for DGP analysis. The results of this analysis show that for all rooms under the 'overcast' and 'intermediate' sky conditions, the DGP value is below the 'perceptible' level (0.30 < DGP < 0.35). The DGP values under the clear sky condition for the rooms on the north-east, south-west, and south-east side are in the range of 'intolerable glare' (DGP > 0.45). For more details, see Figure 9.

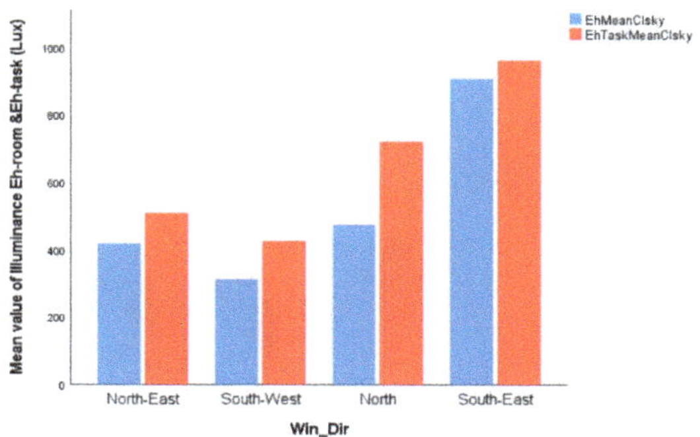

Figure 6. Overview of the results for the mean value of horizontal illuminance of the calculated sensor points of the room areas at work plane height.

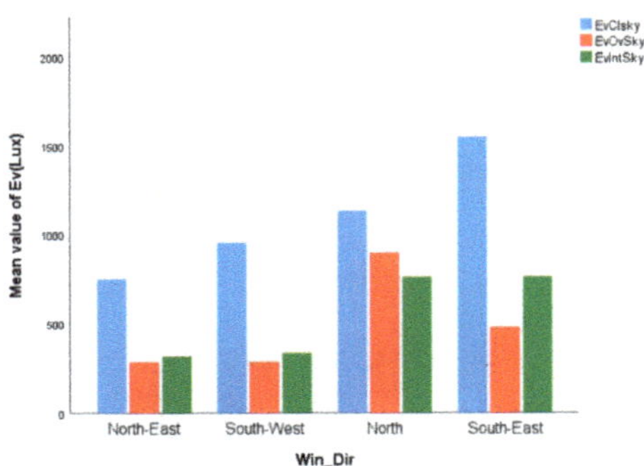

Figure 7. Comparison of the amount of daylight incident at the task area against the mean value of $E_{h\text{-room}}$.

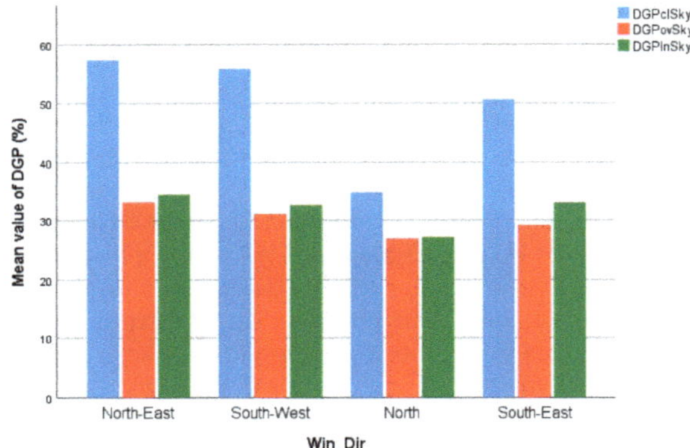

Figure 8. The overview of the results for the mean value of vertical illuminance in the center point of each room with viewing direction towards the daylight opening of the ten rooms under three different sky conditions on 4 June at 10 a.m.

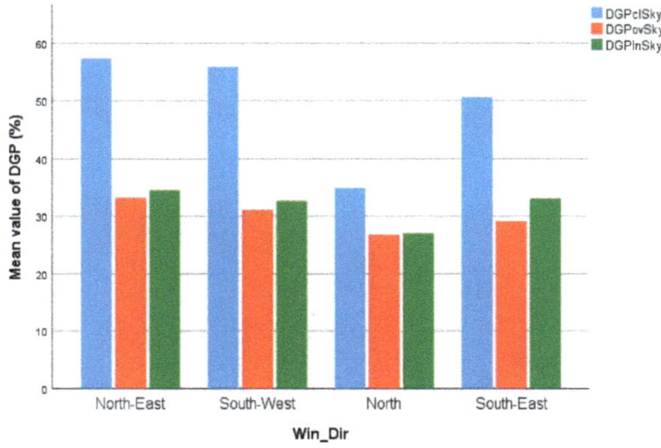

Figure 9. DGP values under clear sky (**blue**), overcast sky (**red**), and intermediate sky (**green**) conditions: imperceptible level (DGP < 0.3), perceptible level (0.3 < DGP < 0.35), disturbing glare (0.35 < DGP < 0.4), intolerable glare (DGP > 0.45).

3.2.2. Subjective Evaluation Using Questionnaires

Descriptive statistics—The sample of all participants in June (N = 46) shows a similar trend to the smaller sample. This means that the highest satisfaction in June belonged to satisfaction with 'A-Job' (M = 2.4, SD = 1.87). After that, daylight (both annually: 'A-Natural', and moment: 'Natural') with (M = 2.9, SD = 1.75 and M = 2.8, SD = 1.72) and 'Glare artificial' (M = 2.8, SD = 1.6) had the highest satisfaction rates. For more details, see Table 8.

Table 8. Descriptive statistics analysis of all 15 satisfaction variables (1 = very satisfactory, 7 = very unsatisfactory).

Variable	Mean ± Std. Deviation (JUNE, N = 46)
A-Light Desk	3.0 ± 1.56
A-Natural	2.9 ± 1.75
A-Artificial	3.3 ± 1.71
A-Glare Sun	3.4 ± 1.76
A-Glare Artificial	3.0 ± 1.72
A-View	3.3 ± 2.23
A-Lighting Quality	3.3 ± 1.42
A-Overall IEQ	3.4 ± 1.60
A-Job	2.4 ± 1.87
Light Desk	3.0 ± 1.63
Natural	2.8 ± 1.72
Artificial	3.1 ± 1.66
Glare Sun	3.3 ± 1.72
Glare Artificial	2.8 ± 1.59
Lighting Quality	3.0 ± 1.61
Valid N (listwise)	46(42)

Correlation analysis—The correlation of all fifteen satisfaction variables was explored by calculation of Pearson correlation coefficients, and only significant correlations are reported here. The results of correlation analysis for the larger samples (N = 46 and N = 61) show that there were 'very strong' ($0.8 < p < 1$) degrees of correlations for all the variables between annual and at the moment variables. Similarly, for the smaller samples (N = 10 for both studies in June and October), the highest correlation was found for 'Glare artificial light'. For daylight (A-Natural vs. Natural), a correlation was found only for June in both small and large samples. For more details, see Table 9.

Table 9. Correlation analysis for variables between 'annual' and 'at the moment'.

Annual/Moment Variables	POE1-Oct. (N = 10)	POE2-Jun. (N = 10)	Jun. (N = 46)	Oct. + Jun. (N = 61)
A-Light Desk vs. Light Desk	86 **	68 *	84 **	84 **
A-Natural vs. Natural	not significant	78 **	87 **	80 **
A-Artificial vs. Artificial	88 **	84 **	86 **	86 **
A-Glare Sun vs. Glare Sun	81 **	95 **	87 **	85 **
A-Glare Artificial vs. Glare Artificial	97 **	92 **	89 **	89 **
A-Light Quality vs. Light Quality	94 **	71 *	86 **	87 **

* $p \leq 0.05$; ** $p \leq 0.01$.

All the samples consistently showed the strongest correlation between 'Light quality' and 'Light desk' for both annual and moment situations. For the large sample in June (N = 46), all correlations of the variables for the moment situation were stronger than the same annual situation. For example, the correlation of 'Light desk' with 'Light quality' was (r = 0.87, $p < 0.01$), and the correlation of 'A-Light desk' with 'A-Light quality' was (r = 0.85, $p < 0.01$).

Both daylight and artificial light seem to contribute to the user assessments of the light quality in the room, with the correlation between 'Artificial light' and 'Lighting quality' ($r = 0.81$, $p < 0.01$) being slightly higher than the correlation between 'Natural light' and 'Lighting quality' ($r = 0.77$, $p < 0.01$). Both sources of glare also contributed to the assessments of the lighting quality in June with the same weight ($r = 0.71$, $p < 0.01$).

Unlike in October, when the assessments of the amount of light on the desk area ('Light desk') seem to only be linked to the amount and quality of artificial lighting, in June, both light sources seem to have contributed to the assessments of the 'Amount of light on desk area'. For both light sources, the correlations were stronger for at the moment assessments compared to the annual assessments. 'Amount of artificial light (moment)' and 'Amount of daylight (moment)' showed strong correlations with the 'Amount of light on desk area' ($r = 0.80$, $p < 0.01$ and $r = 0.71$, $p < 0.01$, respectively). 'Glare from artificial light' and 'Glare from sunlight' showed a lower correlation with 'Light desk' ($r = 0.76$, $p < 0.01$ and $r = 0.64$, $p < 0.01$, respectively).

3.3. Objective–Subjective Correlation Analysis

The correlation between performance metrics and perceived visual comfort for the sample in June ($N = 46$) shows that, in total, eight subjective variables including satisfaction with 'Glare sun', daylight (both annual and at the moment: 'A-Natural' and 'Natural' variables, respectively), 'A-View', total light at the desk ('A-Light desk' and 'Light desk'), and lighting quality ('A-Light quality' and 'Light quality') had a moderate ($0.3 < r < 0.5$, $p < 0.001$) relationship with at least one of the simulated performance metrics. All variables except glare had positive correlations with user satisfaction. Note that it is shown as negative numbers since user satisfaction was defined in inverse order, with 1 being the highest satisfaction level and 7 being the lowest satisfaction level.

The variable 'Glare sun' had the highest correlation with point-in-time horizontal illuminance (momentary) at task area ($E_{h\text{-}task}$) for mean values under the clear sky condition ($r = 0.38$, $p < 0.01$) and the intermediate sky condition ($r = 0.33$, $p < 0.01$).

Satisfaction with daylight for both the annual and momentary situations ('A-Natural' and 'Natural') showed a significant correlation with $E_{h\text{-}room}$, ASE, and DGP. The highest correlations for $E_{h\text{-}room}$ with 'A-Natural' were for the mean value of illuminance under clear sky and intermediate sky conditions ($r = -0.30$, $p < 0.01$), both classified as moderate correlations. Additionally, 'Natural' and 'A-Natural' showed a moderate correlation with DGP ($r = -0.41$, $p < 0.01$).

Since the view to the outside is often inextricably linked to daylight, it was included in the analysis as well. The 'A-View' variable correlated with vertical illuminance ($E_{v\text{-}eye}$), horizontal illuminance at task area ($E_{h\text{-}task}$), and horizontal illuminance for the room area ($E_{h\text{-}room}$) as well as with DA, sDA, and ASE. The highest correlation was found for the mean value of $E_{h\text{-}room}$ under 'clear' sky conditions ($r = -0.46$, $p < 0.01$).

'A-Light desk' and 'Light desk' showed correlations with the horizontal illuminance at task area ($E_{h\text{-}task}$) and room area ($E_{h\text{-}room}$) as well as with DA and sDA. Additionally, 'Light desk' correlated with vertical illuminance ($E_{v\text{-}eye}$) and 'A-Light desk' with ASE. The highest correlation for 'A-Light desk' was found for the minimum value of $E_{h\text{-}room}$ under intermediate sky conditions ($r = -0.47$, $p < 0.01$). The highest correlation for 'Light desk' was found for the minimum value of $E_{h\text{-}room}$ under 'intermediate' sky conditions ($r = -0.47$, $p < 0.01$).

Finally, 'Light quality' and 'A-Light quality' had significant correlations with $E_{h\text{-}room}$, sDA, and ASE. The highest correlation for 'Light quality' was found for the minimum value of $E_{h\text{-}room}$ under 'clear' sky conditions ($r = -0.36$, $p < 0.01$).

4. Discussion

This follow-up study, conducted in June, is the second validation test of the EBD-SIM framework to provide new evidence to obtain better insights about the lit environment, specifically concerning visual comfort in office rooms. The effects of a larger sample size and having two POE studies in two different seasons (October, June) were investigated to find out if previous findings are confirmed (research questions 1 and 2), and to elucidate the usefulness of having two continuous POE studies for better analysis of visual comfort in office environments (research question 3). The results are categorized into two groups: within-subject study and between-subject study.

The main purpose of conducting a between-subject study in June with a larger sample (N = 46) was to verify previous findings in October (N = 15) and provide new or updated evidence for the EBD-SIM framework. Comparison between the two measurement moments showed similarities as well as differences.

The correlation analysis of subjective variables for both studies shows a 'strong' to 'very strong' degree of correlations for all annual and momentary variables except daylight (A-Natural vs. Natural), for which a correlation was not found in October. These strong to very strong correlations can be interpreted in that it is difficult for people to remember a lighting situation throughout the year, and the current situation dominates their feeling regarding the lit environment. Additionally, for daylight (A-Natural vs. Natural), it seems that occupants can better distinguish the difference between annual and momentary situations during dark seasons. From a subjective point of view, for all variables, it was consistently observed that the overall lighting quality perceived by the occupants had the highest correlation with the amount of light on the task area. While in October, the assessment of 'light quality at the task area' seemed to only be linked to the 'amount and quality' of artificial lighting, in June, in addition to artificial lighting, a strong correlation was found with 'Natural light' and 'Glare sun'.

Regarding the first question related to the correlation between visual comfort metrics and perceived occupant visual comfort, similar to the results obtained in October, the results confirm the previous finding that point-in-time-horizontal illuminance had the highest correlation with perceived visual comfort by occupants. Moreover, in the larger study (June), annual performance metrics showed some degrees of correlation. This means that it is worth calculating sDA, DA, and ASE to provide a moderate forecast on occupant perception of lit environments, especially for 'Light desk', 'Light quality', and 'Natural'. This analysis is in agreement with other studies, e.g., [15,24]. Since the results of vertical eye illuminance (E_V) and DGP are sensitive to the point of view of the occupants [25], the calculation of data specifically for the occupants' position and viewpoint improves the accuracy of the data.

Regarding the second question related to instantaneous and annual visual comfort perception and simulated comfort assessment metrics, it was found that for the large sample in June, all correlations of the subjective variables for the moment situation were stronger than the equivalent annual situation. Additionally, E_h, which measures the instantaneous situation, showed the highest number of correlations with perceived variables compared to the annual performance metrics. This could be interpreted in that human visual comfort perception for instantaneous situations and modeled comfort performance metrics have a higher agreement than human perception for annual estimations and modeled performance. In the future, POEs can be integrated into, e.g., (artificially) intelligent building control systems, providing direct feedback to the control agent so that each occupant is provided with their preferred lighting and, in a broader sense, with other desired IEQ conditions. Additional input from occupants such as user characteristics (e.g., age, gender, light sensitivity), user behavior, and user preferences can be collected to analyze the effects of the built environment on occupants in greater depth. Logging this feedback data and storing it in a database can provide a set of valuable evidence from the instantaneous feedback of occupants, which in turn can help with better prediction of human comfort and improvement of lighting design. For example, in an innovative study conducted by Newsham et al. [26], along with lighting simulation tools, a humanoid robot was used to attract the attention of the occupant about their real environmental situation and provide them with personalized suggestions to improve their well-being. If a building is responsive to the requirements and behavior of occupants and organizations, either via a POE and/or via continuous monitoring, it can become a truly intelligent building. Conducting metadata analysis on all data collected from evidence and designing building interfaces or (self-)learning systems based on this evidence-based knowledge would make the interaction of occupants with buildings more convenient.

Regarding the third research question related to the usefulness of a POE with an increased frequency, the results show that in June, people were generally more satisfied with the lit environment compared to answers given in the original study performed in October. As daylight levels are higher in June, it seems that having more daylight has a positive impact on user satisfaction in general. The importance of daylight on user satisfaction was also shown by other researchers, e.g., [27–30]. For example, a study was conducted by Day, Futrell, Cox, and Ruiz [28] that measured physical data and surveys to assess occupants' subjective visual comfort. The results of their survey study ($N = 1068$) showed that occupants who were more pleased with (their access to) daylight were also more likely to have a higher level of satisfaction. This finding also indicates that, depending on the time of year, a single POE study can overestimate or underestimate different lighting quality metrics. The paired sample t-test analysis shows a significant difference for the daylight 'Natural' variable and a marginally significant difference for the daylight 'A-Glare sun' and 'Light desk' variables, which means at least for these variables, it is worth running a POE at least twice at different times of the year.

5. Conclusions

This study addressed implementing subjective–objective results in the evidence-based design process using lighting simulation tools in a POE step of the EBD-SIM framework. The POE focused on assessing occupant visual comfort in an individual office space to provide a systematic approach to repeatedly gather evidence in this field and build a knowledge database that can help improve how the results are analyzed, presented, and interpreted. As this study shows, running the EBD-SIM framework each time can provide new evidence. In the meantime, it helps to tailor the next run based on lessons learned in the previous run.

The results confirm the previous finding that the overall lighting quality perceived by the occupants had the highest correlation with the amount of light on the task area. In parallel, E_h (point-in-time horizontal illuminance) showed a consistently positive correlation with the highest number of subjective variables. Moderate correlations between annual performance metrics and some of the subjective variables were also observed during June, which were non-existent in the October study.

This study suggests that, at least for daylight-related variables (e.g., 'Natural', 'A-Glare sun', and 'Light desk'), it is worth running the POE more than once at different times of the year to obtain a better insight into user perception of the lit environment.

As described in the EBD-SIM framework [6], it is preferable to use the simulation output in the POE step so that the situation present at the time of conducting a POE study can be approximated using lighting metrics, and so that the result of the survey can be analyzed in greater depth.

The scope of this research was limited to the visual aspects of lighting quality, and the simulation results included only daylight aspects. In the future, the study can be extended to include both electric lighting and daylight as well as visual effects and light effects beyond vision through a long-term evaluation.

Author Contributions: Conceptualization, A.D., P.J., and M.A.; methodology, A.D., P.J., and M.A.; software, A.D.; validation, A.D., P.J., and M.A.; formal analysis, A.D.; investigation, A.D., P.J., and M.A.; resources, A.D. and M.A.; data curation, A.D.; writing—original draft preparation, A.D.; writing—review and editing, A.D., P.J., and M.A.; visualization, A.D.; supervision, P.J. and M.A.; project administration, M.A.; funding acquisition, P.J. All authors have read and agreed to the published version of the manuscript.

Funding: This research was funded by the Region Jönköpings Län's FoU-fond Fastigheter and the Bertil and Britt Svenssons Stiftelse för Belysningsteknik.

Institutional Review Board Statement: Ethical review and approval were not requested for this study as collection and analysis of data could not be used to identify participants, did not collect any sensitive personal data, did not include physical contact with participants, would not provide any risk of discomfort, inconvenience, or psychological distress to participants or their families, did not recruit from vulnerable groups, and did not include data collection undertaken overseas. Participants were asked to give informed consent.

Informed Consent Statement: Informed consent was obtained from all subjects involved in the study.

Data Availability Statement: Data not available. The authors do not have permission to share data.

Acknowledgments: The authors would like to acknowledge the survey participants' cooperation in this study as well as the valuable comments by the reviewers and editors of the journal of *Applied Sciences*.

Conflicts of Interest: The authors declare no conflict of interest. The funders had no role in the design of the study; in the collection, analyses, or interpretation of data; in the writing of the manuscript, or in the decision to publish the results.

References

1. Hamilton, D.K.; Watkins, D.H. *Evidence-Based Design for Multiple Building Types*; John Wiley & Sons: Hoboken, NJ, USA, 2009.
2. Levin, D.J. Defining evidence-based design. *Healthc. Des. Mag.* **2008**, *8*, 1.
3. Pati, D. A framework for evaluating evidence in evidence-based design. *HERD* **2011**, *4*, 50–71. [CrossRef] [PubMed]
4. Jakubiec, J.A.; Reinhart, C.F. A concept for predicting occupants' long-term visual comfort within daylit spaces. *Leukos* **2015**, *12*, 1–18. [CrossRef]
5. Chong, G.H.; Brandt, R.; Martin, W.M. *Design Informed: Driving Innovation with Evidence-Based Design*; John Wiley & Sons: Hoboken, NJ, USA, 2010.
6. Davoodi, A.; Johansson, P.; Henricson, M.; Aries, M. A conceptual framework for integration of evidence-based design with lighting simulation tools. *Buildings* **2017**, *7*, 82. [CrossRef]
7. Davoodi, A.; Johansson, P.; Aries, M. The use of lighting simulation in the evidence-based design process: A case study approach using visual comfort analysis in offices. *Build. Simul.* **2020**, *13*, 141–153. [CrossRef]
8. Attia, S.; Gratia, E.; De Herde, A.; Hensen, J.L.M. Simulation-based decision support tool for early stages of zero-energy building design. *Energy Build.* **2012**, *49*, 2–15. [CrossRef]
9. Kheybari, A.G.; Hoffmann, I.S. Data-driven model for controlling smart electrochromic glazing: Living lab smart office space. In Proceedings of the National Conference on Knowledge-Base Urbanism, Architecture, Civil Engineering and Arts, Tehran, Iran, 11 November 2019.
10. Preiser, W.F.; White, E.; Rabinowitz, H. *Post-Occupancy Evaluation*, 1st ed.; Routledge: London, UK, 1988.
11. Li, P.; Froese, T.M.; Brager, G. Post-occupancy evaluation: State-of-the-art analysis and state-of-the-practice review. *Build. Environ.* **2018**, *133*, 187–202. [CrossRef]
12. Dam-Krogh, E.P.; Toftum, J.; Rupp, R.F.; Clausen, G. Post-occupancy evaluation with focus on ieq and productivity metrics in office buildings. In Proceedings of the 16th Conference of the International Society of Indoor Air Quality & Climate, Seoul, Korea, 1–4 November 2020.
13. Hansen, E.K.; Bjørner, T.; Xylakis, E.; Pajuste, M. An experiment of double dynamic lighting in an office responding to sky and daylight: Perceived effects on comfort, atmosphere and work engagement. *Indoor Built Environ.* **2021**. [CrossRef]
14. Suk, J.Y.; Rashed-Ali, H.; Martinez-Molina, A. Daylighting and electric lighting poe study of a leed gold certified office building. In Proceedings of the EAAE—ARCC 2020 International Conference: The Architect and the City, Valencia, Spain, 10–13 June 2020.
15. Shafavi, N.S.; Tahsildoost, M.; Zomorodian, Z.S. Investigation of illuminance-based metrics in predicting occupants' visual comfort (case study: Architecture design studios). *Solar Energy* **2020**, *197*, 111–125. [CrossRef]
16. Doulos, L.T.; Tsangrassoulis, A.; Madias, E.-N.; Niavis, S.; Kontadakis, A.; Kontaxis, P.A.; Kontargyri, V.T.; Skalkou, K.; Topalis, F.; Manolis, E. Examining the impact of daylighting and the corresponding lighting controls to the users of office buildings. *Energies* **2020**, *13*, 4024. [CrossRef]
17. Liu, S.; Ning, X. A two-stage building information modeling based building design method to improve lighting environment and increase energy efficiency. *Appl. Sci.* **2019**, *9*, 4076. [CrossRef]
18. Joseph, A.; Quan, X.; Keller, A.B.; Taylor, E.; Nanda, U.; Hua, Y. Building a knowledge base for evidence-based healthcare facility design through a post-occupancy evaluation toolkit. *Intell. Build. Int.* **2014**, *6*, 155–169. [CrossRef]
19. Hygge, S.; Löfberg, H.A. *Poe Post Occupancy Evaluation of Daylight in Buildings: A Report of IEA SHC TASK 21/ECBCS ANNEX 29*; Centre for Built Environment: Gävle, Sweden, 2000.
20. Ochoa Morales, C.E.; Aries, M.; Hensen, J. State of the art in lighting simulation for building science: A literature review. *J. Build. Perform. Simul.* **2012**, *5*, 209. [CrossRef]
21. CIE. 011/E:2003, C.S.C.S. In *Spatial Distribution of Daylight—Cie Standard General Sky*; CIE: Vienna, Austria, 2003.

22. Likert, R. A technique for the measurement of attitudes. In *Archives of Psychology*; Woodworth, R.S., Ed.; Science Press: New York, NY, USA, 1932; Volume 22.
23. Council, U.S.G.B. Leed bd+c: New Construction | v4-Leed v4 Daylight. Available online: https://www.usgbc.org/credits/new-construction-commercial-interiors-schools-new-construction-retail-new-construction-ret-1 (accessed on 27 January 2021).
24. Newsham, G.R.; Aries, M.B.; Mancini, S.; Faye, G. Individual control of electric lighting in a daylit space. *Light. Res. Technol.* **2008**, *40*, 25–41. [CrossRef]
25. Wienold, J.; Christoffersen, J. Evaluation methods and development of a new glare prediction model for daylight environments with the use of ccd cameras. *Energy Build.* **2006**, *38*, 743–757. [CrossRef]
26. Bonomolo, M.; Ribino, P.; Vitale, G. Explainable post-occupancy evaluation using a humanoid robot. *Appl. Sci.* **2020**, *10*, 7906. [CrossRef]
27. Turan, I.; Chegut, A.; Fink, D.; Reinhart, C. The value of daylight in office spaces. *Build. Environ.* **2020**, *168*, 106503. [CrossRef]
28. Day, J.K.; Futrell, B.; Cox, R.; Ruiz, S.N. Blinded by the light: Occupant perceptions and visual comfort assessments of three dynamic daylight control systems and shading strategies. *Build. Environ.* **2019**, *154*, 107–121. [CrossRef]
29. Cuttle, C. People and windows in workplaces. In Proceedings of the People and Physical Environment Research Conference, Wellington, New Zealand, 8–11 June 1983; pp. 203–212.
30. Galasiu, A.D.; Veitch, J.A. Occupant preferences and satisfaction with the luminous environment and control systems in daylit offices: A literature review. *Energy Build.* **2006**, *38*, 728–742. [CrossRef]

Article

An Adapted Model of Cognitive Digital Twins for Building Lifecycle Management

Ibrahim Yitmen [1,*], Sepehr Alizadehsalehi [2], İlknur Akıner [3] and Muhammed Ernur Akıner [4]

1 Department of Construction Engineering and Lighting Science, Jönköping University, 551 11 Jönköping, Sweden
2 Project Management Program, Department of Civil and Environmental Engineering, Northwestern University, Evanston, IL 60208-3109, USA; sepehralizadehsalehi2018@u.northwestern.edu
3 Department of Architecture, Faculty of Architecture, Akdeniz University, Antalya 07070, Turkey; ilknurakiner@akdeniz.edu.tr
4 Vocational School of Technical Sciences, Akdeniz University, Antalya 07070, Turkey; ernurakiner@akdeniz.edu.tr
* Correspondence: ibrahim.yitmen@ju.se

Abstract: In the digital transformation era in the Architecture, Engineering, and Construction (AEC) industry, Cognitive Digital Twins (CDT) are introduced as part of the next level of process automation and control towards Construction 4.0. CDT incorporates cognitive abilities to detect complex and unpredictable actions and reason about dynamic process optimization strategies to support decision-making in building lifecycle management (BLM). Nevertheless, there is a lack of understanding of the real impact of CDT integration, Machine Learning (ML), Cyber-Physical Systems (CPS), Big Data, Artificial Intelligence (AI), and Internet of Things (IoT), all connected to self-learning hybrid models with proactive cognitive capabilities for different phases of the building asset lifecycle. This study investigates the applicability, interoperability, and integrability of an adapted model of CDT for BLM to identify and close this gap. Surveys of industry experts were performed focusing on life cycle-centric applicability, interoperability, and the CDT model's integration in practice besides decision support capabilities and AEC industry insights. The evaluation of the adapted model of CDT model support approaching the development of CDT for process optimization and decision-making purposes, as well as integrability enablers confirms progression towards Construction 4.0.

Keywords: cognitive; digital twins; building lifecycle management; artificial intelligence; IoT; decision support; self-learning; optimization

Citation: Yitmen, I.; Alizadehsalehi, S.; Akıner, İ.; Akıner, M.E. An Adapted Model of Cognitive Digital Twins for Building Lifecycle Management. *Appl. Sci.* **2021**, *11*, 4276. https://doi.org/10.3390/app11094276

Academic Editor: Andrea Carpinteri

Received: 19 April 2021
Accepted: 7 May 2021
Published: 9 May 2021

1. Introduction

Computerization and digitization are beginning to significantly affect how physical/engineering properties are handled during their life cycles [1,2]. The capture, exchange, use, and control of data and information during an asset's entire life (design, construction, Operation and Maintenance (O&M), and disposal/renewal) are among the most challenging aspects of implementing Building Information Modeling (BIM), so-called BIM in asset management [3]. Intelligent, innovative asset life cycle management has arisen during the last years in the Architecture, Engineering, and Construction (AEC) industry [2]. Digital twins (DT), the blockchains, and the Internet of Things (IoT) draw interest because of their synergistic and information management functionality [4].

Cognitive computing is machines' ability to mimic the human capacity to sense, think, and make optimal decisions in a given situation [5]. While the path reaching fully cognitive systems is still in its early stage, there are several application areas where the technology has already been implemented in many applications such as chatbots by the service sector to provide optimal responses to customer feedback [6].

The DT is already in the early stages, mainly used for prototyping, and includes modeling, simulation, verification, evaluation, and confirmation of the physical artifact using a simulated replica [7]. Analysis emphasis has been heavily placed on simulations and what-if analyses to advise implementation and eventual physical product refinement by continuous monitoring and data assimilation. According to Zhang et al. [8], encompassing intelligence and cognition in a DT is a requirement to realize disruptive technology's potential accurately and to produce integration, calibration, and symbiotic connectivity in the environments, the physical and virtual replica. According to intelligence and cognition, mental abilities and mechanisms that utilize complex information management and synergy across physical and digital settings will manipulate and strengthen the "twining" structure. Dimensions can be listed as stimuli, interaction, aims, time, and situation switching. The main goal is to foster self-adaptive assessment and smart, proactive decision-making through the two realms in an info-symbiotic way and work on the more wealthy and finer-grained information base. Cyber-Physical Systems (CPS) and socio-technical environments, for example, may benefit from this view because their activities are marked by ambiguity, dynamism, and confusion. Cyber Foraging will represent intelligent analysis and planning simulation difficult and costly computational to achieve this in reality with limited computing resources.

In the digital transformation era in the AEC industry, Cognitive Digital Twins (CDT) are introduced as part of the next level of process automation and control towards Construction 4.0 [9]. CDT incorporates cognitive abilities to detect complex and unpredictable actions and reason about dynamic process optimization strategies to support decision-making in building lifecycle management (BLM). Nevertheless, there is a lack of awareness of the real impact of CDT integration, Machine Learning (ML), CPS, Big Data, Artificial Intelligence (AI), and Internet of Things (IoT), all connected to self-learning hybrid models with proactive cognitive capabilities for different phases of the building asset lifecycle. This study investigates the applicability, interoperability, and integrability of an adapted model of CDT for BLM to identify and close this gap. Four research questions are raised in line with the study's goals:

(1) What functionalities do industry professionals allocate a CDT for BLM?
(2) What are achievable interoperability levels between CDT, IoT, Big data, and AI with current BLM technologies?
(3) What integrability enablers are necessary for implementing the CDT for BLM?
(4) How and what information should be retrievable and assignable to CDT?

As the immense contribution, the knowledge domain understands how a CDT model operates and how it connects to most BLM fields. Besides, the study of integrability enablers and professionals' perceptions of the technical ecosystem's accessibility is facilitated by synthesizing industry professionals' questionnaire perspectives. Understanding DT's interoperability value, IoT, AI, big data, and sophisticated building management systems could also help build life cycle management.

As shown in Figure 1, this article is structured as follows: in section two (theoretical background), BLM's essential concepts and CDT are depicted. Section three describes the adapted model of CDT for BLM. In section four, an evaluation of the CDTsBLM model is presented. Section five offers the discussions on decision support capabilities, integrability enablement, and practical implications. In section six, the conclusions, recommendations, and future road map are presented.

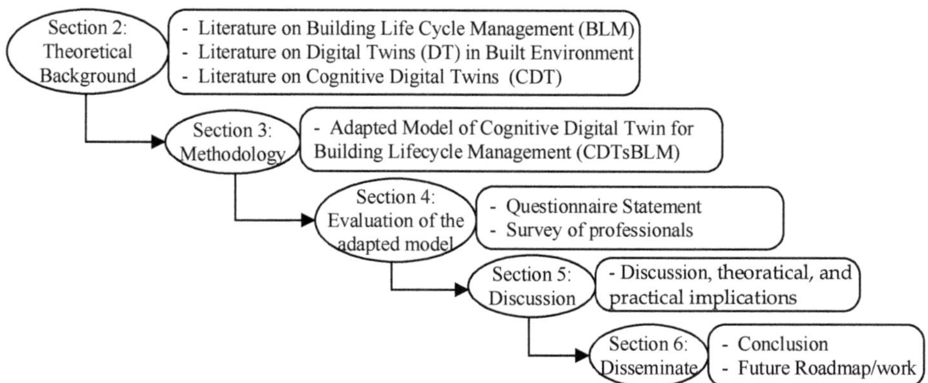

Figure 1. Research framework of the adapted model of Cognitive Digital Twins for Building Lifecycle Management.

2. Theoretical Background

This section outlines key related works in the three main areas relevant to this article: (1) Building Lifecycle Management (BLM)'s essential concepts; (2) Digital Twins (DTs) in build environment and; (3) Cognitive Digital Twins (CDT) are presented.

2.1. Building Lifecycle Management (BLM)

The building lifecycle mainly includes the design, construction, operation, maintenance, and end-of-life stages. Each step can be separated into superimposed information layers that entail efficient data/information exchange strategies for interoperability throughout all lifecycle phases [10]. BLM refers to a method of integrating and handling the different stages of a construction project's lifecycle [11]. BLM is a strategic planning process that supports the development, operation, and maintenance of buildings and their associated infrastructure, including building planning, design, construction, operation, and maintenance. It aims to reduce costs and improve efficiency by ensuring that buildings are built, operated, maintained, and replaced in the most cost-effective and timely manner. The BLM is an integrated approach to building management that considers all the building activities, the building's surroundings, and the impact of these activities on the environment. The BLM, directly and indirectly, affects many aspects, such as buildings or infrastructures' operation and efficiency, operational risks, the environmental impact of buildings, people's quality of life, safety, and businesses. Such a complex and complicated process needs to be real-time, accurate, intelligent, and automated to monitor, detect, learn, analyze, simulate, validate, and operate. There are disintegrated data and information in every phase of the construction project, which contains a significant amount of design and cost information in the design process and decision-making steps; in the implementation step, a considerable amount of material consumption beyond the data generated during the design process and decision-making steps; and a vast data and information in operation and maintenance step.

As a result, to enforce BLM, a management mechanism must be developed that communicates each participant's expertise and phase of the development project to avoid a lack of sufficient and timely information connection and dissemination at different life cycle stages [12]. BLM sometimes starts with a physical analysis of the structure to generate a numerical representation (e.g., CAD documents). In this sense, at the beginning of its lifecycle, developing an effective BLM system is more than enough to avoid data loss during the building's construction, use/maintenance, and disposal [13]. By offering an interactive IT environment to handle the whole construction lifecycle, BLM seeks to migrate and enhance knowledge exchange in all phases of the building process [14]. Energy management, facilities management, maintenance management, and product/information traceability management are part of a scalable BLM scheme that allows users to incorporate and reuse building knowledge, domain expertise across a building's life cycle [13–16]. A scalable networking infrastructure that offers uniform interfaces for sharing all forms of data consumed or generated by the participants, corporations, or information systems, in general, participating in the building lifecycle is a vital component of a functional BLM framework. BLM must interact with any intelligent items/systems that are part of the building lifecycle (sensors, actuators, RFID, databases) [10]. According to BIM, a technique that seeks to manage a building's entire life cycle in a particular data environment, proper data digitalization may optimize knowledge management and share within the multidisciplinary team [17].

2.2. Digital Twins (DTs) in the Built Environment

The CPS is realized through the DT for visualization, modeling, simulation, analysis, prediction, and optimization. DT contains three main components to create a practical loop: a physical entity, a virtual entity, and a data link [18]. Usually, there are two approaches to dynamic mapping in the DT. Inspection data are gathered in the physical world and subsequently transmitted to the virtual world for further analysis. Simulation, prediction, and optimization are achieved in the virtual model by learning data from multiple sources, offering prompt solutions to guide the realistic process and adapt to the changing context.

Based on Alizadehsalehi and Yitmen [19], DTs have various features in the AEC industry such as Real-time (gather and present real-time data of physical assets), Analytics (store data, run continuous analytics from historical data, and provide helpful insight), Simulations (utilize to run various data-driven simulations), visualization (overlay real-life and live 3D BIM models, images, and videos of the physical asset and also the foundation for immersive visualizations), Automation (a bi-directional system that can manage the behavior of physical assets), and Predictions (provide predictions of assets' future behaviors using historical data and analytics of various scenarios assets). As a comprehensive summary, Table 1 presents DT applications in the AEC industry that appeared in the recent literature (2019–2021).

Table 1. Digital Twins applications in the AEC industry that appeared in the recent literature.

n	Author(s)	References	Year	Applications
1	Alizadehsalehi and Yitmen	[19]	2021	Developed and evaluated a DT-based construction progress monitoring system called DRX.
2	Deng et al.	[20]	2021	The transition from BIM to DTs in built-environment applications was studied.
3	Pan and Zhang	[18]	2021	A data-driven DT architecture based on data mining, BIM, and IoT was developed for comprehensive project management.
4	Bosch-Sijtsema et al.	[21]	2021	Examined the digital Technology applications in the AEC industry.
5	Hasan et al.	[22]	2021	Investigated construction machinery operation and work tracking through AR and DT.
6	Camposano et al.	[23]	2021	Examined how AEC/FM professionals describe built asset DTs.
7	Meža et al.	[24]	2021	Devoted BIM-based DT for road constructed using secondary raw materials (SRMs)
8	Hou et al.	[25]	2021	Reviewed the applications and challenges of DTs in construction safety.
9	Borowski	[26]	2021	Reviewed the contemporary actions utilized and challenged in the energy sector through the enterprises.
10	Del Giudice and Osello	[27]	2021	Investigated DT-based approaches, tools, and implementations that can be adapted for achieving smart city objectives.

Table 1. *Cont.*

n	Author(s)	References	Year	Applications
11	Tagliabue et al.	[28]	2021	Proposed leveraging DT for Sustainability Assessment of an Educational Building.
12	Boje et al.	[29]	2020	Examined the many uses and limitations of BIM, as well as the need for Construction DT.
13	Liu et al.	[30]	2020	Investigated building indoor safety management.
14	Austin et al.	[31]	2020	Presented the smart city DT challenges and proposed approaches regarding the architectural and operational stages.
15	Lu et al.	[32]	2020	Detected anomalies by DT for developed asset tracking in service and maintenance.
16	Greif et al.	[33]	2020	Developed the concept of a lightweight DT for non-high-tech sectors such as construction.
17	Lu et al.	[34]	2020	Proposed moving BIM to DT for operation and maintenance.
18	Rausch et al.	[35]	2020	Implemented a computational algorithm to support DTs in construction.
19	Dawood et al.	[36]	2020	Reviewed, developed, and implemented DT, VR, AR, and BIM in AECO.
20	Götz et al.	[4]	2020	Researched asset lifecycle management.
21	Alonso et al.	[37]	2019	Presented the SPHERE platform for improving the building's energy performance.
22	Mathot et al.	[38]	2019	Developed and discussed the next-generation parametric system Packhunt.io with BIM, DT, and Mixed Reality (XR) technologies.
23	Khajavi et al.	[39]	2019	Discussed DT for building lifecycle management.
24	Kan and Anumba	[40]	2019	Presented a comprehensive review of DT applications in the construction domains.
25	Lu et al.	[41]	2019	Proposed the DT-based smart asset management framework.
26	Kaewunruen and Lian	[42]	2019	Recommended using DT to maintain the lifecycle of railway turnout systems sustainably.
27	Lydon et al.	[43]	2019	Conducted simulations of thermally active building systems to assist DT.

2.3. Cognitive Digital Twins (CDT)

The DT concept allows the physical equivalent to be mirrored in virtual space, including exchanging data between them [7]. CDT expresses an evolution of the DT concept. It has been crafted to fit the requirement of monitoring complex industrial processes and apply the same trade model, shadow, and thread of DTs [6]. The balance between rapidity, resolution, and exception handling is crucial from any industry's economic perspective [44]. Virtualization in a dynamic, run-time process allows the digital counterpart's behavioral model to be constantly modified to mimic the physical element's actions, resulting in the CDT [45]. Virtualization is a dynamic design-time process involving computational approaches to model the physical feature, evolving into the complex, run-time process that allows the digital counterpart's behavioral model to be constantly adjusted to mirror the physical element's actions, resulting in the CDT. The CDT is a DT with cognitive abilities, including detecting anomaly and behavioral learning, and the power to determine physical twin actions to improve measures defining its state or function [46]. Therefore, a CDT uses optimization approaches to aid decision-making and data from the physical twin analyzed using analytics or ML.

To put it another way, the CDT is envisioned as a robust monitoring and control mechanism and an essential part of the decision-making action that leads to system optimization. Using optimization techniques inside the heart of the cognitive twin and its impact is the primary crucial differentiating point instead of currently available DT solutions [46]. To make the transition from physical assets in the form of digital replicas to cognitive advancement, Abburu et al. [6] used a three-layer structure to describe the types of twins needed: digital, hybrid, and cognitive. The need to build isolated models of systems for anomaly detection, connect the models for predicting unusual behavior, and problem-solving skills to deal with uncertain situations constitutes the three-layer separation. CDT is characterized in DT through advanced semantic abilities to detect the mechanisms of virtual model evolution, enhance DT-based decision-making, and foster the interpretation of virtual model interrelationships [47]. The CDT ensures that assets are adequately managed and that problems outside technical stakeholders are resolved by implementing Internet of Things (IoT) systems [48].

CDTs may have a high degree of intelligence, allowing them to mimic human cognitive processes and perform conscious acts with little or no human intervention [8]. The Knowledge Core of CDT has semantic-driven recognition, learning, inference, estimation, and decision qualifications consisting of a series of prediction and ML models developed using the data from multiple sources such as physical equivalents and sensors from all aspects of operational conditions of the industrial systems. Besides, it incorporates temporal supply chain data and processes as well as experts' domain knowledge. As a result, the CDT can train and improve to represent and depict the physical asset's current state and operating conditions in real-time. Furthermore, in both the digital and physical worlds, the CDT can identify, analyze, deduce, forecast the twinned physical system's present and potential actions, and produce decisions by interrelating machines and humans.

Lu et al. [49] suggested a new cognitive twins so-called CT definition and a knowledge graph-centric framework for the CT process. Du et al. [50] explored how to build individualized information systems for future smart cities using a human-centered DT simulation approach of cognitive behaviors. Eirinakis et al. [46] suggested an Enhanced Cognitive Twin so-called ECT introducing advanced cognitive skills to the DT asset that allow assisting choices to allow DTs to respond to internal or external stimulation in the context of process industries. The ECT can be used at varying levels of the supply chain hierarchy, including sensor, device, process, workforce, and manufacturing stages, and can be integrated to allow lateral and vertical interaction.

The concepts of the Hybrid and Cognitive Digital Twin (COGNITWIN) toolbox were developed by Abburu et al. [6] to cover cognitive skills for efficient management and operation of processing equipment, for lowering production costs, and efficiency improvements in the process industry. A sensor network can constantly track and capture data from different plant processes and properties stored in a standard setup database. The COGNITWIN project mainly aims at adding the cognitive component to process control systems, thus enabling them to self-organize and provide solutions in case of unexpected behaviors. Figure 2 shows the different stages of DT to CDT. A DT is a formal digital representation of an asset, process, or system that captures any systems' attributes and behaviors through IoT-based various reality capturing sensors suitable for communication, storage, interpolation, and processing to measure, simulate, and experiment with the digital replica to understand its physical counterpart. A DT for monitoring, diagnostics, and prognostics to optimize asset performance and utilization uses sensory data combined with historical data, human expertise, and fleet and simulation learning to improve prognostic outcomes. A DT gets data from physical entities and applies them to the model.

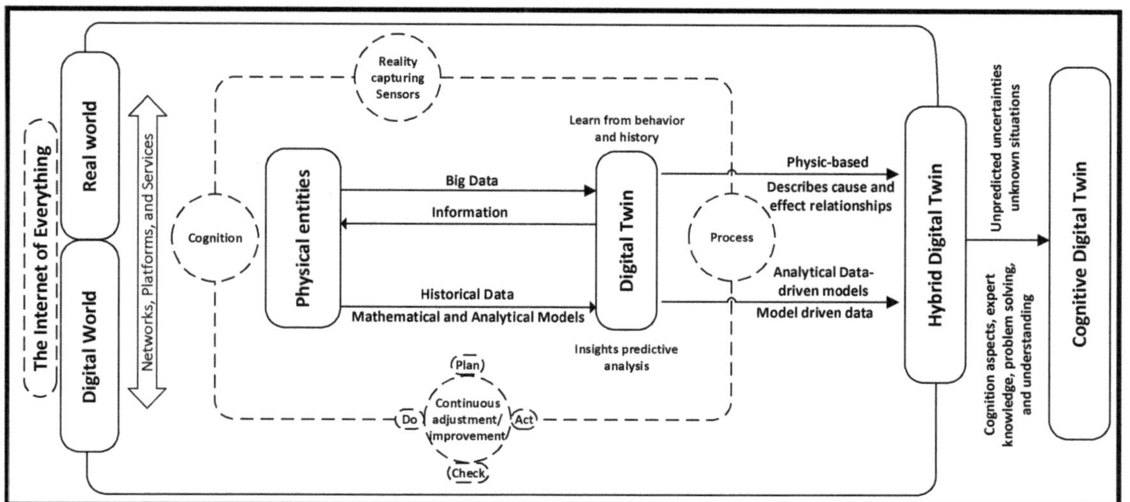

Figure 2. Different stages of Digital Twins to Cognitive Digital Twins.

A Hybrid Digital Twin is usually defined as the DT comprised of combined multiple models. A Hybrid Twin (HT) extends the DT by intertwining different models to take advantage of both physics space and data-driven modeling. HT gets the data from the physical entities and uses them in several models jointly. The way to increase the degree of influence and scope of DT is to have cognitive features, such as reasoning, planning, and learning. Digital twins based on data analytics require immense amounts of data for accuracy, and while physics-based simulation models are highly accurate, they take an incredible amount of time to run. New hybrid systems are combining the best of both worlds for a digital twin that is both quick and exact.

Although HT has a lot of different models, there are so many parameters that influence the processes that, in some situations, are not covered by existing models. CDT represents the next step in evolving the DT concept in the AI era, incorporating cognition aspects to deal with unforeseen situations effectively. Revolutionary DTs will arise as a result of intertwining distinct models to accomplish advanced predictive capabilities and finding solutions to problems to be encountered by integrating expert knowledge. CDT gets data from physical entities and compares them with models, including models of expert knowledge.

Table 2, as a comprehensive summary, exhibites diverse CDT applications in various fields of industry based on the latest research (2019–2021).

Table 2. Diverse Cognitive Digital Twins applications in various fields of industry.

n	Author(s)	References	Year	Industry	Applications
1	Rožanec et al.	[51]	2021	Manufacturing	To capture specific knowledge related to demand forecasting and production planning.
2	Berlanga et al.	[52]	2021	Computer Science	Proposed a platform for social networks.
3	Abburu et al.	[6]	2020	Engineering	Proposed a framework for the implementation of Hybrid and Cognitive Twins as part of the COGNITWIN software toolbox.
4	Kalaboukas et al.	[47]	2021	Manufacturing	Implementation of CDT in Connected and Agile Supply Networks.

Table 2. *Cont.*

n	Author(s)	References	Year	Industry	Applications
5	Zhang et al.	[8]	2020	Computer science and Engineering	Discussed how the different levels of self-awareness can be harnessed for the design of CDTs.
6	Du et al.	[50]	2020	AEC industry	Established methods and tools for the intelligent information systems of smart cities.
7	Eirinakis et al.	[46]	2020	Management	Proposed enhanced cognitive capabilities to the DT artifact that facilitate decision making.
8	Albayrak and Ünal	[53]	2020	Engineering	Smart Steel Pipe Production Plant via CDT-based systems.
9	Abburu et al.	[54]	2020	Engineering	Proposed the CT control system for automation in the process control system.
10	Essa et al.	[55]	2020	Computer Science	Introduced the automation of defect detection.
11	Saracco	[56]	2019	Computer Science	Proposed to bridge Physical Space and Cyberspace.
12	Fernández et al.	[57]	2019	Engineering	Introduced the concept of Associative CDT, which explicitly includes the associated external relationships of the considered entity for the considered purpose.

3. Methodology

3.1. Adapted Model of aCognitive Digital Twin for Building Lifecycle Management (CDTsBLM)

This paper reviews previous work on BLM, DT in the built environment, and CDT and presents an adapted framework developed by Lu et al. [48] and Abburu et al. [6] to improve BLM with CDT in the AEC industry. The adapted framework in this research is referred to as the CDTsBLM Model of the framework. This framework's processes, as shown in Figure 3, are discussed in detail in this section. The CDT is a capabilities-driven digital representation of its physical twin. It should be a capability augmentation and an intelligent digital companion cycle and evolution phases. CDT facilitates cognition towards improving the behavior of the complex process systems inherent in planning, design, construction, and operations. An ML pipeline automates the ML workflow by facilitating data to be converted and associated into a model that can be processed to automate the ML model's outputs and input data completely. As shown in Figure 3, the conceptual framework facilitates the implementation and evaluation of consistent CDT in BLM by integrating various pipelines of ML and analytical tools at various stages from planning to the whole operations through the processing phases during operations.

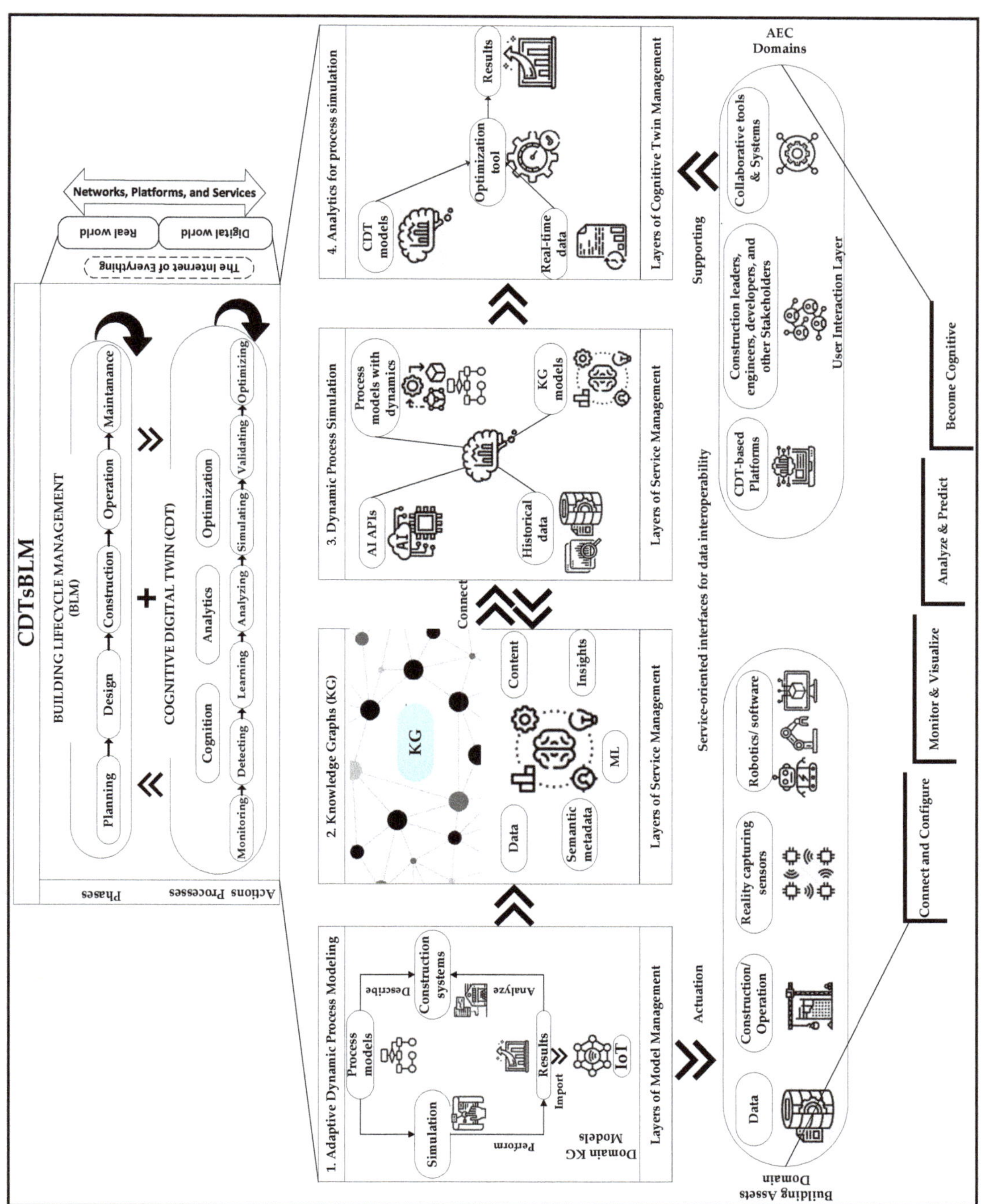

Figure 3. Knowledge Graph-centric Conceptual Framework of Cognitive Digital Twin for Building Lifecycle Management (CDTsBLM) (adapted from the developed framework by Lu et al. [48] and Abburu et al. [6]).

3.1.1. CDTsBLM Framework

The first section of CDTsBLM is CDT and adaptive dynamic process modeling. In this section, all IoT-based systems, consisting of reality capturing sensors, other construction-related technologies, networks, and computational composition, are considered hybrid systems, including continuous systems and discrete systems. DT is an incorporated structure of mathematical models and data ensuring that real physical systems and their virtual entities are synchronized in real-time. Such a method can be characterized as whole workflows where the computing composition and other plant nodes are connected. A process modeling and simulation approach is applied to enact these workflows and simulate the hybrid system behaviors in this arrangement.

Knowledge Graph (KG) helps to represent the data that can achieve cognitive learning by machines. Knowledge is awareness or familiarity, someone or something gained by the experience of a fact or situation. On the other side, a Graph represents how any data are stored in the form of associations. KG is a term of how the engine builds relationships between people, technology, and facts. The KG models are focused on topological relationships between physical and cognitive entities. Ontologies for KG models are created before designing KG models to describe semantics and syntax. KGs will serve as the core mechanism for ML flows, extending data manipulation to enable practical consumption through CDTsBLM. KGs and ML techniques provide the required abstraction layer to clarify better (a) the context of each method and (b) the complex interactions that represent machine-understandable data and ML algorithms to make it easier for data and information extraction tools to communicate.

Artificial Intelligence (AI) APIs, historical data, process models with dynamics, and KG models are integrated to produce CDT models. CDT models aim to support decision-making for dynamic processes of physical entities. The use of dynamic process simulation has been developed as a reliable and effective tool to examine the transient behavior of process systems.

In the CDT and analytics for the process simulation stage, optimization tools will support process optimization through real-time data and CDT models. The result of this optimization is implemented to make decisions for physical entities manipulating.

A service-oriented interface for the data interoperability approach is offered to develop interfaces for heterogeneous data, and for that reason, all the assets and business domain data should be converted into integrated formats through the established interfaces. It means that all generated and captured data at any stage of projects need to be converted to a common data environment.

3.1.2. Layers in CDT

The architecture of CDTsBLM has essentially four layers with each of them providing a set of services as follows.

Model Management Layer is in charge of three different kinds of models: (1) first-principle models for processes based on underlying physics; (2) analytical models based on various AI methods and ML; and (3) information-driven models focused mainly on their detailed work experience based on tacit knowledge of the domain and human operating experts. This *Layer's* primary role is to ensure that various services, including modeling, data-driven, and human experts, provide efficient storage and access for multiple models.

The Service Management Layer makes effective use of all available services to solve the fundamental domain issues. It is focused on a complicated organization of services, combining data-driven model-based services to create value-added pipelines. It contains a registry service, enabling the rapid discovery of the orchestral services required. Service results should be made public, and practical and scalable communication of service can be ensured.

The User Interaction Layer is a digital definition of a physical device simulating its actions. It's critical to assist a user in discovering a CDT's data and models, as well as its characteristics. To put it another way, intuitive yet exploratory user interaction should be possible.

A twin represents a dynamic framework that should be handled effectively, and the Cognitive Twin Management Layer models a physical system's behavior. A Twin can define the system's actions as a digital description of a physical system by offering a standard behavior model. Contrarily, a Twin is a digital entity whose life cycle is affected by a physical system; in other words, physical design behavior changes should be replicated in double structure models as soon as the physical environment's corresponding data become apparent.

3.1.3. CDT realization within Cognitive Building Lifecycle Environment (CBLE)

Figure 4 depicts the CDTsBLM conceptual architecture built on service pipelines from which accessible data flow. The use of data streams implements the cognitive center of the CDT through one pipeline, which provides learning, event identification, and prediction and reasoning skills. A second pipeline for each CDT allows analysis and justification of vast quantities of raw data from different sources. A meta-structure improves the CDT by allowing multi-source data flow interoperability, higher reasoning, cognition pipelines since they interconnect through KGs.

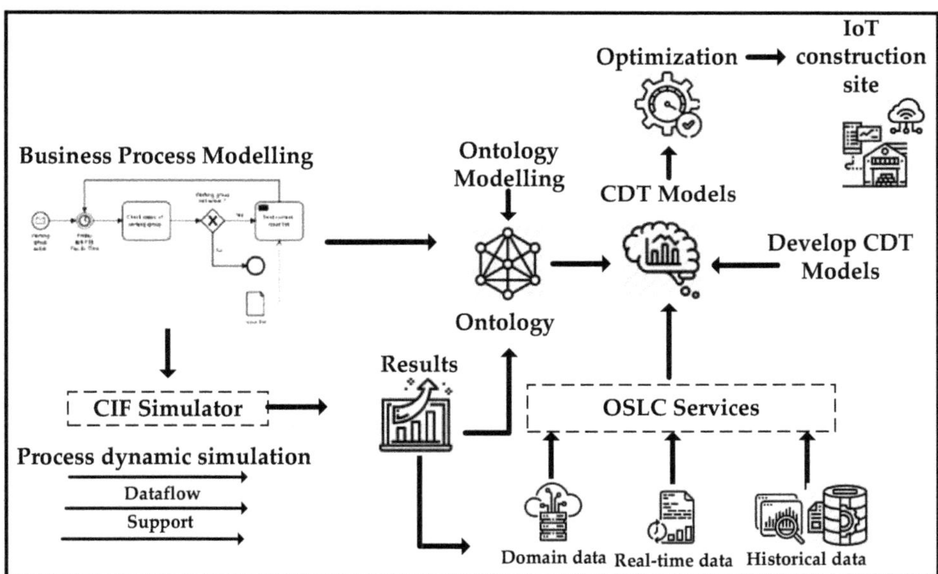

Figure 4. Outline of the process flow of Cognitive Digital Twin for Building Lifecycle Management (CDTsBLM) (adapted from the developed framework by Lu et al. [48]).

KGs enrich and direct the relationship between these two data-driven modeling approaches. ML algorithms, data analytics, and KGs form the foundation of a robust cognitive computing framework that allows for fine-tuned outcomes and increased process and reasoning abilities. The semantic models augment a set of data with features that enable cognitive processes to be far more agile. Combining quantitative-driven ML, qualitative KGs, and data analytics combines machines' computing power with the human intuition and experiences needed to solve various Construction 4.0 use cases. Optimization can be exploited, related to the scope of the activity, the time horizon, providing CDTs with the capability to resolve optimized production issues, such as short-term real-time reorganization and reconfiguration of entire systems, mid-term timing, and lots for individual activities or whole construction and long-term capacity planning.

In the transition from CDTsBLM Conceptual Architecture to Technical Architecture, there is a data communication framework for collecting, processing, integrating, and managing multi-source, multi-scale, and multivariate data from production assets. CDT module interaction and API-based communication with the business domain can also be supported by a messaging and operation bus.

Actuators can be implemented to execute real-time decisions dynamically and communicate them directly from CDT to physical twin. Several enablers must characterize the CDT definition: (1) A profile describes the twinned asset and descriptive details, as well as models; (2) the connections between CDT and other CDT construction, as a network of linked CDTs, a factory or process or different architecture; (3) facets of cognition such as thinking, modeling, estimation, and optimization; (4) aspects of confidence to ensure correct knowledge transmission; (5) status and future notification visualization for end-users; (6) computation requirements as well as implementation aspects; (7) Identifying the CDT lifecycle.

These enablers tackle various stages of a cognitive factory model. The first is to simulate the construction or even other development contexts as a network of interconnected CDTs, for example, the workstation, process, and machine. Data from different sources, including ERP, Physical Twins, Human Operators, were initially added. Detection services (CEP), which are combined with Simulation and Optimization Services in the Cognitive Core functionality, use data-driven process models to allow CDTs to (1) detect a natural anomaly, such as an impending system malfunction, (2) forecast possible response steps with ideal outcomes, (3) simulate the optimized outputs and future consequences gradually, (4) return a well-thought-out proposal for the future course of action, which will be submitted to the appropriate stakeholders or actuaries for approval or denial.

3.1.4. CDT and Cognition

The definition, which is data-oriented, resides at the heart of the CDT. Construction 4.0 needs a greater cognitive increase in assets to allow continuous improvement of the data-driven process. CDTsBLM uses a modern architecture for generation construction data analytics to integrate cognition into DTs and as a meta-platform to help create and implement a range of building applications, such as quality management and predictive maintenance. The CDTsBLM approach depends on a new DT data analytics, in which an innovative CDT-driven metaphor represents a system model, improving the DT base and the integrated CDT structures to understand and solve situations that cannot be modeled, for example, by design allowed in numerical models or experience in the context of numeric models. Based on the current process data in real-time, this integrated approach will explain the issue, including tool deterioration for each particular machine and product type. This cognitive function is assisted in rare cases by the just-in-time process status simulation to measure an anomaly's assurance that needs to be resolved in an extraordinary circumstance. This data-driven simulation would also indicate whether an anomaly is triggered by a particular scenario, meaning how long current process settings will remain unchanged. This novel approach describes data-driven model simulations from twins using novel predictive clustering methods and advanced inductive database mining rules.

3.1.5. CDT and Data

For the analytical models implemented in CDTsBLM, extensive data sources are required. The well-known critical issue in developing analytical models is processing and modeling various types of data in real-time. A complex architecture must adhere to the other methods to construct a universal analytical structure in Construction 4.0 for real-time data stream processing. It is also possible to synchronize and optimize data until they are fed into the analysis models. CDTsBLM seeks to provide an exhaustive and modern, multi-level model of uncertainties and causal relationships that include the following submodels at multiple levels: model content fluxes, statistical capability models, technological process models, deviation models between optimal technical simulations, observations, and logistics demand projects. Material flow pattern models consider the lead time of input materials and generate uncertainties for understanding product transport and logistics and the various technical procedures to be used. Machinery availability, breakdown model, and models for employee insecurity will be considered for statistical capability models. Models of technical processes can incorporate domain awareness. Due to its state-of-the-art data-guided online processing algorithms for broad re-in-time data streams, the CDTsBLM uses the framework as its analytical tool to handle data requirements for CDTs. It aims to address multimodal data fusion, data preparation, optimization, and the analytical design of a manufacturing process as an entity that generates a typical analytical structural model for intrinsic, interrelated process variables. The architecture allows CDTs, when introduced, to seamlessly use many multimodal data types.

3.1.6. CDT and Optimization

When converting a DT to a CDT, the implemented optimization that allows the CDT to generate optimization functions is a critical enabler and differentiator. The vast majority of batch processing, construction planning optimization issues are NP-hard, which applies to Construction 4.0. Consequently, using conventional algorithmic and mathematical programming approaches to generate a proper solution to a real-life issue is computationally intractable. It is not always possible to have precise values of optimization criteria such as inventory supply, production times, costs, human resource efficiency, equipment durability, and construction industry specifications, or to be mindful of future diversities in material order preference, equipment failures due to a lack of information or the changing existence of actual construction sites. As a result, an optimum solution for approximate parameters could be inadequate until the parameters are realized. This complexity is present, especially in the process industries, where the quality of a given material inside a construction cannot be calculated with certainty before the component is processed. As a result, dealing with complexity is almost as critical as making the model itself, as it can be used to verify mathematical models and maintain production viability during operations.

The proposed solution dealing with decision circumstances involving complexity treats all potential realizations of parameters as part of the feedback. This collection is referred to as a scenario set, and each parameter completion is referred to as a scenario. As a result, a scheme reflects a possible condition of the universe. Since the cost of a solution is determined by a situation, its value is therefore unknown. Ex-post research compares a solution to an optimal solution that might have been obtained if the parameters had been realized in their original form. Decision-makers who do not want to take risks are more interested in avoiding the worst situations in the real world. A robust optimization methodology, under a discrete or intermittent uncertainty and the regret criteria of max–min or min–max, is a critical modeling approach for meeting the above requirements.

Another important aspect is the manufacturing process's performance. Its capability primarily determines the control system's capability to adjust schedules to changing conditions, especially at the short-term decision support level for real-time adaptive optimizing. These dynamic problems with re-optimization can be handled via reactive and proactive frames, in which the optimization process is progressively conducted at some intervals and dynamically evolves into integrated new or old knowledge. These methods can address complex optimization problems effectively with input parameters and variables that have not been completed, uncertain or unknown, that are modified simultaneously by the development of the real-time solution process. A suggested architecture for this path utilizes vigorous optimizations of different models (1) builds on well-established concepts such as negligence in which unique input parameters are not known to provide a universal solution which is efficient to optimize the worst-case solution and optimize overall actuality in all realizations of hidden parameters, (2) implements the adaptive, efficient, multi-stage optimization technique for planning where optimized decisions on unknown criteria and action on recourse depend on the realization of insecurity.

CDTsBLM is designed to deliver a complete CDT optimization toolkit based on a local hybrid search, evolutionary calculation, and data-driven techniques to scalable resource-aware planning and optimization algorithms that can be utilized to solve complicated planning issues with a variety of constraints, including utilities, renewable resources, and machinery service restraints. They also possess the potential to hierarchically address various schedule targets, including time and energy-conscious combinations. They often have a high degree of precision by strengthening their forecasting capabilities by utilizing (1) multiple design processes as decision variables to help monitor construction site factors such as efficiency and length while controlling several scheduling parameters such as processing time, energy usage, and operational expense, (2) a variety of execution types, including, for example, alternate routings and resource demand variations for each construction operation. A modified algorithm is designed to complement the prior algorithm set in the Optimization toolkit to endorse the CDT for rigorous online preparation problems, easily extended to resource-conscious purpose multifunctional optimization alternatives.

4. Evaluation of the Proposed CDTsBLM Model

Testing the proposed CDTsBLM model aims to recognize its effectiveness in practice and thus validate it. In this assessment, a digital survey was established with the literature review as a basis and distributed to industry professionals across countries. The survey's core theme was to provide practitioners with an insight into the life cycle-centric applicability and integrality of CDTs with existing BLM practices.

4.1. Sampling

AEC increasingly involves multiple stakeholders ranging from Design Manager to Design Coordinator, Designer, BIM Manager, BIM Coordinator, Digitalization specialist, Project Manager, Construction Manager, Asset Manager, Asset Administrator, and Asset Controller. The longevity of assets may mean that the stakeholders or even the type of usage may change over time; this poses challenges in how these assets are managed over their life and specific challenges to the way data and information about them are handled. Therefore, the study focused on private organizations dealing with building projects operating in the United States (USA), United Kingdom (UK), and Sweden. The sample includes only large firms.

4.2. Data Collection

The data collection was confined to actors that have vital roles in capturing, delivering, and using the information in the building life cycle and technology domain innovation projects. Design, project management, contracting, and facilities management firms were compiled by searching for geographic position cataloging enterprises. The survey included owners and consultants for asset management. The national inclusions improve the validity of questions as they represent the different cultures, experiences, and ways of working of corporate and national groups. The proportions of company positions, sizes, and regions are shown in Table 3.

Table 3. Company region, size, and role in percentage.

Company Type		Design Firm		Project Management Firm		Contracting Firm		Facility Management Firm	
Role		Design Manager	9%	BIM Manager	9%	Project Manager	8%	Asset Manager	10%
		Design Coordinator	8%	BIM Coordinator	9%	Construction Manager	7%	Asset Administrator	9%
		Designer	8%	Digitalization Specialist	7%	BIM Manager	7%	Asset Controller	9%
Company size	Large (>250 employees)	10%		10%		9%		12%	
	Medium (50–250 employees)	9%		6%		7%		10%	
	Small (<50 employees)	8%		5%		6%		8%	
Region	USA	8%		5%		6%		8%	
	UK	9%		6%		7%		10%	
	Sweden	10%		10%		9%		12%	

LinkedIn contacted 271 businesses, and a single representative from each was requested to participate in the questionnaire. Contributors were apprised about the search's aims, and their answers were kept private and anonymous. A total of 45 percent of completed queries were collected. Experts were asked to talk regarding their work, observations, and organizations. Participant experts used a five-point Likert scale to rate their agreement with BLM digitalization-central statements, with one being the most disagreeable and five being the most agreeable.

4.3. Descriptive Statistics

Descriptive statistics reporting the mean values and standard deviations of questionnaire responses are presented in Table 4. The summarized statistics speficies interesting comprehensions as an overview of the AEC industry's perception of the concepts. According to the results, the mean scores for 16 of the 20 questions were higher than 3.65 out of 5.00. The proposed model's overall mean rating was 4.06, which means that industry professionals support approaching CDT development for process optimization and decision-making purposes and that integrability enablers confirm progression towards Construction 4.0.

The argument that sought respondents' opinions on real-time analytics for data-driven models enhanced with cognitive resources was conducted to support decision-making and aid learning, optimization, and reasoning had the highest mean of 4.45 in the relative importance of the variables. Through reason, learning, and optimization, CDTs can monitor, project, modify, and make better choices in real-time. CDT is a robust monitoring and tracking method, and the overall system is optimized with a mean value of 4.38. CDT covers existing process control systems with cognitive elements that allow them to organize themselves and provide a so-called indication of unanticipated actions at an average of 4.34. Overall, respondents agree that CDT should provide cognitive features that enable it to sense complex and unpredictable movements and reason about dynamic process optimization strategies to aid decision-making in BLM.

Table 4. Descriptive statistics, factor analysis and reliability test.

	Questionnaire Statement	Mean	SD	Factor Loading	Cronbach α	Rank
Building Lifecycle Management (BLM)	BLM employs a CDT approach, which allows for a highly effective expanded collaborative process built on AEC industry best practices.	3.26	1.23	0.703	0.710	20
	Using a BLM framework, users can proactively fix real-time problems. RFIs, submittals, and change orders may be minimized or withdrawn.	3.49	1.25	0.707		19
	With BLM, designers can make more intelligent choices in a richer data context while maintaining greater control over the final product output.	3.54	1.19	0.718		17
	BLM is intended to minimize waste by forecasting results correctly, defining possible tension points, and improving procedures.	3.65	1.14	0.713		18
Cognitive Digital Twin (CDT)	CDT offers live data feeds for primary metrics, visualizations, models, and scenario generation applications.	4.21	0.95	0.842	0.869	8
	CDT integrates cognitive components into current process management structures, helping them self-organize and respond to unpredictable activities.	4.34	0.92	0.868		3
	CDT models aid in decision-making for complex systems, including physical actors.	4.32	1.16	0.864		4
	CDT is a valuable monitoring and control mechanism that helps in overall system optimization.	4.38	0.94	0.876		2
Internet of Things (IoT)	The connectivity of real-time data allows for fast reporting and data explosion, enabling deep data analytics.	4.12	0.92	0.824	0.829	13
	In the IoT lifecycle, virtual model assets are needed to identify, detect, and address dependencies across domains in the system, subsystems, and components.	4.04	0.89	0.808		14
	IoT system architecture allows for simple connectivity, communication, and control across domain-specific applications.	4.19	0.93	0.838		10
	As a hybrid architecture, IoT connects the physical and virtual worlds.	4.23	0.91	0.846		6
Self-Learning	Learning introduces new expertise to current data, models, and approaches to learn more reliable models from existing datasets.	3.96	1.12	0.792	0.813	16
	Cognitive aspects help benefit from past process data and incidents to predict and provide the best feasible solutions for unwanted events.	4.01	1.15	0.802		15
	Hybrid models that self-learn and have proactive cognitive skills.	4.13	0.95	0.826		12
	The real and the virtual space can reason and learn about stimuli, interaction, aim, and time.	4.16	0.96	0.831		11
Process Optimization	Real-time analysis for data-driven models augmented by cognitive resources is conducted to facilitate decision-making and improve learning, optimization, and reasoning.	4.45	0.96	0.892	0.856	1
	Dynamic process optimization techniques contribute to an environment in which digital structure and behavior are continually evolved.	4.22	1.18	0.844		7
	Process optimization is conducted to support and manipulate physical structures based on CDT models and real-time data.	4.25	1.19	0.850		5
	Assessing optimization scenarios in the virtual environment before bringing them into effect in the real world.	4.20	1.20	0.840		9

4.4. Factor Analysis

Functionalities of the BLM, CDT, IoT, and Process optimization and achievable levels of interoperability ad integrability of the proposed model as rated by the industry professionals are presented in Table 4. Confirmatory factor analysis boosts trust in the assessment's precision and quality. Table 4 lists the items that were used to calculate each element. A five-point Likert scale was applied to measure all objects, and they were found perceptual. All factor loadings between 0.703 and 0.892, as well as all Cronbach's coefficients less than 0.70, were considered to be adequate.

According to Table 4, CDT was ranked as the highest, Process Optimization as the second, IoT as the third, Self-learning as the fourth, and BLM as the fifth factor for life cycle-centric applicability and integrability of CDTs with current BLM practice, exploring decision support capabilities and AEC industry insights.

4.5. Correlation Analysis

Spearman's rank-order correlation was used to validate the relationships, and the evaluation of the matrix shows a correlation. A positive linear relationship exists within BLM for improved productivity and sustainability, CDT for enhanced decision-making, IoT for real-time connectivity, and self-learning by applying new knowledge on the existing data, models, methods, and optimization simulation decision support. The highest correlation occurs between CDT for improved decision making and IoT for real-time connectivity in $\rho < 0.01$ ($r = 0.812$). The second significant positive correlation exists between CDT for enhanced decision making and optimization and simulation for decision support in $\rho < 0.01$ ($r = 0.799$). The correlation calculations of respondents' perception of CDT decision support abilities are depicted in Table 5.

Table 5. Correlational analysis of Cognitive Digital Twins' perception of decision support capabilities.

		Spearman's Matrix of Correlation Rank				
		BLM for improved productivity and sustainability	CDT for improved decision making	IoT for real-time connectivity	Self-learning by applying new knowledge to the existing data, models, and method	Optimization and simulation for decision support
Spearman'sRho (ρ)	BLM for improved productivity and sustainability	1.000				
	CDT for improved decision making	0.776	1.000			
	IoT for real time connectivity	0.695	0.812	1.000		
	Self-learning by applying new knowledge to the existing data, models, and methods	0.687	0.797	0.790	1.000	
	Optimization and simulation for decision support	0.707	0.799	0.789	0.781	1.000

Notes: $N = 85$. Correlations have a (2-tailed) level of significance "Sig. < 0.000". Correlation is significant at the 0.01 level.

5. Discussion

The motivation for this research came from the novelty of the DT concept and its future applications, which will establish the adaption of CDTs in the AEC industry. Besides, the lack of attention paid in the literature to CDT in AEC project management led the authors to investigate this research. The adapted model in this study provides a viable solution to the identified problem. Process modeling has been used to explain the steps and significant aspects of the CDTsBLM framework. This study presents a novel adapted model that integrates CDT and BLM concepts and allows all project stakeholders to identify and collect the right data sets and implement them properly to optimize the system. The proposed model attempts to improve the BLM performance compared to the traditional and current methods.

In BLM, the CDT can be used to represent any physical unit. Buildings, process phases, total procedures, and ultimately an entire construction operation can be virtualized using CDT. CDTs can be elicited at various hierarchical levels, with CDTs combining horizontally and vertically to form an aggregated structure. The Cognitive Building Lifecycle Environment (CBLE), built by combining CDTs, shares significant knowledge horizontally. Only important decision-making material, on the other hand, is transmitted vertically to upper levels. A mission-critical building's CDT (monitoring and managing its condition and actions) supplying input to a particular process phase that feeds the building design process's CDT is an example. These CDTs will act and respond when sharing data with the various exchanged data sets and their semantics. Hence, the respective CDTs must be coordinated by a supervisory check, resulting in the CBLE, with market requirements, time horizons, and the essence of various activities that must be processed at any given time determined. This study examined the implementation, integrability, and interoperability of CDT in existing BLM practices in the life cycle, exploring decision-making skills and AEC industry insights. It is anticipated that the CDTsBLM model will promote the qualifications mentioned to allow better knowledge, analysis, optimization, and decision-making, which will concentrate on evaluation. The CDTsBLM model will enable re-evaluation, projection, and re-evaluation in a dynamic and complex world, with the possible planning, design, structuring, and operating. The operational processes' environmental effect must be reduced in the AEC industry by optimizing the building lifecycle processes by the CDTsBLM model.

One path forward to achieving new operational efficiencies depends on the reality that much of the human-dependent work activity can be significantly reduced by automating repetitive activities such as data acquisition, base data analysis, and the need for physical presence at physical locations, yielding a faster and safer approach to gathering data as well as reducing the time it takes to correlate and analyze that information. This rich collection of information is accessed, maintained, and controlled by humans for three primary activity streams. Analytics involves various analytical models, technologies, and approaches providing historical, current, and predictive insights from the data gathered. The workflow requires information about and procedures to inspect, maintain, modify and repair the physical asset. Visualization involves information, including the spatial geometry used for primarily planning and engineering.

The cognition, interpretation, and optimization of decision-making skills are fundamental to CDTsBLM. The connections between the ecosystem capacity perceived as a collective framework and all of the capabilities contribute to establishing a crucial cycle-centric application with inclusive aspects that contribute to explain the value of technology integrity from a professional inducible usability perspective. The CDTsBLM uses its models to evaluate data from the current framework to provide feedback and support decision-making. Depending on the study, the data and intelligence displayed are performed by the CDTsBLM.

The quantitative analysis of the data collected from Design Managers, Design Coordinators, Designers, BIM Managers, BIM Coordinators, Digitalization specialists, Project Managers, Construction Managers, Asset Managers, Asset Administrators, and Asset Controllers indicated that there is a willingness to use this type of CDT technology and related models. This analysis justified that the CDTsBLM model framework helps to provide a real-time analysis for data-driven models augmented by cognitive resources, which was conducted to facilitate decision-making and improve understanding, optimization, and thinking. Further, it shows that CDTsBLM is a valuable monitoring and control mechanism that helps in overall device optimization. It helps managerial levels of projects self-organize and respond to unpredictable activities and aid in decision-making for complex systems, including physical actors. The AEC industry can revolutionize how to design, build truly, and operate in a complex project environment. The AEC industry will inevitably adapt cognition, analytics, self-learning, and optimization techniques due to the emergence of DT, Cognitive computing, AI, ML, and cloud-based systems. Table 6 shows such a system's process, opportunities, and challenges. This research indicates that the CDTsBLM is an intelligent system that seamlessly connects engineering operational data, information, and models utilized over the whole building asset lifecycle with self-learning and predictive capabilities. It then makes the results readily available in real-time and the proper context for all related stakeholders to prevent or solve potential issues proactively. The findings from this research could serve as a base to pave the way for promoting progression towards Construction 4.0.

Table 6. Sample of Cognitive Digital Twins cognition, analytics, and optimization processes in Building Lifecycle Management.

	Building Lifecycle Management		
	Process	**Opportunities**	**Challenges**
Cognition	Sensing complex and unpredicted behavior, and reasoning and insights from real-time processing, where cases, knowledge, and experience interoperate to facilitate to comprehend and control the progress	Creating cognitive artificial intelligence from raw data and maximizing monitoring accuracy.	IoT network in terms of scalability, security, data loss, competent human resources, lack of enabling technologies
Analytics	Monitoring, refining, and utilizing the flow of incoming real-time data from various sources (the physical counterparts and sensors)	Applying cognitive analytics through data-enriched simulations enhanced by cognitive computing insights and predictive analytics	Lack of fully automated DT platform, various types of captured data, experienced staff, IT infrastructure, trust with respect to data, privacy and security, lack of historical data
Self-Learning	Extracting knowledge from aggregated data, automatically learning from data, identifying patterns, and making decisions.	Applying intelligent and self-learning planning and control to improve the accuracy of monitoring through iterative updating.	Integration of transfer learning algorithms, lack of comprehensive modeling language, data availability, validation of data
Optimization	For schedule design, task allocation, and workflow optimization of the relevant construction process and resource allocation	Combining reasoning and optimization for establishing planning and design, construction schemes based on analytic algorithms	Uncertainty quantification algorithms, multi-objective algorithms, complex environment modeling, large-scale computation

6. Conclusions

Construction projects and their data from the first stage to the last day of AEC projects are becoming huge and more complex to gather and manage. It is becoming exceedingly difficult, if not impossible, to identify and collect the right data sets and put them in the proper context to enable the optimization of the system. However, with help from a DT that can sense, reason, and act, such intelligent systems will help projects' stakeholders make the right decisions or autonomously trigger the right actions in the digital or physical world. Increasing complexity in terms of the DT becomes apparent when looking at the various application streams and their need for precise and real-time data. The DTs' highest level is the CDT connected with the top-level cognitive engineering maturity, including AI and ML. This article introduced the newly adapted CDT paradigm, including BLM's capabilities in cognition, analytics, and optimization for construction 4.0. The most significant advantage of cognition is the ability to solve problems preventively unknown. The CDT provides a toolkit for optimizing based on its cognitive components that enables CDT to carry out optimization tasks and delivers valuable results that other CDTs or process actors consume. Industry practitioners have examined the technical framework of CDTsBLM and taken full advantage of additional CDT capabilities, including construction schedules, preventive maintenance, and other goals, in the traditional DT sense. The benefits of implementing the CDT concept in the construction industry are intended by opening the optimization tool kit inside the CDT and enhancing real-time or almost real-time choices by interplaying optimization and simulation. Finally, the CDT description and conceptualization formalities will be further evolved alongside this implementation and assessment by tailoring every application scenario specific to the technology, ability, and KPIs that show this CDT effect.

The findings demonstrate the applicability of the CDTsBLM integration for a variety of AEC analysis scenarios. Future research directions could focus on investigating the processes and sub-processes of CDTsBLM applications in various AEC projects. Utilizing this system's legal and financial aspects will also lead to future research opportunities of CDTsBLM. Researchers might want to explore the processes and integrability of various construction technologies with CDT for various purposes.

Author Contributions: Conceptualization, I.Y., S.A.; methodology, I.Y., S.A., İ.A., M.E.A.; software, I.Y., S.A.; validation, I.Y., S.A., İ.A., M.E.A.; formal analysis, I.Y., S.A., İ.A., M.E.A.; investigation, I.Y., S.A., İ.A., M.E.A.; resources, I.Y., S.A., İ.A., M.E.A.; data curation, I.Y., S.A., İ.A., M.E.A.; writing—original draft preparation, I.Y., S.A., İ.A., M.E.A.; writing—review and editing, I.Y., S.A., İ.A., M.E.A.; visualization, I.Y., S.A., İ.A., M.E.A.; supervision, I.Y. All authors have read and agreed to the published version of the manuscript.

Funding: This research received no external funding.

Institutional Review Board Statement: not applicable.

Informed Consent Statement: not applicable.

Data Availability Statement: not applicable.

Acknowledgments: The authors would like to thank all the survey respondents in the AEC industry.

Conflicts of Interest: The authors declare no conflict of interest.

References

1. Alizadehsalehi, S.; Yitmen, I. A Concept for Automated Construction Progress Monitoring: Technologies Adoption for Benchmarking Project Performance Control. *Arab. J. Sci. Eng.* **2018**, *44*, 4993–5008. [CrossRef]
2. Salehi, S.A.; Yitmen, I. Modeling and analysis of the impact of BIM-based field data capturing technologies on automated construction progress monitoring. *Int. J. Civ. Eng.* **2018**, *16*, 1669–1685. [CrossRef]
3. Pärn, E.; Edwards, D.; Sing, M. The building information modelling trajectory in facilities management: A review. *Autom. Constr.* **2017**, *75*, 45–55. [CrossRef]
4. Götz, C.S.; Karlsson, P.; Yitmen, I. Exploring applicability, interoperability and integrability of Blockchain-based digital twins for asset life cycle management. *Smart Sustain. Built Environ.* **2020**. [CrossRef]
5. Lytras, M.; Visvizi, A. Artificial Intelligence and Cognitive Computing: Methods, Technologies, Systems, Applications and Policy Making. *Sustainability* **2021**, *13*, 3598. [CrossRef]
6. Abburu, S.; Berre, A.J.; Jacoby, M.; Roman, D.; Stojanovic, L.; Stojanovic, N. Cognitive Digital Twins for the Process Industry. In Proceedings of the The Twelfth International Conference on Advanced Cognitive Technologies and Applications (COGNITIVE 2020), Nice, France, 25–29 October 2020.
7. Hartmann, D.; Van der Auweraer, H. Digital Twins. In *Progress in Industrial Mathematics: Success Stories: The Industry and the Academia Points of View*; Springer International Publishing: Cham, Switzerland, 2021; pp. 3–17.
8. Zhang, N.; Bahsoon, R.; Theodoropoulos, G. Towards Engineering Cognitive Digital Twins with Self-Awareness. In Proceedings of the 2020 IEEE International Conference on Systems, Man, and Cybernetics (SMC), Toronto, ON, Canada, 11–14 October 2020; p. 3891.
9. Alizadehsalehi, S.; Hadavi, A.; Huang, J.C. From BIM to extended reality in AEC industry. *Autom. Constr.* **2020**, *116*, 103254. [CrossRef]
10. Vanlande, R.; Nicolle, C.; Cruz, C. IFC and building lifecycle management. *Autom. Constr.* **2008**, *18*, 70–78. [CrossRef]
11. Di Biccari, C.; Mangialardi, G.; Lazoi, M.; Corallo, A. Configuration Views from PLM to Building Lifecycle Management. In *IFIP International Conference on Product Lifecycle Management*; Springer: Berlin/Heidelberg, Germany, 2018; pp. 69–79.
12. Wu, S.; Liu, Q. Analysis on the Application of BIM and RFID in Life Cycle Management of Prefabricated Building. In Proceedings of the IOP Conference Series: Materials Science and Engineering, Chengdu, China, 13–15 December 2019; Volume 780.

13. Kubler, S.; Madhikermi, M.; Buda, A.; Främling, K.; Derigent, W.; Thomas, A.; Thomas, A. Towards data exchange interoperability in building lifecycle management. In Proceedings of the Proceedings of the 2014 IEEE Emerging Technology and Factory Automation (ETFA), Barcelona, Spain, 16–19 September 2014; pp. 1–8.
14. Ustinovičius, L.; Rasiulis, R.; Nazarko, L.; Vilutienė, T.; Reizgevicius, M. Innovative research projects in the field of Building Lifecycle Management. *Procedia Eng.* **2015**, *122*, 166–171. [CrossRef]
15. Puķīte, I.; Geipele, I. Different Approaches to Building Management and Maintenance Meaning Explanation. *Procedia Eng.* **2017**, *172*, 905–912. [CrossRef]
16. Kubler, S.; Buda, A.; Robert, J.; Främling, K.; Le Traon, Y. Building lifecycle management system for enhanced closed loop collaboration. In *IFIP International Conference on Product Lifecycle Management*; Springer: Cham, Switzerland, 2016; pp. 423–432.
17. Malagnino, A.; Mangialardi, G.; Zavarise, G.; Corallo, A. Business process management and building information modeling for the innovation of cultural heritage restoration process. In Proceedings of the IMEKO International Conference on Metrology for Archaeology and Cultural Heritage, Lecce, Italy, 23–25 October 2017.
18. Pan, Y.; Zhang, L. A BIM-data mining integrated digital twin framework for advanced project management. *Autom. Constr.* **2021**, *124*, 103564. [CrossRef]
19. Alizadehsalehi, S.; Yitmen, I. Digital twin-based progress monitoring management model through reality capture to extended reality technologies (DRX). *Smart Sustain. Built Environ.* **2021**. [CrossRef]
20. Deng, M.; Menassa, C.C.; Kamat, V.R. From BIM to digital twins: A systematic review of the evolution of intelligent building representations in the AEC-FM industry. *J. Inf. Technol. Constr.* **2021**, *26*, 58–83. [CrossRef]
21. Bosch-Sijtsema, P.; Claeson-Jonsson, C.; Johansson, M.; Roupe, M. The hype factor of digital technologies in AEC. *Constr. Innov.* **2021**. [CrossRef]
22. Hasan, S.M.; Lee, K.; Moon, D.; Kwon, S.; Jinwoo, S.; Lee, S. Augmented reality and digital twin system for interaction with construction machinery. *J. Asian Arch. Build. Eng.* **2021**, 1–12. [CrossRef]
23. Camposano, J.C.; Smolander, K.; Ruippo, T. Seven Metaphors to Understand Digital Twins of Built Assets. *IEEE Access* **2021**, *9*, 27167–27181. [CrossRef]
24. Meža, S.; Pranjić, A.M.; Vezočnik, R.; Osmokrović, I.; Lenart, S. Digital Twins and Road Construction Using Secondary Raw Materials. *J. Adv. Transp.* **2021**, *2021*, 1–12. [CrossRef]
25. Hou, L.; Wu, S.; Zhang, G.; Tan, Y.; Wang, X. Literature Review of Digital Twins Applications in Construction Workforce Safety. *Appl. Sci.* **2020**, *11*, 339. [CrossRef]
26. Borowski, P. Digitization, Digital Twins, Blockchain, and Industry 4.0 as Elements of Management Process in Enterprises in the Energy Sector. *Energies* **2021**, *14*, 1885. [CrossRef]
27. Del Giudice, M.; Osello, A. *Handbook of Research on Developing Smart Cities Based on Digital Twins*; IGI Global: Hershey, PA, USA, 2021.
28. Tagliabue, L.; Cecconi, F.; Maltese, S.; Rinaldi, S.; Ciribini, A.; Flammini, A. Leveraging Digital Twin for Sustainability Assessment of an Educational Building. *Sustainability* **2021**, *13*, 480. [CrossRef]
29. Boje, C.; Guerriero, A.; Kubicki, S.; Rezgui, Y. Towards a semantic Construction Digital Twin: Directions for future research. *Autom. Constr.* **2020**, *114*, 103179. [CrossRef]
30. Liu, Z.; Zhang, A.; Wang, W. A Framework for an Indoor Safety Management System Based on Digital Twin. *Sensors* **2020**, *20*, 5771. [CrossRef] [PubMed]
31. Austin, M.; Delgoshaei, P.; Coelho, M.; Heidarinejad, M. Architecting Smart City Digital Twins: Combined Semantic Model and Machine Learning Approach. *J. Manag. Eng.* **2020**, *36*, 04020026. [CrossRef]
32. Lu, Q.; Xie, X.; Parlikad, A.K.; Schooling, J.M. Digital twin-enabled anomaly detection for built asset monitoring in operation and maintenance. *Autom. Constr.* **2020**, *118*, 103277. [CrossRef]
33. Greif, T.; Stein, N.; Flath, C.M. Peeking into the void: Digital twins for construction site logistics. *Comput. Ind.* **2020**, *121*, 103264. [CrossRef]
34. Lu, Q.; Xie, X.; Parlikad, A.K.; Schooling, J.M.; Konstantinou, E. Moving from Building Information Models to Digital Twins for Operation and Maintenance. In *Proceedings of the Institution of Civil Engineers—Smart Infrastructure and Construction*; Thomas Telford Ltd.: London, Greater London, UK, 2020; pp. 1–9.
35. Rausch, C.; Sanchez, B.; Esfahani, M.E.; Haas, C. Computational Algorithms for Digital Twin Support in Construction. In *Construction Research Congress 2020*; American Society of Civil Engineers (ASCE): Tempe, AZ, USA, 2020; pp. 191–200.
36. Dawood, N.; Rahimian, F.; Seyedzadeh, S.; Sheikhkhoshkar, M. Enabling the Development and Implementation of Digital Twins. In *Proceedings of the 20th International Conference on Construction Applications of Virtual Reality*; Tesside University Press: Middlesbrough, UK, 2020.
37. Alonso, R.; Borras, M.; Koppelaar, R.H.E.M.; Lodigiani, A.; Loscos, E.; Yöntem, E. SPHERE: BIM Digital Twin Platform. *proceedings* **2019**, *20*, 9. [CrossRef]
38. Mathot, M.; Hohrath, B.; Rolvink, A.; Coenders, J. Design Modelling with Next Generation Parametric System Packhunt.io. In *Impact: Design With All Senses*; Springer: Cham, Switzerland, 2019; pp. 277–287.
39. Khajavi, S.H.; Motlagh, N.H.; Jaribion, A.; Werner, L.C.; Holmstrom, J. Digital Twin: Vision, Benefits, Boundaries, and Creation for Buildings. *IEEE Access* **2019**, *7*, 147406–147419. [CrossRef]

40. Kan, C.; Anumba, C.J. Digital Twins as the Next Phase of Cyber-Physical Systems in Construction. In *Computing in Civil. Engineering 2019: Data, Sensing, and Analytics*; American Society of Civil Engineers (ASCE): Atlanta, GA, USA, 2019; pp. 256–264.
41. Lu, Q.; Xie, X.; Heaton, J.; Parlikad, A.K.; Schooling, J. From BIM Towards Digital Twin: Strategy and Future Development for Smart Asset Management. In *International Workshop on Service Orientation in Holonic and Multi-Agent Manufacturing*; Springer: Cham, Switzerland, 2019.
42. Kaewunruen, S.; Lian, Q. Digital twin aided sustainability-based lifecycle management for railway turnout systems. *J. Clean. Prod.* **2019**, *228*, 1537–1551. [CrossRef]
43. Lydon, G.; Caranovic, S.; Hischier, I.; Schlueter, A. Coupled simulation of thermally active building systems to support a digital twin. *Energy Build.* **2019**, *202*, 109298. [CrossRef]
44. Alizadehsalehi, S.; Yitmen, I. The Impact of Field Data Capturing Technologies on Automated Construction Project Progress Monitoring. *Procedia Eng.* **2016**, *161*, 97–103. [CrossRef]
45. Rozanec, J.M.; Jinzhi, L. Towards actionable cognitive digital twins for manufacturing. In Proceedings of the International Workshop on Semantic Digital Twins, co-located with ESWC, Heraklion, Greece, July 2020.
46. Eirinakis, P.; Kalaboukas, K.; Lounis, S.; Mourtos, I.; Rozanec, J.M.; Stojanovic, N.; Zois, G. Enhancing Cognition for Digital Twins. In Proceedings of the 2020 IEEE International Conference on Engineering, Technology and Innovation (ICE/ITMC), Cardiff, UK, 15–17 June 2020; pp. 1–7.
47. Kalaboukas, K.; Rožanec, J.; Košmerlj, A.; Kiritsis, D.; Arampatzis, G. Implementation of Cognitive Digital Twins in Connected and Agile Supply Networks—An Operational Model. *Appl. Sci.* **2021**, *11*, 4103. [CrossRef]
48. Lu, J.; Zheng, X.; Gharaei, A.; Kalaboukas, K.; Kiritsis, D. Cognitive Twins for Supporting Decision-Makings of Internet of Things Systems. In *Proceedings of the 5th International Conference on the Industry 4.0 Model for Advanced Manufacturing*; Springer International Publishing: Cham, Switzerland, 2020; pp. 105–115.
49. Lu, V.Q.; Parlikad, A.K.; Woodall, P.; Ranasinghe, G.D.; Heaton, J. Developing a Dynamic Digital Twin at a Building Level: Using Cambridge Campus as Case Study. In *International Conference on Smart Infrastructure and Construction 2019 (ICSIC) Driving Data-Informed Decision-Making*; ICE Publishing: London, UK, 2019.
50. Du, J.; Zhu, Q.; Shi, Y.; Wang, Q.; Lin, Y.; Zhao, D. Cognition Digital Twins for Personalized Information Systems of Smart Cities: Proof of Concept. *J. Manag. Eng.* **2020**, *36*, 04019052. [CrossRef]
51. Rožanec, J.M.; Lu, J.; Rupnik, J.; Škrjanc, M.; Mladenić, D.; Fortuna, B.; Kiritsis, D. Actionable Cognitive Twins for Decision Making in Manufacturing. *arXiv* **2021**, arXiv:2103.12854.
52. Berlanga, R.; Museros, L.; Llidó, D.M.; Sanz, I.; Aramburu, M.J. Towards Semantic DigitalTwins for Social Networks. 2021.
53. Albayrak, Ö.; Ünal, P. Smart Steel Pipe Production Plant via Cognitive Digital Twins: A Case Study on Digitalization of Spiral Welded Pipe Machinery. In *Cybersecurity Workshop by European Steel Technology Platform*; Springer: Cham, Switzerland, 2020.
54. Abburu, S.; Berre, A.J.; Jacoby, M.; Roman, D.; Stojanovic, L.; Stojanovic, N. COGNITWIN—Hybrid and cognitive digital twins for the process industry. In Proceedings of the 2020 IEEE International Conference on Engineering, Technology and Innovation (ICE/ITMC), Cardiff, UK, 15–17 June 2020.
55. Essa, E.; Hossain, M.S.; Tolba, A.S.; Raafat, H.M.; Elmogy, S.; Muahmmad, G. Toward cognitive support for automated defect detection. *Neural Comput. Appl.* **2019**, *32*, 4325–4333. [CrossRef]
56. Saracco, R. Digital Twins: Bridging Physical Space and Cyberspace. *Computer* **2019**, *52*, 58–64. [CrossRef]
57. Fernández, F.; Sánchez, Á.; Vélez, J.F.; Moreno, A.B. Symbiotic Autonomous Systems with Consciousness Using Digital Twins. In *International Work-Conference on the Interplay Between Natural and Artificial Computation*; Springer: Cham, Switzerland, 2019.

MDPI

Article

Towards an Occupancy-Oriented Digital Twin for Facility Management: Test Campaign and Sensors Assessment

Elena Seghezzi [1,*], Mirko Locatelli [1], Laura Pellegrini [1], Giulia Pattini [1], Giuseppe Martino Di Giuda [2], Lavinia Chiara Tagliabue [3] and Guido Boella [3]

1 Department of Architecture, Built Environment and Construction Engineering, Politecnico di Milano, via Ponzio 31, 20133 Milan, Italy; mirko.locatelli@polimi.it (M.L.); laura1.pellegrini@polimi.it (L.P.); giulia.pattini@polimi.it (G.P.)
2 Department of Management, Università Degli Studi di Torino, Via Verdi 8, 10124 Turin, Italy; giuseppemartino.digiuda@unito.it
3 Department of Computer Science, Università Degli Studi di Torino, Corso Svizzera 185, 10124 Turin, Italy; laviniachiara.tagliabue@unito.it (L.C.T.); guido.boella@unito.it (G.B.)
* Correspondence: elena.seghezzi@polimi.it

Abstract: This study focuses on calibration and test campaigns of an IoT camera-based sensor system to monitor occupancy, as part of an ongoing research project aiming at defining a Building Management System (BMS) for facility management based on an occupancy-oriented Digital Twin (DT). The research project aims to facilitate the optimization of building operational stage through advanced monitoring techniques and data analytics. The quality of collected data, which are the input for analyses and simulations on the DT virtual entity, is critical to ensure the quality of the results. Therefore, calibration and test campaigns are essential to ensure data quality and efficiency of the IoT sensor system. The paper describes the general methodology for the BMS definition, and method and results of first stages of the research. The preliminary analyses included Indicative Post-Occupancy Evaluations (POEs) supported by Building Information Modelling (BIM) to optimize sensor system planning. Test campaign are then performed to evaluate collected data quality and system efficiency. The method was applied on a Department of Politecnico di Milano. The period of the year in which tests are performed was critical for lighting conditions. In addition, spaces' geometric features and user behavior caused major issues and faults in the system.Incorrect boundary definition: areas that are not covered by boundaries; thus, they are not monitored

Keywords: Building Management System; Digital Twin; Post-Occupancy Evaluations; facility management; asset management

Citation: Seghezzi, E.; Locatelli, M.; Pellegrini, L.; Pattini, G.; Di Giuda, G.M.; Tagliabue, L.C.; Boella, G. Towards an Occupancy-Oriented Digital Twin for Facility Management: Test Campaign and Sensors Assessment. *Appl. Sci.* **2021**, *11*, 3108. https://doi.org/10.3390/app11073108

Academic Editor: Jorge de Brito

Received: 28 February 2021
Accepted: 23 March 2021
Published: 31 March 2021

Publisher's Note: MDPI stays neutral with regard to jurisdictional claims in published maps and institutional affiliations.

1. Introduction

The operation and maintenance (O&M) phase of buildings and civil infrastructures ranges between 20–30 years for buildings, but it can cover more than 50 years of the whole lifecycle [1]. It is essential to ensure an actual and efficient management of buildings during the O&M phase. Occupancy and actual use of spaces strongly affect the organizational effectiveness and functioning during the operational phase [2,3]. Typically, standardized and fixed values of occupancy are considered during design phases, e.g., maximum occupancy values from fire regulations or scheduled occupancy for energy models [4]. Consequently, actual occupancy and space use levels may significantly vary from and rarely correspond to the values considered during the design phase. Occupancy strongly influences use and cleanness of spaces, which in turn are related to well-being, satisfaction, and productivity of users [5,6]. In recent years, a consistent number of studies investigated the segment of the performance gap between expected energy consumptions, defined during the design phase, and actual consumptions, due to human-building interaction and variable occupancy [6–17]. However, other promising fields in building management include security, safety, cleanness, and space management. These aspects can have a crucial role, especially in light of current sanitary emergencies related to the spread of the COVID-19 pandemic: space monitoring is a key aspect to guarantee safety in existing buildings [18].

In this context, the aim of the ongoing research project here presented is to define a Building Management Systems (BMS) based on an occupancy-oriented Digital Twin (DT), evolving from and enriching the Building Information Model (BIM) and integrating occupancy levels and additional relevant data from Post-Occupancy Evaluations (POEs). Analyses and simulations of the occupancy-oriented DT would support the decision-making processes during the O&M phase.

The case study for the application of the methodology is an existing office building hosting the Department of Architecture, built environment, and construction engineering (DABC) at Politecnico di Milano, Italy, used by people working at the university and performing their research and administrative activities in the indoor spaces of the building. The maintenance and cleanness of the distribution spaces and offices is a very important aspect in the facility management of the building and department business plan; strong variations in occupants' flows are experienced by the users and particularly during the pandemic.

The IoT network of sensors that represents the source of data for the occupancy-oriented DT and that was tested and calibrated as described in this article was provided and installed by an external consulting company (Laser Navigation srl). They provided the hardware part of the system that is the camera-based sensors with an embedded deep learning algorithm, the installation, and the technical settings of the sensors. They also provided an online platform named SophyAI and integrated with the IoT system, that allows to visualize, store, and download collected data.

This paper focuses on the preparatory phases for the definition of the DT, i.e., sensor system calibration and collected data quality validation. In fact, a fundamental characteristic of a DT is the connection, alignment, and reciprocity between the physical and virtual part [19]. Therefore, a key aspect is the data collection process, ensuring data quality on which the correct digital representation of the physical phenomenon depends [20,21], since, in order to obtain satisfactory results, is essential to ensure the quality of input data [22]. In this perspective, fundamental steps are the selection of sensor types that are most suitable for the specific application [23], the spatial distribution of sensors in the indoor spaces [24], and the setting and calibration of the IoT sensor system [25,26], to allow a correct detection and collection of data.

Given the importance of data quality for the proper digital representation of the building occupancy phenomenon, the objectives of the research are: optimization of spatial distribution and orientation of sensors for system planning and installation, identification of issues and faults of the detection system, and resolution of issues and faults by performing an assessment of the detection system through test campaigns. This study proposes method and evaluation criteria for system calibration and data quality validation, also defining parameters for occupancy analysis. Two test campaigns were performed until all major faults have been checked and solved, allowing for the verification and validation of collected data quality to monitor building occupancy. The study also describes and tests the use of the platform SophyAI for real-time visualization of data during the test campaigns.

2. Literature Review

2.1. *Evolution, Main Applications, and Features of Post-Occupancy Evaluations*

Post-Occupancy Evaluations (POEs) aim at assessing building performances, users' behavior, and feedback regarding existing buildings during the operational phase and once the building has been occupied for some time [27–31]. POEs were first introduced in the UK and US in the 1960s in order to assess building performances from user perspective, by means of interviews, questionnaires, photographic surveys, and walk-through surveys [27,28]. The major developments of POEs were during the 1980s, aiming at analyzing and optimizing the facility management and design [29]. POEs had been performed in the US, mainly in the public sector, UK, New Zealand, and Canada [32], and, since a correlation between workplace features and worker productivity was proposed in 1985, they have been also applied in the private sector to improve costumer and worker satisfaction and to optimize the workplaces [30]. In the mid-90s, the interest moved from analyses during the operational phase alone to an entire building life cycle process, i.e., Building Performance Evaluation (BPE) [33,34]. Insights and findings from POEs could be applied in the subsequent design and building life cycle process [34–36].

In the last two decades, POEs have mainly been applied to assess and optimize building energy performances, and to reduce the building environmental impacts [37]. A less investigated but promising research field is the optimization of occupancy patterns, and cleaning activities and contracts. Space features and workplace cleanness have been classified as basic factors affecting user satisfaction [5], and, consequently, user productivity. The variable "interior use of space" can account for around 43% of the variance in employees' enjoyment at work, well-being, and perceived productivity [6].

As above mentioned, POEs are analyses of the built environment, aiming at defining the effectiveness and functionality of spaces for users, building performances, and user satisfaction and perception regarding facilities in general and workplaces in particular [27,38]. There are three levels of POEs depending on accuracy, time needed to be performed, tools, and levels of invasiveness of user privacy [7,30]:

- Indicative POEs enable to perform overall non-invasive analyses of the building, with selected interviews and photographic surveys to detect critical areas of the building.
- Investigative POEs are more in-depth analyses, and more invasive, with questionnaires, video recordings, and measurement, and they are meant to find causes and consequences of the building performances.
- Diagnostic POEs are the most in-depth analyses, with high levels of user privacy invasiveness and high costs, since they can imply the use of sensor systems to monitor the building, providing data to analyze and optimize building performances and future designs.

Despite being expensive and invasive for user privacy, especially when performing diagnostic analyses, POEs can have several benefits, ranging from the optimization of the operational phase in terms of performances and user satisfaction, to an increased facility adaptation to organizational change and growth over time [27,30], and to the definition of design criteria and requirements based on actual user and space needs for similar buildings [39].

2.2. From Building Information Models to Digital Twins for Asset Management

In recent years, a major evolution of Building Information Modelling (BIM) occurred in the construction sector. BIM models are parametric models, centralized sources of information mostly for the design and construction phases, and instruments to improve collaboration among specialists and document management [40]. The application of BIM for facility management can result in several benefits: customer services improvement, time and cost reduction resulting from better planning capabilities, and higher consistency of data [41,42]. The integration of POEs in a BIM approach enables the connection between POE data and the digital model [8,12,43], with the advantages of defining a single source and storage of POE and building data, integrating structured data into the BIM, and identifying POE data and related issues in a visual representation of the building space [8,44,45]. Despite the advantages of adopting BIM during the operational phase, a BIM approach for asset management lacks of information richness, analysis, and simulation capability, which are usually manually implemented and time-consuming when using a BIM model [20]. In addition, an effective and efficient management of buildings during the operational phase strongly rely on continuous flows of real time data regarding the building, its performances, and conditions [20,46]. However, BIM models present limitations for the integration with different data sources and systems, e.g., sensor data, and lack of automatic updating and evolution over time [20]. Therefore, in order to overcome these limitations, the definition of a Digital Twin is investigated.

2.3. Evolution of the Digital Twin Concept

The Digital Twin (DT) concept dates back in 2002 when the idea of a virtual space containing the information of and linked to the real space emerged in the field of study of complex systems, in particular regarding the Product Lifecycle Management (PLM) [19]. When the concept emerged, it was not referred to as DT, but it was presented as the "Conceptual Ideal for PLM" [19], evolved then to Mirrored Spaces Model in 2005 [47] and to Information Mirroring Model in 2006 [48] and 2011, when also the term Digital Twin was first used to describe the model [49]. In recent years, the concept of a DT has been studied also in the aerospace sector: the DT represents an ultra-realistic digital replica of real flying vehicles, considering one or more interconnected systems allowing for probabilistic simulations that take into account physical characteristics and models, sensor data, and history of previous flights and vehicles [50–52]. Recent definitions of DTs can be found in various sectors, with a wide use and diffusion of the concept of a virtual replica of physical entities whose purpose is to manage, optimize, and control the physical asset itself. In the infrastructure sector, DT was defined as a realistic virtual representation of the corresponding infrastructure, adding the built or natural context in which the object is contained and to which it is connected [53]. In the manufacturing sector, the idea of the connection between physical components and virtual models is widened, adding the necessary mono- or bi-directional flow of data between the physical asset and its virtual counterpart in order to real-time monitoring the actual object, supporting simulations, analytics, and control capabilities of the dynamic virtual model [54]. The construction industry can be still considered in its beginning regarding the definition of a DT for buildings. Despite the various attempts to define a DT in construction industry [21,24,53,55,56], a comprehensive definition was proposed by Al-Sehrawy and Kumar [57]: "an approach for connecting a physical system to its virtual representation via bidirectional communication (with or without human in the loop) using temporally updated Big Data [...] to allow for exploitation of Artificial Intelligence and Big Data Analytics by harnessing this data to unlock value through optimization and prediction of future state". This definition includes all the fundamental parts of a DT, which are described in detail in the following paragraph.

2.4. Elements and Characteristics of a Digital Twin

As stated, a DT is composed by some elements. A list of components for DTs in construction industry is provided as follows:

- A physical asset and its virtual counterpart, and data connecting them [23];
- Platform to visualize and manage sensor data, e.g., data and virtual model visualization, analysis, and simulation, which is a key aspect for real-time remote monitoring [23]. The platform should return insights, alerts, or predictions regarding the physical object, thus supporting the decision-making process for the definition of O&M objectives and plans [40,58];
- An acquisition layer such as an IoT system [40,46,53,59], since sensing is a vital component of a DT [60–63], allowing for continuous monitoring of the physical asset. The virtual component enriched with real-time data regarding the real object represents a dynamic digital replica of the physical asset [46,53];
- BIM model as a starting point, especially as regards the geometrical virtual replica of the building, allowing for the evolution and information enrichment of the BIM model itself [20,23];
- Artificial intelligence (AI) tools to analyze data and provide predictions, simulations, and data analytics [20,46].

In addition, some characteristics are fundamental for the correct definition of a DT:

- Synchronization between physical and virtual component [40], with data flowing at least in one direction allowing for analyses, control, and simulation on the virtual model [20,46,64]. Any change in the monitored characteristics or conditions of the asset is detected and, through data flow, is reflected in the virtual counterpart [20,21];

- Bidirectional communication between physical and virtual part, either with or without humans in the loop, defining a Passive DT or an Active DT, respectively [57]. The knowledge regarding the asset provided by the virtual part results in either human intervention of direct actuation in the real asset [21];
- A DT represents specific and selected aspects of the physical asset, i.e., the subjects of monitoring, simulating, and analyzing, so it does not represent an exact duplication of reality [57];
- Data or status visualization capabilities in order to support the monitoring and decision-making processes by the actors that are in charge of the asset O&M phase [46].

As previously specified, one of the main characteristics of a DT is the direct connection between physical and virtual entity, with the concept of twinning as alignment and reciprocity between the two components [19]. Therefore, a fundamental aspect for the definition of a DT is the data collection process, i.e., data quantity, quality, and granularity, on which depends the correct detection of changes of the real object over time, and thus the correspondence of the digital object to the real one and its continuous evolution through the building lifecycle [20–22]. A first fundamental step is the selection of sensor types that are most suitable for each specific application [23]. In addition, the spatial distribution of sensor network, i.e., the spatial distribution of sensors in the indoor spaces, is another theme that should be faced [24]. Furthermore, in order to allow a correct detection and collection of data, the IoT sensor system should be properly set and calibrated [25,26] to ensure the quality of data collected. In fact, the output of an analysis strongly depends on the data that are used as input for the system or algorithm; therefore, to obtain satisfactory results, data quality is essential [22]. Nonetheless, existing studies tend to focus on different phases and aspects of the DT definition and creation, while IoT sensor system definition is a less investigated aspect, almost taken for granted [23].

As stated above, a fundamental preliminary step is a detailed analysis of sensor types in order to identify the most suitable ones for the research objectives. Such analysis is described in the following paragraph, in which existing studies regarding types of sensors for occupancy detection are analyzed, highlighting features, pros, and cons. In addition, a brief review of the concept of occupancy detection is provided. The investigation supported the selection of the sensor type for the case study, as explained in the following methodology section.

2.5. Occupancy Detection: Analysis of Occupancy Monitoring Systems

Occupancy detection consists in the definition of occupancy levels and patterns of buildings during the operational phase. Occupancy patterns consist of occupancy values at room-level and user movements inside the building [65]. Monitoring occupancy patterns and optimizing the use of spaces and cleaning activities, based on occupancy data, can increase user satisfaction and productivity at work. In fact, occupancy levels of buildings have a strong influence on cleanness and use of spaces that, in turn, are strongly related to well-being, satisfaction, and productivity of users [5,6]. Table 1 focuses on IoT monitoring sensor systems studies, highlighting main features, pros, and cons.

As shown in Table 1, camera-based sensors and PIR (Passive Infra-Red) sensors present the best accuracy levels, followed by CO_2 sensors, but they are also affected by detecting and privacy issues [9], such as the Hawthorne effect for camera-based sensors. It mainly causes alterations of behavior when users are aware of being observed and, if ignored, can affect the reliability of collected data [26]. One strategy implies the combination of more types of sensors, some of which may already exist in the building, having been previously installed for other purposes [9]. Additionally, system implementation costs can be reduced by previously analyzing the building with Indicative POE analyses in order to identify the most critical areas to be further analyzed [27,32] by means of sensor systems and other techniques.

Table 1. Sensor systems and related features to monitor occupancy.

Sensor Type	Main Aspects	Pros	Cons
Camera-based sensors [25,26]	Average accuracy of 97%	High accuracy Security and safety applications	Users detection only within field-of-view Privacy issues and Hawthorne effect
CO2 concentration change sensors [25,26,66]	Average accuracy of 94%	Often used in buildings No privacy issues	Less reliable than other type of sensors
Visual light and infrared (PIR) technologies [9,16,67,68]	High accuracy of 97% (unoccupied–occupied scenarios) Accuracy 93% (stationary and moving occupants)	High accuracy No privacy issues	Issues in detecting stationary occupants Users' presence/absence detection only within the field-of-view
Radio frequency identification (RFID) sensors [9,15,69]	Accuracy of 88% (stationary occupants)Low accuracy 65% (moving occupants)	No privacy issuesAccess-control system applications	Low accuracy compared with other sensor systems
Wi-Fi connections [8,9,70–74]	Average accuracy of 80%	Available in most buildings	Privacy issues in visualizing and analyzing users' connections

The highlighted advantages and disadvantages of existing sensor types supported the selection of the type of sensors for the methodology and case study, as described in the following section.

3. Research Project Stages

This paper presents some stages of an ongoing research project. The aim of the research project is to define a Building Management Systems (BMS) based on an occupancy-oriented Digital Twin (DT), evolving from and enriching the Building Information Model (BIM) and integrating occupancy levels and additional relevant data from Post-Occupancy Evaluations (POEs). The expected results of the research project are: monitoring occupancy and defining building occupancy patterns, optimizing current O&M management, building space use and organization, cleaning activities, and, as possible future implementations, applying Smart Contract to cleaning and maintenance services and extending the IoT network with other kind of sensors for safety and quality control. The research project stages are presented in Figure 1.

The first two stages, "definition of BIM guidelines" and "BIM model creation", have been previously analyzed in a publication by Di Giuda et al. [75]. "Preliminary analyses", "system installation", and "test campaigns" are presented in this paper, as they are critical to provide the foundations upon which the occupancy-oriented DT should be based.

The "occupancy-oriented DT" set of activities is currently under development. In future steps of the research, collected data will be analyzed to identify occupancy patterns of the building spaces, and evaluate current management of spaces in terms of people permanence and cleaning frequency. In addition, benchmarks to evaluate optimization strategies will be defined together with the subjects in charge of O&M in the case study building, a fundamental step to evaluate advantages and results of the methodology [76]. The defined occupancy-oriented DT will be the base for the subsequent phase, i.e., "FM scenario definition and optimization", that will allow the optimization of cleaning activities and contracts that are currently based on the building floor areas, and to reach a better organization and planning of space usage.

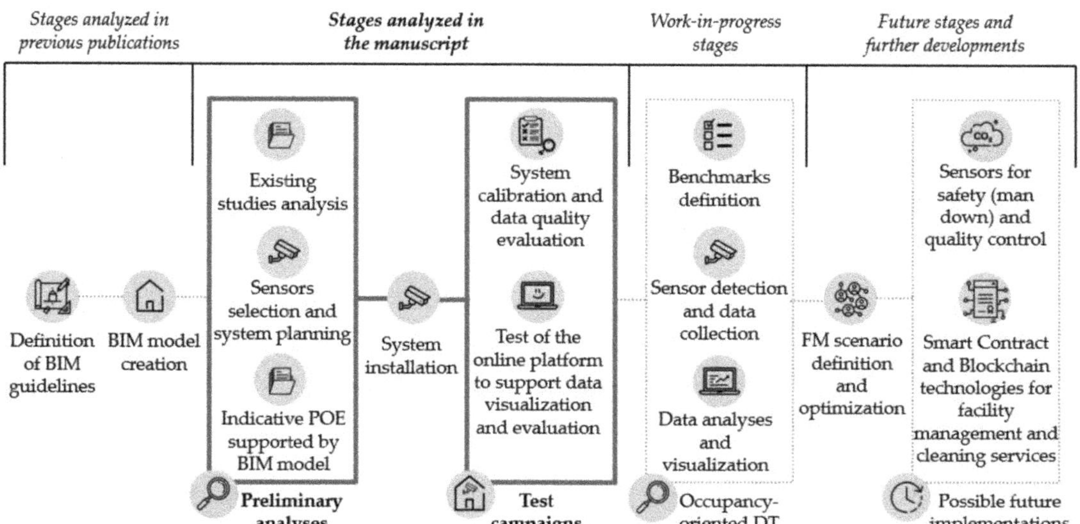

Figure 1. Stages of the research project.

Figure 1 also provides "possible future implementations" of the research project. The integration of other kinds of sensors, such as "sensors for safety (man down) and quality control" will allow the monitoring and optimization of different aspects of the building management, resulting in a complete report of building conditions and indoor environmental quality. In addition, a possible future implementation of the system will be the definition of "smart contracts for facility management and cleaning activities" that would be based on the actual need of cleaning defined in previous stages. Smart Contract based on Blockchain technologies and on the occupancy-oriented DT data will provide relevant advantages, i.e., increased network security, reliable data storage, traceability [77], and the possible automation of payments for cleaning activities [78,79].

4. Method

This section provides the methodology applied for the "preliminary analyses" and "test campaigns" stages, analyzed in detail in this article. The "system installation" task was performed by an external consulting company that provided and installed the IoT sensor network, and the platform SophyAI for visualization, storage, and download of collected data.

4.1. IoT Network of Camera-Based Sensors

The "sensors analysis and selection" phase relies on the proposed literature review. As previously stated, most analyzed recent applications aimed at optimizing energy performances and consumptions rather than building operation and use [9,12,16]. Nonetheless, existing studies allowed objectively comparing several available sensor types, supporting the selection of the most suitable type for occupancy monitoring.

Camera-based sensors were selected considering their high accuracy and the possibility to perform other kind of analyses, such as security and safety monitoring, thus allowing for further implementations of new features in the system, increasing the scalability of the system itself.

The limitations of camera-based sensors that have been presented in the literature review section, and how they have been overcome, are described as follows:

- Detection only within field-of-view of the sensors: the BIM model was used to ensure the best positioning and orientation of sensors and to maximize the area covered by the sensors' field-of-view;
- Privacy issues and Hawthorne effect: the system was set to anonymously monitor users and not to store any images. The user is recognized as a human by the deep learning algorithm embedded in the camera-based sensors and translated into an anonymous agent that cannot be linked to a specific user identity. Consequently, the movements of the user can be anonymously monitored in real time and visualized in the online platform SophyAI, without storing any real image or video recording.

The sensors can detect occupancy; in particular, they can visualize real time movements of users that are instantaneously transformed into anonymous virtual agents.

The detection of anonymous real-time movements of users is limited to common spaces, i.e., circulation areas and corridors, and they can be visualized in the online platform SophyAI, but are not stored in the database (DB), to protect the users' privacy. On the contrary, sensors count and store the number of agents that are entering or leaving rooms, which are the main objects of monitoring.

Two values are recorded by the sensors for each monitored room:

- O: Occupancy values at room-level, i.e., the number of people (p) occupying a room in a certain period of time;
- T: Period of time in which one or more virtual agents occupy a room (minutes/hours).

4.2. *Visualization and Analysis Platform*

A critical theme for real-time monitoring is the possibility of plotting sensors data for visualization, verification, and analyses [23]. Data visualization is a primary subject to support decision-making processes and to help people who are in charge of O&M in reaching management goals, since they may not possess the technical ability to effectively use the indexes and information directly extracted by the sensor system [58].

As shown in Figure 2, the described monitoring system is intertwined with the online platform SophyAI. The platform can:

- Visualize real-time occupancy count, i.e., the instantaneous value of O of each space;
- Visualize real-time movements of anonymous virtual agents in a 2D visualization of spaces;
- Store in a DB data regarding the occupancy count of each space (O) and of each day of the week;
- Store in a DB data regarding the room occupancy time (T) during each day of the week.

Data stored in the online platform DB can then be downloaded as CSV files that contain the values of O and T for all days of a specified period of time, which could be a week, a month, or a year. In addition, data can be processed in graphs and diagrams and visualized through the online platform.

The online platform displays a 2D visualization of the spaces. Each monitored area is contained in a 2D boundary, which defines the contours of the area itself. The check between the area displayed in the 2D visualization and the 3D view of the same area detected by cameras is a key aspect to correct the optical distortion between the 3D view of the camera and the 2D view of the online platform. The check between 2D boundary and 3D view was performed as a part of the study during the system test and calibration, as described in following paragraphs.

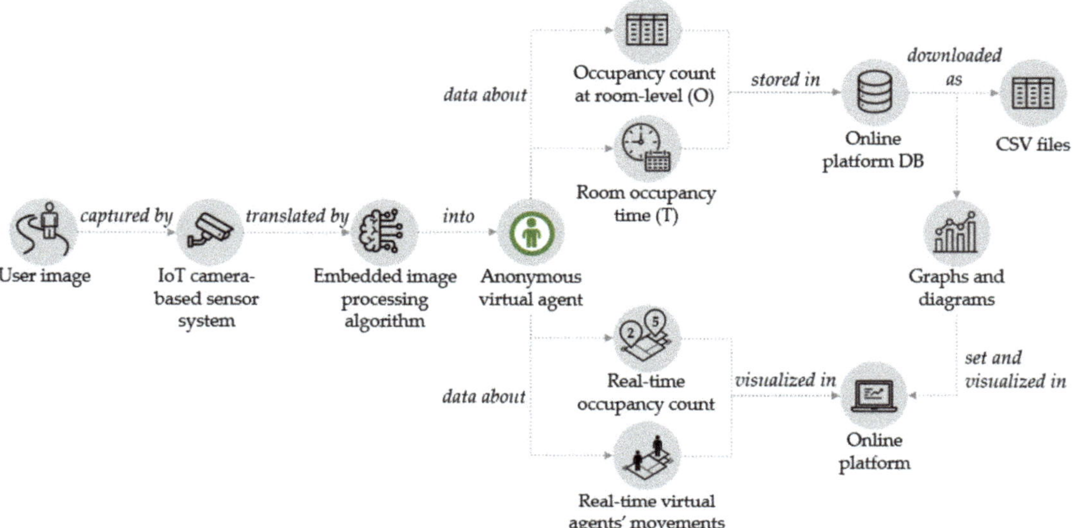

Figure 2. Data and information flow.

4.3. Preliminary Analyses Based on Post-Occupancy Evaluations and Building Information Modelling

This phase applies Indicative POEs to preliminary analyze the building by means of general and low-invasive analyses, using the BIM model:

- Analysis of the geometry of spaces: identification of number of levels of the building and number and geometry of rooms. The geometry of spaces influences the number and position of cameras that are needed to monitor the whole space. In addition, the height of spaces represents the maximum height at which the sensor can be installed, and in turn influences the field-of-view of the sensor;
- Analysis of the functions of spaces: identification of the function of spaces, e.g., bathroom, office, equipment room, etc. The function of spaces influences the definition of the area to be monitored. For example, an equipment room with no variable occupancy, since only technicians can enter the rooms for planned maintenance, does not represent a critical area for occupancy monitoring. As a result, the critical areas whose variable occupancy needs to be monitored are identified. The installation of sensors is limited to the identified critical areas, thus reducing implementation costs of the overall system;
- Analysis of electrical and data and communication systems: analysis of presence, distribution, and equipment of electrical equipment. A non-homogeneous distribution of the electrical and data wiring can in fact represent a limitation for sensors installation;
- Simulation of sensors location and orientation: virtual objects representing the sensors are placed into the BIM model, and each virtual sensor is linked to a field-of-view to simulate the area covered and seen by the sensor itself. The height of installation of the sensor also influences the field-of-view. The simulation of several configurations allows the optimization of number, position, and orientation of sensors, maximizing the area covered by sensors.

The use of the BIM model as a source of information and simulation tool to perform the Indicative POE ensures the minimization of user privacy invasiveness. In addition, the sensor system plan is optimized by comparing different configurations.

4.4. Test Campaign Methodology for Data Quality Evaluation

The preliminary analyses allow for an efficient planning and installation of sensors, selecting critical areas to be monitored, and optimizing spatial distribution, orientations, and fields-of-view of sensors. Nonetheless, after the first phases of data collection, some errors and faults, described in the following paragraph, may occur, and the system needs to be calibrated to ensure data quality. An incorrect calibration would lead to incorrect data collection and to an erroneous modelling of the occupancy patterns of the spaces, with repercussions on the whole BMS.

The iterative process to perform the "test campaigns" (Figure 1) is presented in detail in Figure 3 and described in the following paragraphs.

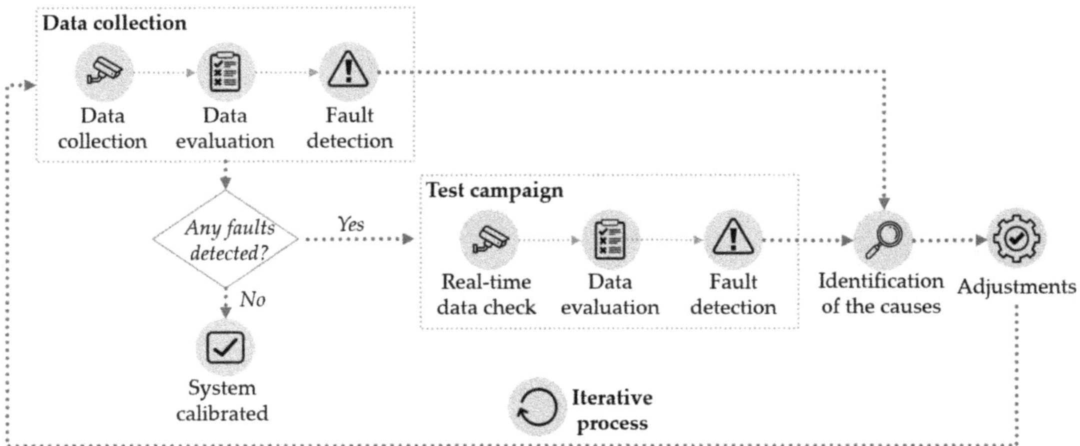

Figure 3. Data collection, test campaign, and adjustments application iterative process.

Once the system is installed as planned with the support of preliminary analyses, data are collected for a representative period of time that should be identified for each case study. Then collected data are downloaded in CSV format from the online platform DB. Collected data are analyzed in order to identify the possible data errors and related system faults, as described in Table 2. If no faults that could compromise the following analyses are detected, the system is properly functioning and calibrated. Otherwise, test campaigns are performed to verify the errors detected in the collected data.

The real time test campaign involves two operators. One operator (operator A) monitors through the online platform the position and movements of the other operator (operator B) inside the building. The two operators are constantly connected via earphones to communicate and coordinate with each other. In particular, operator A guides operator B towards the areas where errors were previously detected in the collected data. Moving inside the building and entering/exiting the rooms, operator B tests the detection of user movements and the room occupancy count (O) by the system. At the same time, operator A monitors the response of the system by checking the real-time displayed user movements and instantaneous values of O of the rooms through the online platform. Consequently, the operators search for detection errors and system faults in order to identify the causes, as described in Table 2. System faults can be classified as missing data, outliers, stuck values, and noise. Each fault can be identified in collected data or during test campaigns according to specific values of O. In addition, noise can be detected only during real-time test campaigns, by comparing the movements of operator B and his anonymous digital counterpart displayed on the online platform. Some examples of the causes of the errors and faults are camera malfunctioning in the case of missing data and extreme lighting contrast in the monitored area, which impedes a correct detection and causes noisy data.

Table 2. System faults [22] and related data errors. Data errors are divided into errors observed in collected data and errors detected during real-time test campaigns.

System Fault	Data Error in Collected Data	Data Error during Real-Time Test Campaign
Missing data: data are not collected	O = 0 p	O = 0 p
Outliers: one or more consecutive anomalous values	O < 0 p Values of O unacceptable for room dimensions, e.g., O = 50 p in a 10-square-meter office	O < 0
Stuck values: those values occur when a sensor fails in detecting and a previously-detected value remains fixed	O > 0 outside the working hours	No correspondence between detected O and actual occupancy values, e.g., O > 0 in an empty room
Noise: it represents corrupted values	Not detectable in collected data	No correspondence between the virtual agent movements detected by the system and displayed in the platform, and the actual movements of operator B

Once the causes of data errors and faults are identified, some adjustments are proposed and applied to the system. Then the system must be verified again, in an iterative process, until no errors are detected and consequently the required data quality level is reached. This iterative process is also useful to check overtime the effectiveness of improvement solutions or to check the system after geometry changes in the building, e.g., in the case of refurbishments.

Regarding the possible adjustments to solve the errors and the related causes, some general rules were identified to define a hierarchy of possible solutions.

Generally, the most preferable solution would be not acting on the hardware of the system: in the case of a recently added physical obstacle that prevents the camera-based sensor from detecting, the most preferable solution would be moving the object before moving the sensor. In addition, before acting on the hardware part of the system (e.g., adding or replacing cameras), the camera settings could be checked, and the software system would be improved. An example of camera setting adjustment is the modification of contrast and luminance settings of the camera in the case of extreme lighting contrast in the monitored area. In addition, modifying the software is faster, less invasive, and cheaper than working on the hardware. Specifically, the deep learning algorithms of the embedded artificial intelligence system of the cameras for image recognition could be improved and optimized with the support of the external consulting company Laser Navigation srl. Consequently, the adjustments are hierarchized based on those general rules using the following symbols: from the most preferable solution, identified with (++), to the least preferable one, identified with (–).

5. Case Study

The building chosen as case study hosts the Department of Architecture, Built Environment and Construction Engineering (DABC) of Politecnico di Milano, and is located in Milan (Italy). It is a four-story building, hosting administrative offices, research spaces, and university staff offices, for a total of 4300 square meters of gross floor area. Rooms have variable dimensions depending on their use. The building has a symmetrical layout, with a common space in the center and two side corridors. The offices and workspaces are located on either sides of the corridors. Each floor houses at least one bathroom. Before the current study, the building has never been monitored. Therefore, neither data regarding the actual occupancy patterns, nor information about actual cleaning and maintenance activities are currently analyzed and optimized. Furthermore, no space optimization has been performed in relation to the use of available rooms and the actual occupancy indexes at room-level in the building. This case study building acts as prototype for a future application of the proposed method to other university's buildings.

The case study section is divided in two subsections: the first one describes the application and results of preliminary analyses on the building that supported the planning and installation of the IoT sensor system; the second subsection describes the two test campaigns with specific focus on the detected system faults and related proposed adjustments.

5.1. Preliminary Analyses: Sensors Spatial Distribution and Orientation

A preliminary study of the building ("Indicative POE supported by BIM model" phase as in Figure 1) was performed to identify critical areas to be monitored and to optimize number, position, and orientation of sensors, which in turn allowed the reduction of implementation costs and proper planning of the IoT sensor system.

As described in the methodology section, the preliminary analyses included the following activities that are analyzed in detail in the following paragraphs:

- Analysis of the functions of spaces;
- Analysis of the geometry of spaces;
- Simulation of sensors location and orientation;
- Analysis of electrical and data and communication systems.

5.1.1. Analysis of the Functions of Spaces

The analysis of the building through the BIM model allowed the identification of number and type of rooms of the building, as shown in Table 3. The BIM model had been previously defined and modeled, as described in Di Giuda et al. [75], who also performed a complete survey to update the as-built documents and to ensure the correspondence between the BIM model and the building. Equipment rooms, storage closets, and archives were excluded, since the only users are cleaning services employees or technicians in charge of maintenance activities. The analyses highlighted that sensors installed in common spaces, i.e., corridors, would be sufficient to monitor room occupancy, i.e., the count of users entering and leaving rooms. Anonymous real-time agent movements are detected only in corridors and can be visualized in the online platform, but are not stored in the DB, to ensure and protect the privacy of users. On the contrary, as regards rooms, the count of number of users (O) and time of occupancy (T) is recorded, as shown in Table 3. The occupancy of critical rooms is monitored to optimize their use, cleanness, and maintenance, while corridors are considered only as circulation areas. As shown in Table 3, 70 rooms out of 87 were selected as critical areas to be monitored.

Table 3. Type and quantity of rooms and necessity to be monitored.

Building Level	Space Type	Quantity	Monitored/Not Monitored
Underground Level	Laboratory/Office	5	Monitored
	Bathroom	2	Monitored
	Classroom	1	Monitored
	Equipment Space	3	Not monitored
	Storage Room	10	Not monitored
Ground Level	Laboratory/Office	22	Monitored
	Bathroom	3	Monitored
	Meeting Room	1	Monitored
	Equipment Space	1	Not monitored
First Level	Laboratory/Office	21	Monitored
	Bathroom	3	Monitored
	Storage Room	1	Not monitored
	Terrace	1	Not monitored
Second Level	Laboratory/Office	10	Monitored
	Bathroom	1	Monitored
	Meeting Room	1	Monitored
	Equipment Space	1	Not monitored

5.1.2. Analysis of the Geometry of Spaces and Simulation of Sensors Location and Orientation

During the preliminary phases regarding the system planning, the BIM model of the building was used to optimize locations and orientations of the camera-based sensors. The geometry analysis highlighted that the building corridors are long, low ceiling, and narrow (length: 32 m; height: 2.40–2.70 m; width: 1.60 m). Three simulations of the interrelated position of the sensors in a corridor have been performed to define the best configuration. Virtual objects representing the sensors were added to the BIM model in different locations according to the three possible configurations. Each virtual sensor was then linked to a field-of-view that allowed to virtually check through the model the area covered by each sensor. The BIM-based simulation analyzed three possible configurations, as shown in Figure 4:

- The first solution considered two standard cameras at the two opposite sides of the corridor. Each corridor would be entirely monitored by two sensors at the same time, but in the central area, the detection could be less precise due to the distance from the sensors. In addition, cameras would have difficulty in monitoring areas near the end of the corridor, i.e., the area close to each sensor. Users passing through a door near the end of the corridor would be extremely distorted in the view of the nearby camera, making recognition difficult.
- The second solution considered two standard camera-based sensors located at 1/3 and 2/3 of the corridor. This solution allows for a better monitoring of the end areas of corridors, but limits the simultaneous monitoring by both sensors to the central area only.
- The third solution implied the use of a single 360-degree camera at the center of the corridor. Those kind of sensors are more expensive than standard cameras, but the total cost would be comparable, since this solution would consider only one sensor instead of two. This solution results in the corridor being entirely monitored by a single camera.

The chosen solution was the second one, since in many cases there are doors near the end of corridors, thus excluding the first solution. In addition, due to the reduced width of the corridors, one single camera could struggle in identifying two people walking lined up. Therefore, the third solution, which involved only one camera, was also less preferable than the second one.

The chosen sensors are High Quality Bullet Pro Camera PoE, with the following features: they provide HD quality images; the Power over Ethernet (PoE) allows to supply power and network connection to the camera with a single cable; a Wide Dynamic Range (WDR) allows to compensate problems due to exposure to light; the view angle of the camera reaches a maximum of 110 degrees. The system is installed in a dedicated Virtual Local Area Network (VLAN), and a static IP is provided for each element of the system. As stated in the Introduction, the sensor system was provided by a third-party organization, Laser Navigation Srl, who operated in full compliance with EU General Data Protection Regulation (GDPR). In fact, the deep learning algorithm does not record images, but only metadata regarding the anonymous movements and count of users are processed by the system, inhibiting the recognition of the observed subjects.

The 20 sensors were installed directly in the ceiling, i.e., at height 2.40/2.70 m depending on the level of the building, ensuring the maximum coverage area. Figure 5 shows the plan of the IoT sensors system in the case study building.

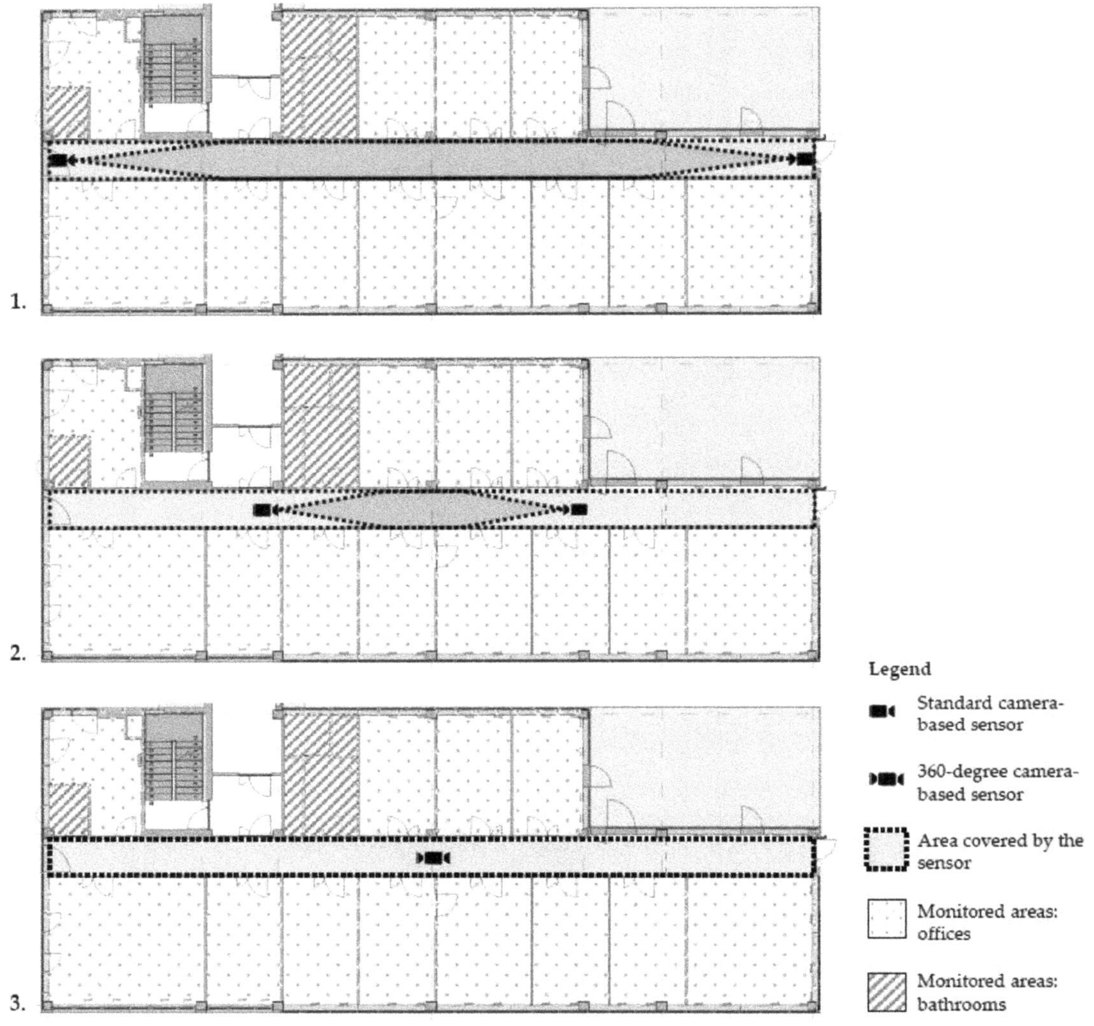

Figure 4. 2D visualization of the three simulations of sensors positioning in a typical corridor through the BIM model.

The BIM model also allowed for an optimization of the field-of-view of the sensors, as shown in Figure 6. The virtual camera field-of-view simulation supported the definition of the best orientation, i.e., the best tilt angle of each camera on x- and y-axis. This ensured that all the offices and bathrooms defined as critical, whose occupancy needed to be monitored, were correctly detected by the sensors.

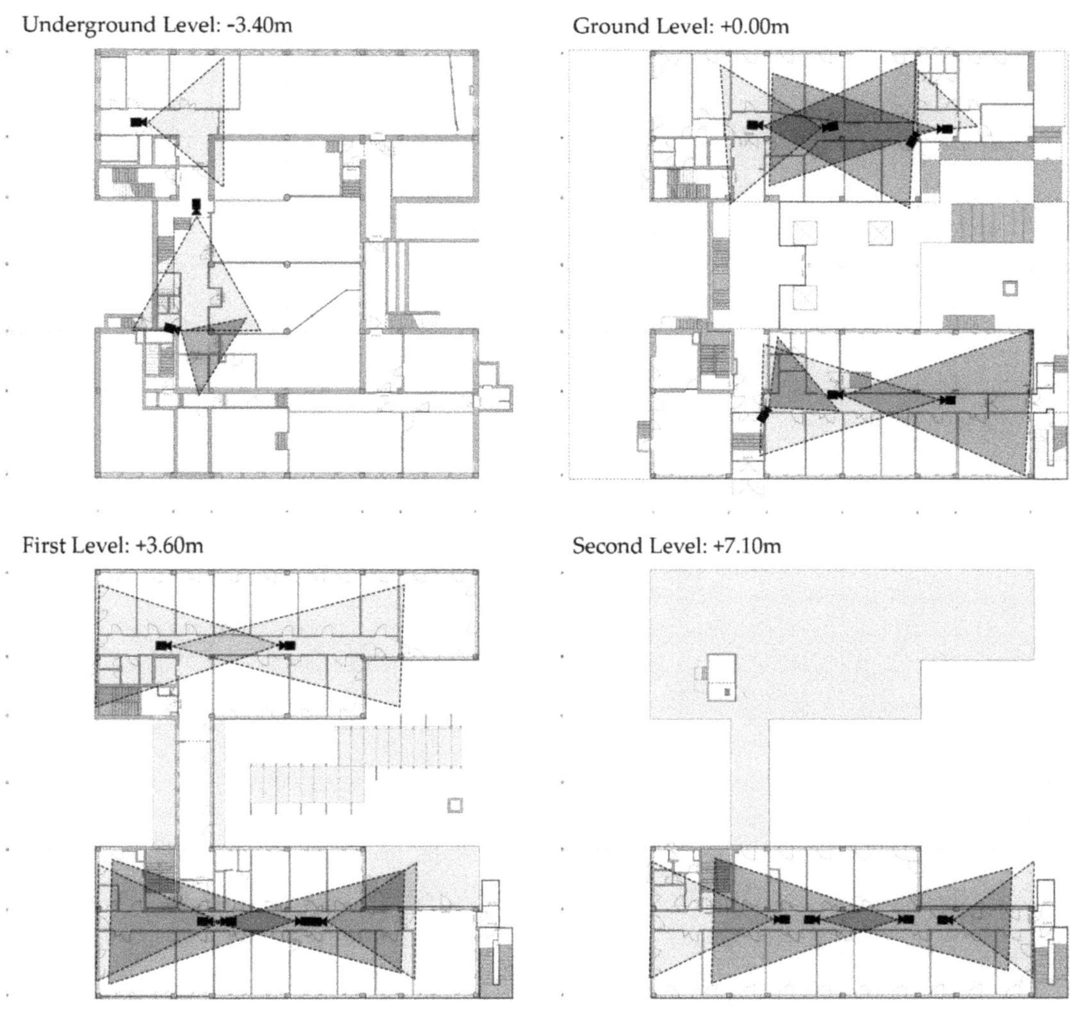

Figure 5. Ultimate spatial distribution of camera-based sensors inside the case study building.

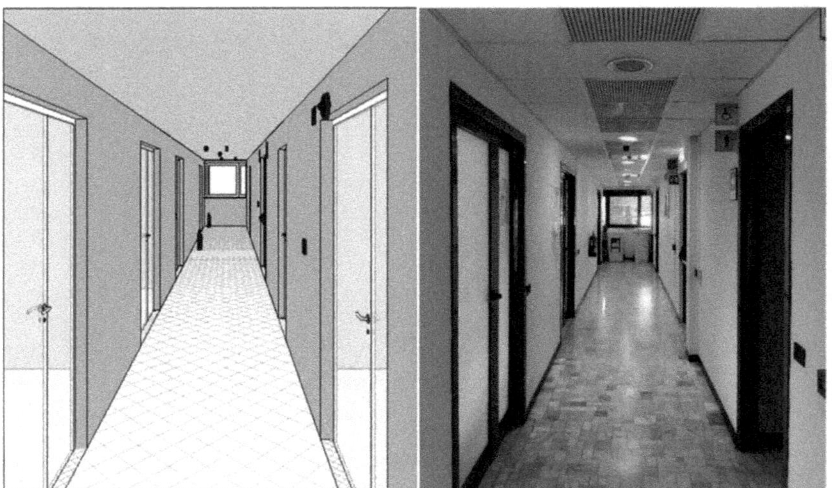

Figure 6. Comparison between the simulation of the virtual sensor field-of-view in the BIM model (**left**) and the actual field-of-view of the installed sensor (**right**).

5.1.3. Analysis of Electrical and Data and Communication Systems

The last preliminary analysis performed with the BIM model was the check of the electrical and data system equipment and wiring distribution already available in the building. The analysis showed that since cameras would be installed in corridors, all the necessary wiring was already available. Therefore, no implementation was needed to install the system.

5.2. Test Campaigns

Once the preliminary analyses had been performed, the system had been installed. First data collection was performed, and collected data were analyzed to identify issues and faults. Data were collected during a three-month period, i.e., the representative period, as it is the minimum period of time to encounter all possible activities conducted by the users of the department. A qualitative analysis was conducted on the collected dataset to identify rough errors.

Figure 7 shows a graph of collected data about a bathroom during one day. Stuck values are identified since 4 p.m., because the occupancy raises but never decreases. It is a stuck value because the bathroom can host only one person at a time; therefore, a continuous occupancy of four people represents without any doubt a blunder in the detection. The solution to this specific problem is provided in Table 4.

After faults of the system were detected, a first test campaign was performed to understand the causes and propose improvements for the system. Detected data were tested in real time by two operators, as described in the method, to identify the causes of system faults. An example of the visualization of real time data in the online platform is shown in Figure 8.

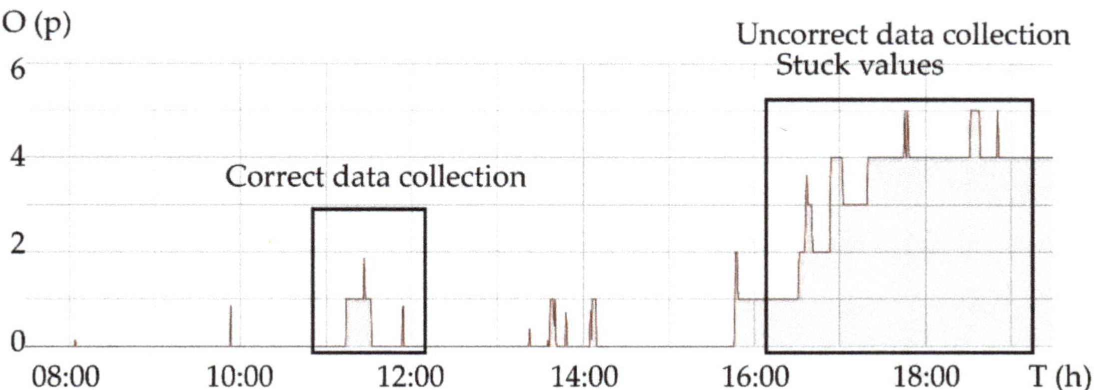

Figure 7. Graph of collected data regarding occupancy overtime in a bathroom (O-T).

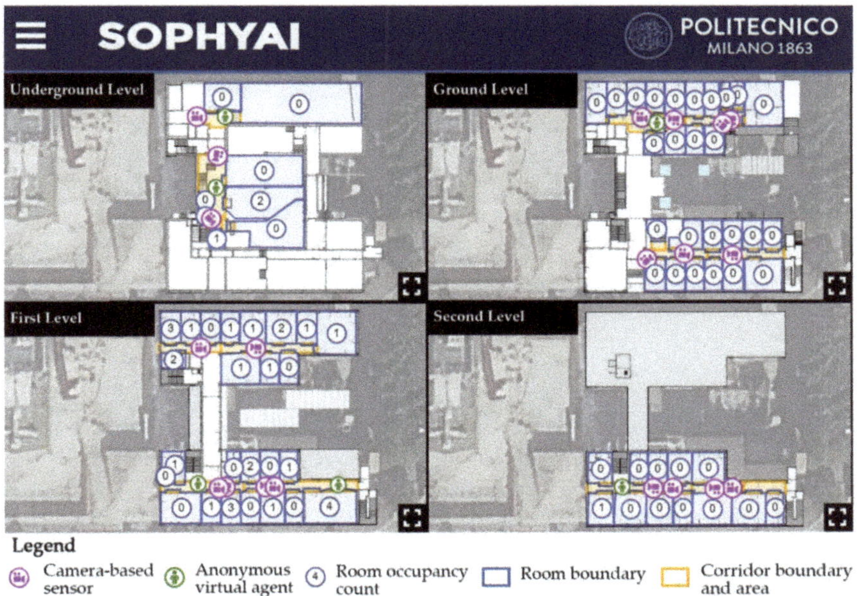

Figure 8. Visualization of real time data in the online platform during test campaign.

This first test campaign was followed by a second test campaign, to properly calibrate the system and ensure detected data quality, as shown in Figure 9. The two test campaigns were carried out during different periods of the year. This represented a key aspect for the recognition of lighting contrast issues. The first test campaign was performed in June 2020, with data collection for a three-month period from November 2019 to January 2020. The first test campaign was performed after the end of the first Italian shutdown period due to COVID-19 pandemic (early March–early June 2020). The second test campaign was performed in November 2020, with data collected for another three-month period from July to October 2020, excluding August, during which the building is usually under-occupied due to summer holidays. After the shutdown period March–June 2020, administrative and research activities have been resumed. Therefore, all the data collected for system test and calibration can be considered reliable.

Table 4. Identified issues and effects on the data collecting system during the test campaigns. Table legend: (a): data collection phase; (b): real-time test campaign phase; O: occupancy values at room-level, number of people occupying the room (p); T: period of time in which users occupy a room (minutes/hours); adjustments hierarchy scale ranging from (++) most preferable system adjustment to (–) least preferable system adjustment.

Detected Issue, Evaluation Criteria, and Fault Classification	Cause Identification and Effects on the System	Proposed Hierarchized Adjustments
Data are not detected and collected. The system detects: (a)–(b): "Data = null". This fault is classified as Missing data (a)–(b).	Camera-based sensors not working	Verification of sensor integrity, functioning, and connection
	Incorrect boundary definition: areas that are not covered by boundaries; thus, they are not monitored	Perform a real time test to identify and verify the optimal boundary definition to minimize optical distortion between the 3D view of the camera and the 2D floor map visualization
	Obstructions or obstacles in corridors that impede users' vision, like printers, waste bins for separate collection of paper, and presence of platforms for people with disabilities	(++) Remove the obstacle, if possible (+) Improve the deep learning algorithms for image recognition (–) Add a new camera and re-verify the system
	Behaviors of users that deceive the detection system: blind zone caused by unexpected doors left open	(++) Verify the possibility to avoid keeping the door open with a communication to the users (–) Add a new camera in a different position and re-verify the system
	Blind zone of the sensor: when two people are walking in corridors towards a sensor, the person further away from the camera generally is not detected	(++) Possibility to ignore the related error, which does not affect the next phase of statistical data analysis for the definition of the occupancy pattern (–) Add a new camera allowing a multiple detection of the same area and re-verify the system
Fast increase/decrease of occupancy values. The system detects: (a)–(b): Negative or too high values of O. This fault is classified as Outliers (a)–(b).	Behaviors of users that deceive the detection system: difficulty in counting users when they are standing in front of the door opening the room or talking right in front of the entry of a room	(++) Add an automatic routine to the algorithm that records the occupancy data after a minimum user presence (+) Add an automatic routine to the algorithm that brings the count back to 0 when the displayed count is negative. (–) Enrich the system with the possibility of manually resetting rooms occupancy values in the presence of a wrong count (only in Administrator mode).
Unexpected length of the period of time the room is occupied (for bathrooms).(An example is shown in Figure 7) The system detects: (a): T > 15 min (b): Irregular real-time user detection This fault is classified as Stuck values (a) and Noise (b).	Too high distance of the camera from the to-be-detected area: irregular detection of users with a continuous detection/disappearance of a moving user, resulting in wrong collected data, as if there were multiple users closely entering the room one after the other	(++) Improve the deep learning algorithms for image recognition (–) Add new cameras and re-verify the system
	Elevated lighting contrast between different areas of corridors: irregular detection of users with a continuous detection/disappearance of a moving user, resulting in wrong collected data, as if there were multiple users closely entering the room one after the other	(++) Review the camera settings regarding lighting and contrast (–) Add a new camera in the brighter zone and re-verify the system

Table 4. *Cont.*

Detected Issue, Evaluation Criteria, and Fault Classification	Cause Identification and Effects on the System	Proposed Hierarchized Adjustments
Unexpected moment of the day in which the room is continuously occupied (for offices) The system detects: (a): O > 0 outside working hours (b): Irregular real-time user detection This fault is classified as Stuck values (a) and Noise (b).	Too high distance of the camera from the to-be-detected area: irregular detection of users with a continuous detection/disappearance of a moving user, resulting in wrong collected data as if there were multiple users closely entering the room one after the other. Due to the higher value of O than the real number of people in the room, when people leave, O does not return to zero, with remaining values of O > 0 even after the end of the working day	(++) Improve the deep learning algorithms for image recognition (–) Add new cameras and re-verify the system
	Elevated lighting contrast between different areas of corridors: irregular detection of users with a continuous detection/disappearance of a moving user, resulting in wrong collected data as if there were multiple users closely entering the room one after the other. Due to the higher value of O than the real number of people in the room, when people leave, O does not return to zero, with remaining values of O > 0 even after the end of the working day	(++) Review the camera settings regarding lighting and contrast (–) Add a new camera in the brighter zone and re-verify the system
	Difficulty detecting two people entering in a room close together and/or quickly. This causes wrong collected data as if there were multiple users closely entering the room one after the other. Due to the higher value of O than the real number of people in the room, when people leave, O does not return to zero, with remaining values of O > 0 even after the end of the working day	(++) Possibility to ignore the related error, which does not affect the next phase of statistical data analysis for the definition of the occupancy pattern (–) Add a new camera allowing a multiple detection of the same area and re-verify the system
	Difficulty detecting cleaning employees due to the presence of the cleaning trolley, which impedes a complete view of the operator. Therefore, often the cleaning employee is detected entering the room (O = +1) but not leaving, so the value O remains unchanged	(++) Optimization and training of the recognition algorithm to identify the cleaning trolley by excluding the cleaning service employee in the occupancy count (–) Add a new camera allowing a multiple detection of the same area and re-verify the system

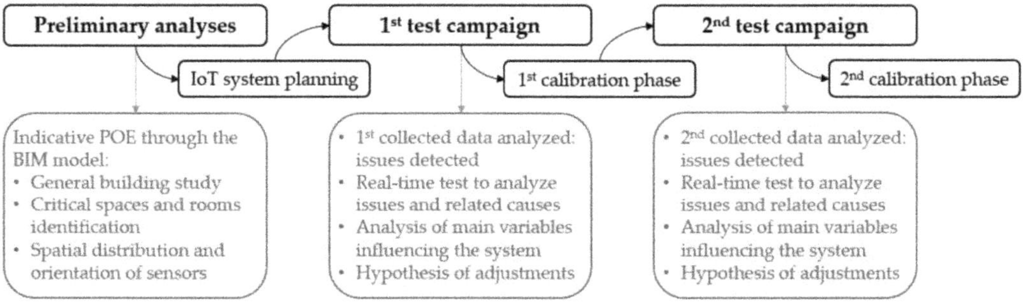

Figure 9. Preliminary analyses and two test campaigns process.

6. Results and Discussion

Table 4 provides a resume of errors, related evaluation criteria, fault identification and classification, causes of the faults, and proposed solutions, hierarchized and listed from the most preferable one (++) to the least preferable one (–) of the two test campaigns.

The first test campaign highlighted the following issues:

- Difficulty of the system in detecting two people entering in a room close together and/or quickly;
- Issues in the detection of two people walking in a corridor towards a sensor, since the person further away from the camera is not detected. The error occurs in all areas not covered by the fields of view of two cameras at the same time;
- Irregular detection of users with a continuous detection/disappearance of a moving user due to the high distance between the camera and the to-be-detected area. This issue was in fact detected mainly in areas far from the sensors.

The identified issues are mainly due to the geometry of corridors, which are low ceiling, long, and narrow. Due to the limited height of the corridor, the cameras struggle in detecting people walking in groups or lined up (Figure 10). The issues led to an incorrect user detection affecting the displayed data in the online platform. However, while the possibility of having people walking lined up or in groups in corridors is relatively high, the probability of two or more people entering a room simultaneously is low, due to the standard dimensions of the doors, that allow the entrance of one person at a time. For this reason, it is possible to ignore these issues. To overcome the issue related to irregular detection of users due to the distance of areas from the camera, improvement in the deep learning algorithms for image recognition were implemented.

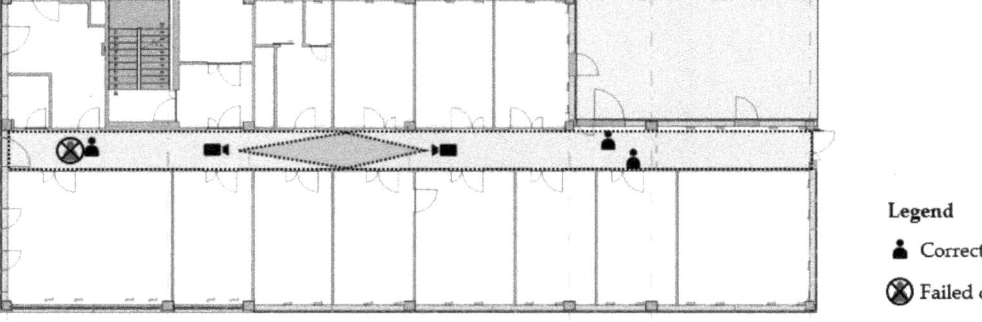

Figure 10. Location and orientation of camera-based sensors in corridor.

Considering the online platform, a major issue was related to the values indicating the presence of people in the rooms showing a negative value or a high positive value. This means that, according to the collected data, many people were entering or leaving the room in a very short time. Automatic routines to solve and mitigate incorrect data have been implemented to the software:

- Automatism that brings the count back to 0 when the displayed count is negative;
- Possibility to manually reset rooms that present a wrong count;
- Automatic routine that records the occupancy data after a minimum user presence (i.e., 5 min) for office spaces only.

The improvement of the automatic routine solves the related problem of users' behavior that deceive the detection system, such as standing in front of the door when opening the office or talking right in front of the entry of a room.

After the modifications and improvements applied, data have been collected for a period of three months from July to October 2020, excluding August for lower building occupancy due to summer holidays, to verify the effectiveness of the strategies adopted.

The qualitative analysis of the second dataset highlighted a general improvement in detection capabilities of the system, since technical issues were not identified anymore, but some faults occurred anyway. Therefore, a second real-time test campaign was carried out in November 2020, resulting in the following sensor-related issues:

- Difficulty in detecting users at the end of the corridors due to the presence of windows. The intense natural light generates a high luminous contrast between the central part of the corridor and the terminal part. The light contrast of the two zones generates an unstable detection of users. The detecting issue related to lighting contrasts of different zones of the building was only discovered in the second test campaign and not during the previous test. Considering the location of the building (Milan, 45°28′46.8″ N 9°13′48.0″ E), the sun is low in the sky during the winter season. This can generate detection issues related to lighting contrast, which cannot be detected during others seasons of the year, which explains the newly emerged detection issue, since the first test campaign had been performed in May. Therefore, conducting several tests during different periods of the day and year is strongly recommended for camera-based sensor systems. A preferable solution to overcome the lighting contrast issue is modifying the settings of the camera to correct the lighting contrast.
- Failure in detecting the users' entrance due to other kind of obstructions such as open doors. Specifically, the doors opening towards the corridor can obstruct the view of the adjacent room entrance, preventing the system from registering users entering the room (Figure 11). The issue can only be managed by adding new cameras to cover the unexpected blind spots. The issue was unexpected, since doors are usually kept closed when offices are occupied.
- Issues in the recognition of cleaning service company employees. The system struggled in detecting the workers due to the presence of the cleaning trolley, which impeded a complete view of the operator. The cleaning trolley provoked an incorrect counting of entries, exits, and occupancy of the rooms. To overcome the issue, the recognition algorithm can be optimized and trained to correctly recognize the cleaning service operator, by recognizing the cleaning trolley.

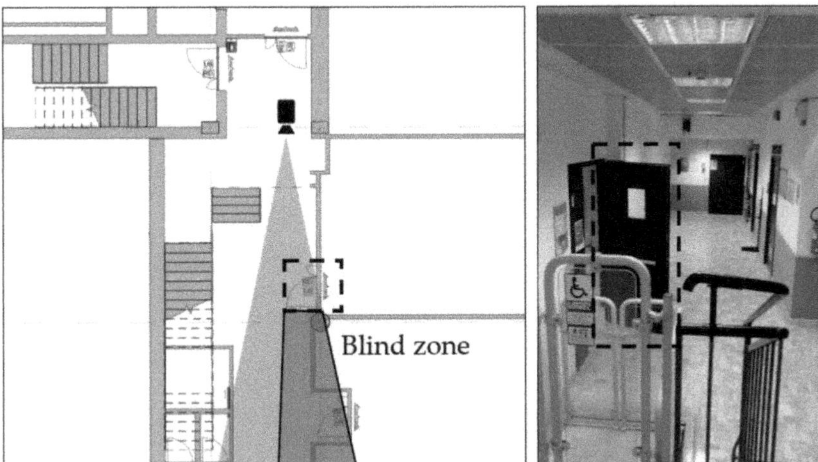

Figure 11. Unexpected blind zone generated by unusual occupants' behavior.

Figure 12 presents the percentages of error types detected during the two test campaigns. During the first test campaign, 30 out of 70 monitored rooms presented detection faults, while during the second test campaign, 38 rooms presented detection issues. The reason of the higher number of faults during the second test campaign is mainly due to the lighting issues that emerged only during the second test campaign.

During the first test campaign, 70% of errors were related to technical issues: 17% of errors due to not-working cameras, and 53% of errors due to difficulty detecting users in areas too far from the camera-based sensors. As shown in Figure 12, these types of technical issues were completely fixed by adjusting the system settings. Once all cameras were properly working and correctly set, those issues did not occur in subsequent analyses.

Another type of issue detected in the first test campaign was related to unexpected user behavior, resulting in 30% of errors. These errors could be adjusted with some improvements in the system. However, a 3% of errors due to unexpected user behavior occurred also during the second test campaign. Despite the error percentage being significantly lower in the second test campaign, this type of error could not be completely avoided because of the unpredictable nature and high variability of user behavior.

Regarding the second test campaign, elevated lighting contrasts between different areas of corridors caused 68% of errors. This kind of error was never detected during the first test campaign because of the different period of the year when the test was performed, a key aspect for proper calibration of a camera-based sensors system.

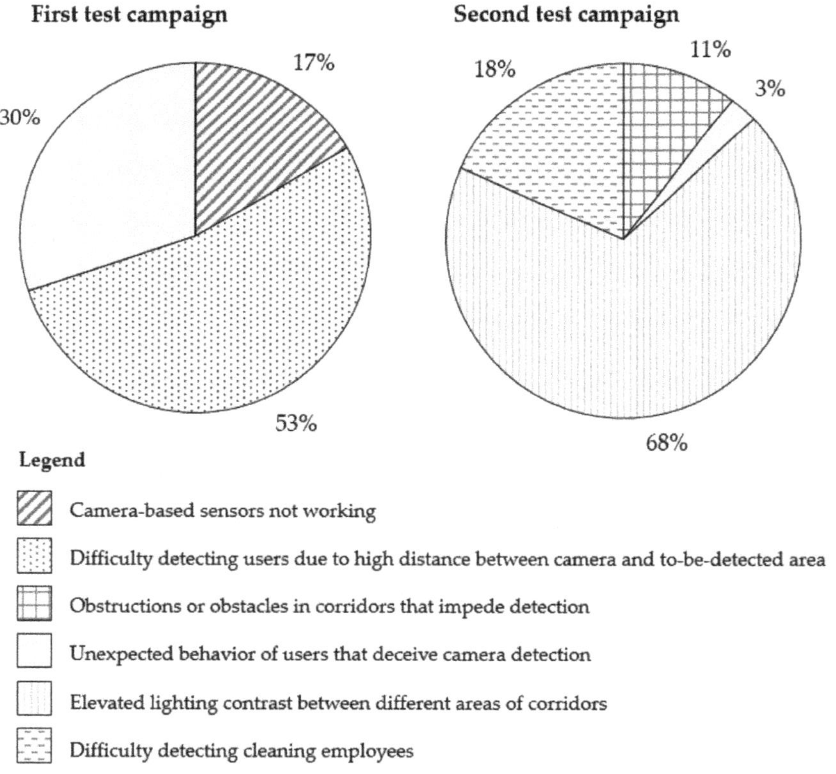

Figure 12. Charts of errors detected during first and second test campaigns.

The remaining 30% of errors of the second test campaign were related to difficulties detecting the cleaning employee (18%) and to obstructions and obstacles that impeded the detection (11%). Cleaning employee detection issues were classified as a technical error in the first campaign. After the resolution of technical issues, the error persisted, and the actual cause was identified, highlighting the importance of a multi-stage testing of the system. Detection issues related to obstacles and obstructions appeared because some pieces of furniture were moved or spaces were reorganized. The frequent check of the correct functioning of the system overtime is fundamental to verify newly appeared issues and consequently adjust and improve the system.

A comparison of these results with other systems could be helpful to provide an assessment of the proposed system. As previously underlined, the setting and calibration of monitoring system has frequently been neglected in existing literature, and the accuracy of IoT systems applied to DTs is often taken for granted. Available data regard, as shown in Table 1, the accuracy of specific sensors' typologies, but do not address the accuracy of systems, which depends on several variables, e.g., building features and use, number of sensors, etc. The test campaigns here presented have been used to explore and improve the efficiency of the entire systems of DT.

For this reasons, these results obtained from the case-study building cannot be compared to other systems, based on different typologies of sensors. The provided case study application is useful to define guidelines to calibrate IoT camera-based sensors system.

7. Conclusions and Further Developments

This work presents the development of first steps of an ongoing research project to define a Building Management System (BMS) for facility management, especially regarding the occupancy and cleaning activities in office buildings that would be based on an occupancy-oriented Digital Twin (DT). The proposed BMS would ensure better space management, organization, and cleaning, since the system would detect actual occupancy levels and related needs for cleaning activities. The advantages result in optimizations of cost and space use, as well as customized cleaning activities and contracts.

In particular, this study presented the IoT system calibration phases, i.e., the preliminary analyses to optimize the planning of the IoT camera-based sensors system, and the test campaigns, in order to ensure the system efficiency and accuracy to monitor occupancy. A key aspect of the definition of a DT has been in fact identified in the data connection between physical asset and virtual counterpart, the main components of a DT. In addition, the data quality is a critical aspect to ensure the quality of the results of the analyses, simulations, and predictions performed on the virtual model.

The case study section highlighted that the preliminary analyses, i.e., Indicative Post-Occupancy Evaluations (POE) supported by the use of the BIM model, were important to plan the IoT system, in particular as regards number, locations, and orientation of the sensors. The analyses allowed the identification of offices and bathrooms as main spaces to be monitored. In addition, the observed configuration of building spaces allowed planning the sensors installation only in corridors, from which it is possible to detect entries and exits from the different rooms. The BIM model allowed for simulations of sensors location and fields-of-view.

As regards the two test campaigns results, some system faults and related causes were identified and solved. The issues generated by user behavior were the least predictable, trivial, and at the same time the most difficult and expensive to solve, requiring the installation of new cameras. The variability of human behavior inside a building is very high; the calibration of the system must cover a sufficient period of time to bring out all problems related to human behaviors. Considering the complexity of the monitoring system and the high dynamicity of the variables involved (e.g., fast-changing spatial conditions and user behavior), a multi-stage test and calibration campaign was fundamental for the correct setting of a camera-based sensor system.

Another interesting aspect resulting from the test campaigns was the influence that the period of the year had on the test itself, due to changing lighting conditions.

Other relevant aspects are the geometric features of the to-be-monitored spaces. For example, the limited width and height of the corridors led to some difficulties in detecting more users moving together. However, those issues did not have critical effects on the collected data. The boundary conditions of the system should be carefully checked, as they could have negative consequences on collected data and on data analyses.

The use of an online platform was useful to real-time check and evaluate data during the test campaigns, as well as to remote controlling the monitoring system.

Once the system is tested and assessed, further developments of the research will regard the proper monitoring of the building. As of now, qualitative analyses have been performed on collected dataset to identify rough errors, and by means of the two test campaigns, the causes of the faults have been identified and solved. During the next phases of the research project, quantitative analyses will be conducted on collected datasets, which will be the basis for the definition of the occupancy-oriented DT. DT analyses and simulations, and resulting optimization scenarios, will be proposed and analyzed to identify real advantages and limitations of the proposed methodology. The proposed method, once completely tested and refined on the case study building, could be extended to large building stock, supporting the decision-making process of building owners and building managers.

Potential applications of the system would entail the integration of other kind of sensors to monitor Indoor Air Quality (IAQ), carbon dioxide, temperature, humidity, and Volatile Organic Compounds (VOC) levels, resulting in a more complete evaluation of the building conditions and Indoor Environmental Quality (IEQ). Sensors could play an important role for safety management purposes. The combined use of the system with Smart Contract and Blockchain technology could ensure increased network security, reliable data storage, traceability of data, and the possible automation of payments for cleaning activities. Cleaning contracts could in fact be customized based on the actual use of spaces, detected by the proposed system.

Author Contributions: Conceptualization: E.S., M.L., and L.P.; methodology: E.S., M.L., and L.P.; software: L.P. and G.P.; validation: E.S, M.L., L.P., G.P., G.M.D.G., L.C.T., and G.B.; formal analysis: E.S., M.L., and L.P.; investigation: M.L., L.P., and G.P.; resources: G.M.D.G.; data curation: M.L. and L.P.; writing—original draft preparation: E.S., M.L., and L.P.; writing—review and editing: E.S, M.L., L.P., G.P., G.M.D.G., L.C.T., and G.B.; visualization: E.S., M.L., and L.P.; supervision: G.M.D.G., L.C.T., and G.B..; project administration, G.M.D.G.; funding acquisition, G.M.D.G. All authors have read and agreed to the published version of the manuscript.

Funding: This research received no external funding.

Institutional Review Board Statement: Not applicable.

Informed Consent Statement: Not applicable.

Data Availability Statement: Restrictions apply to the availability of these data. Data was obtained from Department of Architecture, built environment and construction engineering and are available from the authors with the permission of Department of Architecture, built environment and construction engineering party.

Acknowledgments: The authors thank Eng. Marco Schievano and Eng. Francesco Paleari, and the Department of Architecture, Built Environment, and Construction Engineering for the support in the ongoing research. This research is developed in collaboration with Laser Navigation srl.

Conflicts of Interest: The authors declare no conflict of interest.

References

1. National Research Council. *Stewardship of Federal Facilities: A Proactive Strategy for Managing the Nation's Public Assets*; The National Academies Press: Washington, DC, USA, 1998.
2. Zimmerman, A.; Martin, M. Post-occupancy evaluation: Benefits and barriers. *Build. Res. Inf.* **2001**, *29*, 168–174. [CrossRef]
3. Bento Pereira, N.; Calejo Rodrigues, R.; Fernandes Rocha, P.; Bento Pereira, N.; Calejo Rodrigues, R.; Fernandes Rocha, P. Post-Occupancy Evaluation Data Support for Planning and Management of Building Maintenance Plans. *Buildings* **2016**, *6*, 45. [CrossRef]
4. Dong, B.; Yan, D.; Li, Z.; Jin, Y.; Feng, X.; Fontenot, H. Modeling occupancy and behavior for better building design and operation—A critical review. *Build. Simul.* **2018**, *11*, 899–921. [CrossRef]

5. Kim, J.; de Dear, R. Nonlinear relationships between individual IEQ factors and overall workspace satisfaction. *Build. Environ.* **2012**, *49*, 33–40. [CrossRef]
6. Agha-Hossein, M.M.; El-Jouzi, S.; Elmualim, A.A.; Ellis, J.; Williams, M. Post-occupancy studies of an office environment: Energy performance and occupants' satisfaction. *Build. Environ.* **2013**, *69*, 121–130. [CrossRef]
7. Straka, V.; Aleksic, M. Post-occupancy evaluation: Three schools from Greater Toronto. In Proceedings of the PLEA2009–26th Conference on Passive and Low Energy Architecture, Quebec City, QC, Canada, 22–24 June 2009.
8. Rogage, K.; Clear, A.; Alwan, Z.; Lawrence, T.; Kelly, G. Assessing Building Performance in Residential Buildings using BIM and Sensor Data. *Int. J. Build. Pathol. Adapt.* **2019**. [CrossRef]
9. Wang, W.; Hong, T.; Xu, N.; Xu, X.; Chen, J.; Shan, X. Cross-source sensing data fusion for building occupancy prediction with adaptive lasso feature filtering. *Build. Environ.* **2019**, *162*, 106280. [CrossRef]
10. De Wilde, P. The gap between predicted and measured energy performance of buildings: A framework for investigation. *Autom. Constr.* **2014**, *41*, 40–49. [CrossRef]
11. Marzouk, M.; Abdelaty, A. BIM-based framework for managing performance of subway stations. *Autom. Constr.* **2014**, *41*, 70–77. [CrossRef]
12. Costa, A.A.; Lopes, P.M.; Antunes, A.; Cabral, I.; Grilo, A.; Rodrigues, F.M. 3I Buildings: Intelligent, Interactive and Immersive Buildings. *Procedia Eng.* **2015**, *123*, 7–14. [CrossRef]
13. Yang, J.; Santamouris, M.; Lee, S.E. Review of occupancy sensing systems and occupancy modeling methodologies for the application in institutional buildings. *Energy Build.* **2016**, *121*, 344–349. [CrossRef]
14. Daher, E.; Kubicki, S.; Guerriero, A. Post-occupancy evaluation parameters in multi-objective optimization-based design process. In *Advances in Informatics and Computing in Civil and Construction Engineering*; Springer: Cham, Switzerland, 2018.
15. Demian, P.; Liu, Z.; Deng, Z. Integration of Building Information Modelling (BIM) and Sensor Technology: A Review of Current Developments and Future Outlooks. In Proceedings of the 2nd International Conference on Computer Science and Application Engineering (CSAE), Chicago, IL, USA, 1–3 October 2018.
16. Saralegui, U.; Anton, M.A.; Arbelaitz, O.; Muguerza, J. An IoT sensor network to model occupancy profiles for energy usage simulation tools. In Proceedings of the 2018 Global Internet of Things Summit, Bilbao, Spain, 4–7 June 2018.
17. Ouf, M.M.; O'Brien, W.; Gunay, B. On quantifying building performance adaptability to variable occupancy. *Build. Environ.* **2019**, *155*, 257–267. [CrossRef]
18. Capolongo, S.; Rebecchi, A.; Buffoli, M.; Appolloni, L.; Signorelli, C.; Fara, G.M.; D'Alessandro, D. COVID-19 and cities: From urban health strategies to the pandemic challenge. a decalogue of public health opportunities. *Acta Biomed.* **2020**, *91*, 13–22.
19. Grieves, M.W.; Vickers, J. Digital Twin: Mitigating Unpredictable, Undesirable Emergent Behavior in Complex Systems. In *Transdisciplinary Perspectives on Complex Systems: New Findings and Approaches*; Springer: Cham, Switzerland, 2017; pp. 85–113. ISBN 9783319387567.
20. Lu, Q.; Xie, X.; Parlikad, A.K.; Schooling, J.M.; Konstantinou, E. Moving from Building Information Models to Digital Twins for Operation and Maintenance. In *Proceedings of the Institution of Civil Engineers—Smart Infrastructure and Construction*; ICE Publishing: London, UK, 2020; pp. 1–9.
21. Dawkins, O.; Dennett, A.; Hudson-Smith, A. Living with a Digital Twin: Operational management and engagement using IoT and Mixed Realities at UCL's Here East Campus on the Queen Elizabeth Olympic Park. In Proceedings of the 26th Annual GIScience Research UK Conference: GISRUK 2018, Leicester, UK, 17–20 April 2018; p. 7.
22. Manngård, M.; Lund, W.; Björkqvist, J. Using Digital Twin Technology to Ensure Data Quality in Transport Systems. In Proceedings of the 8th Transport Research Arena TRA, Helsinki, Finland, 27–30 April 2020.
23. Boje, C.; Guerriero, A.; Kubicki, S.; Rezgui, Y. Towards a semantic Construction Digital Twin: Directions for future research. *Autom. Constr.* **2020**, *114*, 103179. [CrossRef]
24. Tomko, M.; Winter, S. Beyond digital twins—A commentary. *Environ. Plan. B Urban Anal. City Sci.* **2019**, *46*, 395–399. [CrossRef]
25. Yang, J.; Pantazaras, A.; Chaturvedi, K.A.; Chandran, A.K.; Santamouris, M.; Lee, S.E.; Tham, K.W. Comparison of different occupancy counting methods for single system-single zone applications. *Energy Build.* **2018**, *172*, 221–234. [CrossRef]
26. Yan, D.; Hong, T.; Dong, B.; Mahdavi, A.; D'Oca, S.; Gaetani, I.; Feng, X. IEA EBC Annex 66: Definition and simulation of occupant behavior in buildings. *Energy Build.* **2017**, *156*, 258–270. [CrossRef]
27. Preiser, W.F.E. Post-occupancy evaluation: How To Make Buildings Work Better. *Facilities* **1995**, *13*, 19–28. [CrossRef]
28. Manning, P. *Office Design: A Study of Environment*; Manning, P., Ed.; The Pilkington Research Unit, Department of Building Science, University of Liverpool: Liverpool, UK, 1965.
29. Hadjri, K.; Crozier, C. Post-occupancy evaluation: Purpose, benefits and barriers. *Balt. J. Manag.* **2009**, *27*, 21–33. [CrossRef]
30. National Research Council. *Learning from Our Buildings: A State-of-the-Practice Summary of Post-Occupancy Evaluation*; National Academies Press: Washington, DC, USA, 2001; ISBN 9780309076111.
31. Preiser, W.F.E.; Rabinowitz, H.Z.; White, E.T. *Post-Occupancy Evaluation*; Van Nostrand Reinhold: New York, NY, USA, 1988.
32. Li, P.; Froese, T.M.; Brager, G. Post-occupancy evaluation: State-of-the-art analysis and state-of-the-practice review. *Build. Environ.* **2018**, *133*, 187–202. [CrossRef]
33. Preiser, W.F.E.; Schramm, U. Building Performance Evaluation. In *Time-Saver Standards for Architectural Design Data*; Watson, D., Crosbie, M.J., Callender, J.H., Eds.; McGraw-Hill Professional Publishing: New York, NY, USA, 1997.

34. Boarin, P.; Besen, P.; Haarhoff, E. Post-Occupancy Evaluation of Neighbourhoods: A Review of the Literature; Building Better Homes Towns and Cities Working Papers; 2018. Available online: https://www.buildingbetter.nz/resources/publicationsand (accessed on 18 July 2020).
35. Preiser, W.F.E.; Nasar, J.L. Assessing Building Performance: Its evolution from Post-occupancy Evaluation. *ArchNet-IJAR* **2008**, *2*, 84–99.
36. Preiser, W.F.E. Building Performance Assessment—From POE to BPE: A Personal Perspective. *Archit. Sci. Rev.* **2005**, *48*, 201–205. [CrossRef]
37. Durosaiye, I.O.; Hadjri, K.; Liyanage, C.L. A critique of post-occupancy evaluation in the UK. *J. Hous. Built Environ.* **2019**, *34*, 345–352. [CrossRef]
38. Zimring, C.M.; Reizenstein, J.E. Post-Occupancy Evaluation: An Overview. *Environ. Behav.* **1980**, *12*, 429–450. [CrossRef]
39. Di Giuda, G.M.; Pellegrini, L.; Schievano, M.; Locatelli, M.; Paleari, F. BIM and Post-occupancy evaluations for building management system: Weaknesses and opportunities. In *Springer Volume Research for Developmen—La Digitalizzazione dei Processi per la Progettazione, Costruzione e Gestione Dell'ambiente Costruito*; Springer International Publishing: Cham, Switzerland, 2020; pp. 319–327, ISBN 9783030335700.
40. Alonso, R.; Borras, M.; Koppelaar, R.H.E.M.; Lodigiani, A.; Loscos, E.; Yöntem, E. SPHERE: BIM Digital Twin Platform. *Proceedings* **2019**, *20*, 6. [CrossRef]
41. Oti, A.H.; Kurul, E.; Cheung, F.K.T.; Tah, J.H.M. A framework for the utilization of Building Management System data in building information models for building design and operation. *Autom. Constr.* **2016**, *72*, 195–210. [CrossRef]
42. Codinhoto, R.; Kiviniemi, A. BIM for FM: A case support for business life cycle. *IFIP Adv. Inf. Commun. Technol.* **2014**, *442*, 63–74.
43. Machado, F.A.; Dezotti, C.G.; Ruschel, R.C. The interface layer of a BIM-IoT prototype for energy consumption monitoring. In *Systems, Controls, Embedded Systems, Energy, and Machines*; CRC Press: Boca Raton, FL, USA, 2017; pp. 9-1–9-10. [CrossRef]
44. Underwood, J.; Isikdag, U. Emerging technologies for BIM 2.0. *Constr. Innov.* **2011**, *11*, 252–258. [CrossRef]
45. Pin, C.; Medina, C.; McArthur, J.J. Supporting post-occupant evaluation through work order evaluation and visualization in FM-BIM. In Proceedings of the ISARC 2018—35th International Symposium on Automation and Robotics in Construction and International AEC/FM Hackathon: The Future of Building Things, Berlin, Germany, 20–25 July 2018; pp. 32–39.
46. Lu, Q.; Parlikad, A.K.; Woodall, P.; Don Ranasinghe, G.; Xie, X.; Liang, Z.; Konstantinou, E.; Heaton, J.; Schooling, J.M. Developing a Digital Twin at Building and City Levels: A Case Study of West Cambridge Campus. *J. Chem. Inf. Model.* **2019**, *53*, 1689–1699. [CrossRef]
47. Grieves, M.W. Product lifecycle management: The new paradigm for enterprises. *Int. J. Prod. Dev.* **2005**, *2*, 71–84. [CrossRef]
48. Grieves, M.W. *Product Lifecycle Management: Driving the Next Generation of Lean Thinking*; McGraw-Hill Education: New York, NY, USA, 2006; ISBN 978-0071452304.
49. Grieves, M.W. *Virtually Perfect: Driving Innovative and Lean Products through Product Lifecycle Management*; Space Coast Press: Cocoa Beach, FL, USA, 2011; ISBN 978-0982138007.
50. Shafto, M.; Conroy, M.; Doyle, R.; Glaessgen, E.H.; Kemp, C.; LeMoigne, J.; Wang, L. *Modeling, Simulation, Information Technology & Processing Roadmap-Technology Area 11*; National Aeronautics and Space Administration: Washington, DC, USA, 2012.
51. Glaessgen, E.H.; Stargel, D.S. The digital twin paradigm for future NASA and U.S. Air force vehicles. In Proceedings of the Collection of Technical Papers—53rd AIAA/ASME/ASCE/AHS/ASC Structures, Structural Dynamics and Materials Conference, Honolulu, HI, USA, 23–26 April 2012; pp. 1–14.
52. Knapp, G.L.; Mukherjee, T.; Zuback, J.S.; Wei, H.L.; Palmer, T.A.; De, A.; DebRoy, T. Building blocks for a digital twin of additive manufacturing. *Acta Mater.* **2017**, *135*, 390–399. [CrossRef]
53. Bolton, A.; Blackwell, B.; Dabson, I.; Enzer, M.; Evans, M.; Fenemore, T.; Harradence, F.; Keaney, E.; Kemp, A.; Luck, A.; et al. *The Gemini Principles: Guiding Values for the National Digital Twin and Information Management Framework*; University of Cambridge: Cambridge, UK, 2018.
54. High Value Manufacturing Catapult Visualization and VR Forum. Feasibility of an Immersive Digital Twin: The Definition of a Digital Twin and Discussions around the Benefit of Immersion. 2018. Available online: hvm.catapult.org.uk (accessed on 18 July 2020).
55. National Infrastructure Commission. Data for the Public Good. 2017. Available online: https://nic.org.uk/ (accessed on 18 July 2020).
56. Batty, M. Digital twins. *Environ. Plan. B Urban Anal. City Sci.* **2018**, *45*, 817–820. [CrossRef]
57. Al-Sehrawy, R.; Kumar, B. Digital Twins in Architecture, Engineering, Construction and Operations. A Brief Review and Analysis. *Lect. Notes Civ. Eng.* **2021**, *98*, 924–939.
58. Love, P.E.D.; Matthews, J. The 'how' of benefits management for digital technology: From engineering to asset management. *Autom. Constr.* **2019**, *107*, 15. [CrossRef]
59. Lu, Q.; Parlikad, A.K.; Woodall, P.; Ranasinghe, G.D.; Heaton, J. Developing a Dynamic Digital Twin at a Building Level: Using Cambridge Campus as Case Study. In Proceedings of the International Conference on Smart Infrastructure and Construction 2019 (ICSIC): Driving Data-Informed Decision-Making, Cambridge, UK, 8–10 July 2019; pp. 67–75.
60. Haag, S.; Anderl, R. Digital twin—Proof of concept. *Manuf. Lett.* **2018**, *15*, 64–66. [CrossRef]
61. Qi, Q.; Tao, F. Digital Twin and Big Data Towards Smart Manufacturing and Industry 4.0: 360 Degree Comparison. *IEEE Access* **2018**, *6*, 3585–3593. [CrossRef]

62. Tao, F.; Zhang, H.; Liu, A.; Nee, A.Y.C. Digital Twin in Industry: State-of-the-Art. *IEEE Trans. Ind. Inform.* **2019**, *15*, 2405–2415. [CrossRef]
63. Ding, K.; Shi, H.; Hui, J.; Liu, Y.; Zhu, B.; Zhang, F.; Cao, W. Smart steel bridge construction enabled by BIM and Internet of Things in industry 4.0: A framework. In Proceedings of the ICNSC 2018—15th IEEE International Conference on Networking, Sensing and Control, Zhuhai, China, 27–29 March 2018; pp. 1–5.
64. Peng, Y.; Lin, J.R.; Zhang, J.P.; Hu, Z.Z. A hybrid data mining approach on BIM-based building operation and maintenance. *Build. Environ.* **2017**, *126*, 483–495. [CrossRef]
65. Barbosa, J.; Mateus, R.; Bragança, L. Occupancy Patterns and Building Performance—Developing occupancy patterns for Portuguese residential buildings. In Proceedings of the Sustainable Urban Communities towards a Nearly Zero Impact Built Environment, Vitória, Brazil, 7–9 September 2016; pp. 1193–1200.
66. Ouf, M.M.; Issa, M.H.; Azzouz, A.; Sadick, A.-M. Effectiveness of using WiFi technologies to detect and predict building occupancy. *Sustain. Build.* **2017**, *2*, 7–17. [CrossRef]
67. Benezeth, Y.; Laurent, H.; Emile, B.; Rosenberger, C. Towards a sensor for detecting human presence and characterizing activity. *Energy Build.* **2011**, *43*, 305–314. [CrossRef]
68. Wu, L.; Wang, Y. A Low-Power Electric-Mechanical Driving Approach for True Occupancy Detection Using a Shuttered Passive Infrared Sensor. *IEEE Sens. J.* **2019**, *19*, 47–57. [CrossRef]
69. Li, N.; Calis, G.; Becerik-Gerber, B. Measuring and monitoring occupancy with an RFID based system for demand-driven HVAC operations. *Autom. Constr.* **2012**, *24*, 89–99. [CrossRef]
70. Martani, C.; Lee, D.; Robinson, P.; Britter, R.; Ratti, C. ENERNET: Studying the dynamic relationship between building occupancy and energy consumption. *Energy Build.* **2012**, *47*, 584–591. [CrossRef]
71. Chen, J.; Ahn, C. Assessing occupants' energy load variation through existing wireless network infrastructure in commercial and educational buildings. *Energy Build.* **2014**, *82*, 540–549. [CrossRef]
72. Wang, W.; Chen, J.; Song, X. Modeling and predicting occupancy profile in office space with a Wi-Fi probe-based Dynamic Markov Time-Window Inference approach. *Build. Environ.* **2017**, *124*, 130–142. [CrossRef]
73. Wang, Y.; Shao, L. Understanding occupancy pattern and improving building energy efficiency through Wi-Fi based indoor positioning. *Build. Environ.* **2017**, *114*, 106–117. [CrossRef]
74. Wang, J.; Tse, N.C.F.; Chan, J.Y.C. Wi-Fi based occupancy detection in a complex indoor space under discontinuous wireless communication: A robust filtering based on event-triggered updating. *Build. Environ.* **2019**, *151*, 228–239. [CrossRef]
75. Di Giuda, G.M.; Giana, P.E.; Schievano, M.; Paleari, F. Guidelines to Integrate BIM for Asset. In *Digital Transformation of the Design, Construction and Management Processes of the Built Environment*; Springer International Publishing: Cham, Switzerland, 2020; pp. 391–400, ISBN 9783030335700.
76. Pärn, E.A.; Edwards, D.J.; Sing, M.C.P. The building information modelling trajectory in facilities management: A review. *Autom. Constr.* **2017**, *75*, 45–55. [CrossRef]
77. Nawari, N.O.; Ravindran, S. Blockchain technology and BIM process: Review and potential applications. *J. Inf. Technol. Constr.* **2019**, *24*, 209–238.
78. Di Giuda, G.M.; Pattini, G.; Seghezzi, E.; Schievano, M.; Paleari, F. The Construction Contract Execution through the Integration of Blockchain Technology. In *Digital Transformation of the Design, Construction and Management Processes of the Built Environment*; Springer International Publishing: Cham, Switzerland, 2020; pp. 309–318, ISBN 9783030335700.
79. Ye, Z.; Yin, M.; Tang, L.; Jiang, H. Cup-of-Water theory: A review on the interaction of BIM, IoT and blockchain during the whole building lifecycle. In Proceedings of the ISARC 2018—35th International Symposium on Automation and Robotics in Construction and International AEC/FM Hackathon: The Future of Building Things, Berlin, Germany, 20–50 July 2018; p. 10.

Review

Building Information Modelling and Internet of Things Integration for Facility Management—Literature Review and Future Needs

Antonino Mannino *, **Mario Claudio Dejaco** and **Fulvio Re Cecconi**

Department of Architecture, Built Environment and Construction Engineering, Politecnico di Milano, Via Ponzio 31, 20133 Milano, Italy; mario.dejaco@polimi.it (M.C.D.); fulvio.rececconi@polimi.it (F.R.C.)
* Correspondence: antonino.mannino@polimi.it

Abstract: Digitisation of the built environment is seen as a significant factor for innovation in the Architecture, Engineering, Construction and Operation sector. However, lack of data and information in as-built digital models considerably limits the potential of Building Information Modelling in Facility Management. Therefore, optimisation of data collection and management is needed, all the more so now that Industry 4.0 has widened the use of sensors into buildings and infrastructures. A literature review on the two main pillars of digitalisation in construction, Building Information Modelling and Internet of Things, is presented, along with a bibliographic analysis of two citations and abstracts databases focusing on the operations stage. The bibliographic research has been carried out using Web of Science and Scopus databases. The article is aimed at providing a detailed analysis of BIM–IoT integration for Facility Management (FM) process improvements. Issues, opportunities and areas where further research efforts are required are outlined. Finally, four key areas of further research development in FM management have been proposed, focusing on optimising data collection and management.

Keywords: Building Information Modelling (BIM); Internet of Things (IoT); facility management; cyber-physical systems; digital twin

Citation: Mannino, A.; Dejaco, M.C.; Re Cecconi, F. Building Information Modelling and Internet of Things Integration for Facility Management—Literature Review and Future Needs. *Appl. Sci.* **2021**, *11*, 3062. https://doi.org/10.3390/app11073062

Academic Editor: Jürgen Reichardt

Received: 27 February 2021
Accepted: 25 March 2021
Published: 30 March 2021

1. Introduction

The construction industry has a relatively low digitisation level compared to other sectors [1,2]. Although it is seen as a major factor in the innovation of the Architecture, Engineering and Construction and Operations (AECO) sector, digitisation in the construction industry still shows a slow growth rate [3]. However, improvements in methodologies and technologies are under development to better manage AECO processes [4].

This article presents a literature review on the integration of Building Information Modelling (BIM) and Internet of Things (IoT) for the Facility Management (FM) of the constructed asset. It is divided into four main parts: (a) an overview of FM and the impact of digitisation in the sector; (b) the description of the research method; (c) an in-depth content analysis of 99 selected journals' articles on BIM-IOT integration for FM, which allows identifying both benefits/opportunities and issues/limits at technical and operational levels; (d) conclusions and a description of a possible future research agenda.

The context is the fourth industrial revolution (Industry 4.0), where several technological changes in many sectors have been made, including in the AECO one [5,6].

There are many studies on the application of digital technologies aiming to promote digitisation in the built environment. However, compared to the design and construction stages, there is a lack of research on applying these new technologies in the operation and use stage of the building life cycle [7], particularly for the FM sector. FM represents up to 85% of the whole life cycle cost of the building [8]. Even though the life cycle cost of a building can and should be controlled in the design phase the adoption of innovative tools and technologies to improve FM in existing buildings is continually increasing. Wong et al. [7] identified and discussed several possibilities for future research into digital technologies like integrating FM with BIM, reality capture technology, IoT, Radio Frequency Identification (RFID), and Geographic Information System.

Among several studies on applying new technologies, a significant solution taken into consideration in the last years by the AECO sector has emerged: the Cyber-Physical Systems (CPS) [6]. CPS, also known as Digital Twins (DT), are systems based on the combination of physical and digital objects. Through simulation of an as-built component (or system), using digital models and several types of data, DT allows mirroring the life of its corresponding real twin to forecast the health of building components, their service life, faults [9] and, in general, the building performances [10].

Even if not risk-free, these digital innovations will enable new dynamics and allow new services that will improve efficiency and sustainability in building management processes [11].

1.1. Facility Management

Facility management is a multidisciplinary topic that requires the collaboration and coordination of different people [6]. ISO 41011:2017 defines FM as an "organisational function which integrates people, place and process within the built environment with the purpose of improving the quality of life of people and the productivity of the core business" [12].

According to International Facility Management Association (IFMA), there are 11 core competencies in FM [13]: Occupancy and Human Factors, Operations and Maintenance (O&M), Sustainability, Facility Information and Technology Management, Risk Management, Communication, Performance and Quality, Leadership and Strategy, Real Estate, Project Management, Finance and Business.

Currently, not all buildings have optimal management [14,15] due to outdated procedures that cause a lack of data and information. In other cases, despite the use of sensors/automatic devices and databases, the information collected is not entirely exploited [16]. An example is given by FM information systems, e.g., Computerised Maintenance Management Systems (CMMS), Energy Management Systems (EMS) and Building Automation Systems (BAS), where data are often fragmented and manually entered after the handover of the building. Fragmentation and data poorness could generate laborious and inefficient processes [7]. Furthermore, FM operators often rely on paper documents in their daily activities. This increases both the time needed and the difficulties of getting accurate information [17]. For these reasons, the improvement of both FM tools and processes is a crucial issue in FM companies [18]. Hence, with increasing industry interest, a review of the current status and a description of a future research agenda on FM is needed.

1.2. Digitisation and FM

New technologies have transformed many people's daily lives and have revolutionised several traditional industry practices aiming to achieve efficiency, accuracy, and precision. This evolution has gained momentum due to advancements in technologies such as the Internet of Things (IoT), big data, cloud computing and cyber-physical systems [19].

The strengths of these innovations 4.0 lie in monitoring, controlling, interoperability, real-time information processing and process self-optimisation [19]. The physical world's connection with the virtual world enables products and components to create a self-adapting and self-managing communication network [20].

In the construction sector, the first attempt at digitisation aiming to increase the sector's efficiency has already been seen with the spread of BIM [21].

1.3. Building Information Modelling for FM

The United States National Institute of Building Sciences (NIBS) defines BIM as "The digital representation of physical and functional characteristics of a facility. As such, it serves as a shared knowledge resource for information about a facility, forming a reliable basis for decisions during its life cycle from inception onwards" [22].

In recent years, BIM has been more and more employed in the AECO sector to improve information management. BIM Models (BIMs) allow integrated management of information throughout the building's entire life cycle, hence improving FM [23]. On the one hand, BIM allows working more efficiently during the design and construction phases by developing a 3D model that avoids project interference and allows project time and cost calculation. On the other hand, it allows acquiring data created during several phases of the building life cycle to use them in operations management, maintenance activities, environmental analysis and energy performance simulations. The latter is related to Building Energy Modelling (BEM), which has become an essential aspect for FM.

Benefits of using BIM in FM include providing "as-is" information and enabling Facility Managers to work on information using a single source of data, overcoming all the issues deriving from the sources' fragmentation.

A BIM model has different Levels of Information Needs [24]. To deal with them, the American BIMForum defined the Level of Development (LOD) Specification. This reference enables practitioners in the AEC Industry to specify and articulate with a high level of clarity the content and reliability of Building Information Models (BIMs) at various stages in the design and construction process. A BIM model has six Levels Of Development (LOD): LOD 100, LOD 200, LOD 300, LOD 350, LOD 400 and LOD 500 [25]. Each LOD defines how much information is included in a building component. The higher the LOD, the greater the clarity and reliability of data and information.

According to Love et al. [26], using the highest LOD is possible in order to enrich the digital model with all the information necessary for assets management and maintenance. In this way, data are more efficiently stored in a single file without fragmentation or loss of information. Moreover, improving the handover process is possible using fewer paper documents or manual transfer of information [26].

As early as 2012, Becerik-Gerber et al. [27] have defined, also through surveys and interviews, how BIM can support FM practices. Their paper assesses the status of BIM implementations in FM, potential applications, level of interest in BIM utilisation, application areas, and data requirements for BIM-enabled FM practices. To date, studies on BIM application in FM confirm momentum (e.g., [18,28–33]).

However, despite all the advantages, BIM is not often used in the FM phase. The most significant causes that hinder this integration are:

(1) Industry perception, which considers BIM models just as 3D models and not as informative models with business value [34].
(2) Lack of involvement of the facility managers in the creation of the BIM model [7,35]. Consequently, less information useful for FM is integrated into the model.
(3) The need for interoperability between BIM and FM technologies and the lack of open systems and standardised data libraries that can be utilised as a bridge between BIM and CAFM technologies [36].
(4) Lack of clear roles, responsibilities, contract and liability framework [36].
(5) Furthermore, the main limitation of BIM methodology is its static information: data are provided during the design phase but not updated during the building's life cycle [37]. This is a relevant issue for buildings management, and research is moving in this direction.

1.4. Internet of Things for FM

Asghari, Rahmani and Javadi [38] define IoT as "an ecosystem that contains smart objects equipped with sensors, networking and processing technologies integrating and working together to provide an environment in which smart services are taken to the end-users". They show how this ecosystem is being applied in healthcare, environmental, smart cities, commercial and industrial contexts. IoT has led to an interconnection between people and objects at an unprecedented scale and pace [39] and will allow new strategies to improve quality of life [40]. Furthermore, connected devices could be programmed to make autonomous decisions and adequately inform users to make the best decisions [41].

Operation and maintenance stages represent 50–70% of the total annual facility operating costs [42], and buildings management requires integrating and analysing different types of data and information generated by various stakeholders. This implies that improved data and information management can have a significant impact on building performance.

In this context, the application and integration of IoT and BIM technologies to gather and store data/information for the entire life cycle of the building have caught wide attention. In recent years, a growing number of innovations have been developed [7].

IoT and smart connection have great potential in optimising FM activities, including inventory and document management, building security, logistics and materials tracking, tracking of building component life cycle and building energy controls [7]. Several studies about the use of data coming from IoT devices have been carried out (e.g., [43–45]), although many of them do not include the integration of BIM.

2. Scope and Aim of the Research

As mentioned before, BIM and IoT- based data sources is a relatively new field. One can consider BIM and IoT data as two complementary entities, where one covers the lack of the other. Researchers have addressed different aspects of BIM, IoT and their use in an integrated way: sustainability, risks, safety and so forth.

In this article, integration is addressed more from the point of view of the information collected/transmitted by sensors and actuators (and used for a specific purpose) than from that of the software or platform used. Hence, studies and research published on BIM and IoT data integration are analysed in this paper. The content is structured as a bibliographic investigation through which an analysis of these technologies' current use is carried out. The aims of this research are:

(1) To provide a detailed bibliographic analysis of the research efforts on BIM–IoT integration for FM processes improvements
(2) To identify limitations in the research and introduce a roadmap for further research on FM improvements through new technologies.

3. Materials and Methods

This study analyses and categorises existing studies on BIM and IoT integration for FM according to the methodology shown in Figure 1. To review BIM-IoT integration comprehensively in the Facility Management context, two electronic databases of peer-reviewed literature have been taken into consideration: Scopus and Web of Science (WoS). The bibliometric analysis presented here aims to analyse academic publications and trends to evaluate the existing research performance and understand patterns. As the first step, keywords to select articles on BIM-IoT integration for FM functions are defined. Table 1 highlights the keywords used to find publications on BIM and IoT. Table 2 shows the set of keywords for each FM core competence.

Table 1. Keywords used for research in the two electronic databases of peer-reviewed literature Scopus and Web of Science (WoS). The asterisk "*" after the keywords tells the search engine to look for all the words beginning with that keywords, i.e., "sensor*" tells the search engine to look for the words "sensor", "sensors", "sensoring", etc … The quotations marks surrounding two or more words tell the search engine to look for the phrase and not the words, i.e., "industry foundation classes" is use to search for the prhase and not for the words industry or foundation or classes.

Tools	Keywords
BIM	BIM or "Building Information Modelling" or "Building Information Modeling" or IFC* or "industry foundation classes"
IoT	IoT or "Internet of things" or sensor* or WSN or "Wireless Sensor* Network*" or "Real Time Data" or "Real-Time Data"

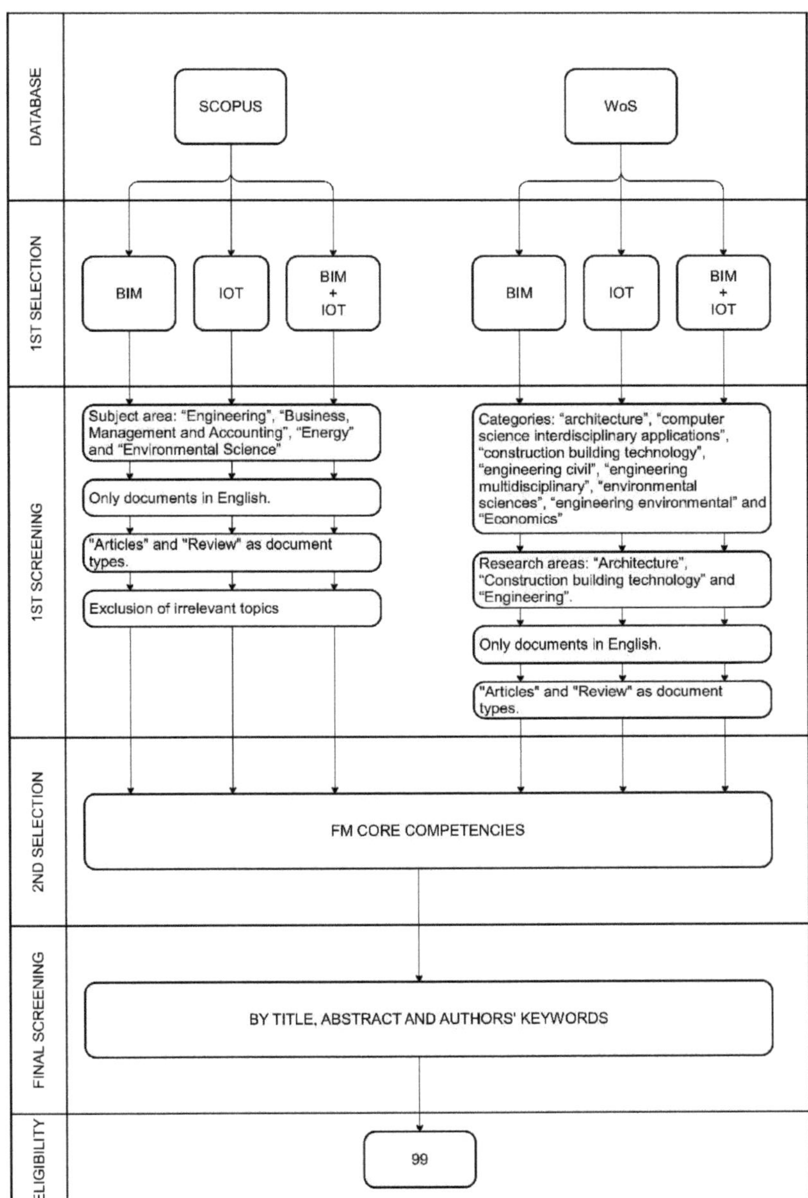

Figure 1. Research methodology.

After the keywords selection, Scopus and WoS databases were queried, using the keywords shown in Table 1, to find publications dealing with: (a) BIM; (b) IoT; and (c) BIM and IoT at the same time. This first-level query investigated how much these topics have been explored by researchers, even outside the FM field.

In a second-level query on the two databases, BIM and IoT keywords (Table 1) were coupled with FM core competencies (Table 2) to measure how deep BIM and IoT permeate FM core competencies.

Table 2. FM and its core competencies keywords used for research in the two electronic databases of peer-reviewed literature Scopus and Web of Science (WoS).

FM and Core Competencies	Keywords
Communication	Communication* or report* or "communication* strategy"
Finance and Business	"Financial management" or budget* or financial or contract* or procurement
Human Factors	Occupancy or "Workplace environment" or "Occupant services" or "Occupant health" or "Occupant safety" or "Occupant security"
Leadership & Strategy	Leadership or "team* management" or "team* organisation" or "conflict management"
O&M	FMM or "Facility Maintenance Management" or "Facilities Maintenance Management" or maintenance or "maintenance management" or operation* or "Physical safety and security" or "Work management systems"
Project Management	"project management" or "project execution" or "project monitor" or "project outcomes" or "project schedule"
Quality and Performance	"Quality management" or "key performance indicators" or KPI or "service level agreements"
Real Estate and Property Management	"real estate management" or "property management" or "Real estate strategies" or "Real estate assessment" or "Asset management"
Risk Management	"Risk*" or "Emergency*"
Sustainability	Sustainab* or "Energy*" or "Water*" or "Waste*" or BEM*
Information and Technology Management	"intelligent building systems" or "automation" or "data collection" or "Information management" or "Information security*" or "Information system" or "cyber-security"
FM	FM or "Facility Management" or "Facilities Management"

A set of filters was applied to the various searches to limit the large number of results. To perform this selection in the WoS database, results were as follows:

- Refined, considering the following WoS categories: "architecture", "computer science interdisciplinary applications", "construction building technology", "engineering civil", "engineering multidisciplinary", "environmental sciences", "engineering environmental" and "economics" (because of Finance and Business FM core function).
- Refined, considering only the following research areas: "architecture", "construction building technology" and "engineering".
- Refined, considering only documents in English.
- Refined, considering only articles and reviews as document types to acquire higher-quality documents.

Articles from "computer science interdisciplinary applications" and "engineering multidisciplinary" domains were included in the review to ensure a comprehensive review of BIM and IoT device integration. Finally, the results were further filtered (by title, abstract and author's keywords) to remove articles not relevant for the research scope.

To perform a similar selection in the Scopus database, results were as follows:

- Refined, considering the following subject area: "engineering", "business, management and accounting", "energy" and "environmental science".
- Refined, considering only documents in English.
- Refined, considering only articles and reviews as document types to acquire higher-quality documents.
- Filtered, excluding irrelevant topics (e.g., medicine, chemistry, biology).

As the last step, the two search results were combined, and duplicated articles were excluded. Finally, a list of 99 articles on the BIM-IOT integration was selected. Five out of ninety-nine articles are general reviews on BIM-IOT for FM and were already discussed in the introduction. To derive patterns and propose future research directions, qualitative data analyses of the 94 articles based on each article's technical aspect were carried out, as discussed in Section 5.

4. Results

The first result of the query using the keywords combination method explained in Section 3 shows a fairly clear gap between the number of publications on the three main topics. More than 95% of the articles deal with IoT (Table 3). A limited number of publications, less than 4%, of articles deal with BIM. Lastly, the integration of BIM and IoT is still at an early stage.

Table 3. Number of journal articles on BIM, IoT and their integration resulting from Scopus and WOS databases research (until February 2021).

	Scopus	WoS
BIM	3316	2717
IoT	157,108	31,088
BIM + IoT	237	166

Considering the number of publications on BIM and IoT integration in the last 30 years, it is possible to see how the first significant increase is registered after 2013 (Figure 2). Furthermore, publications grew almost simultaneously in both citation databases. Interestingly, 80% of the articles on Scopus and 85% of the articles on WoS were published during the last five years; this means that BIM-IoT integration is a new domain with increasing interest, especially during 2019 and 2020. Accordingly, the implementation of BIM-IoT integration for FM is also a new domain, with a limited number of publications (red line in Figure 2).

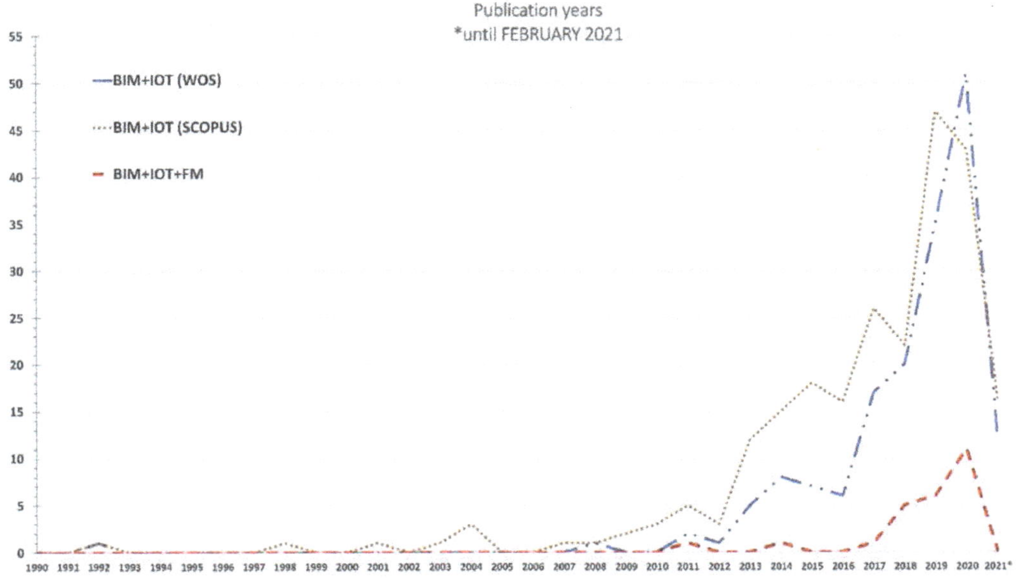

Figure 2. Number of publications per year (* until February 2021) dealing with BIM and IoT in Scopus (yellow dotted), in Web of Science (blue dash-dotted) and number of product dealing with BIM and IoT for Facility Management (red dashed).

A further query using FM core functions keywords (Table 2) was carried out to find as many articles as possible on BIM and IoT integration in the FM field. Results are shown in Table 4.

Table 4. Journals' articles on BIM, IoT and their integration for FM and its core functions from WoS (W) and Scopus (S) databases (until February 2021).

	Functions	Tools		
		BIM	**IoT**	**BIM + IoT**
FM Core Functions	Communication	426 (W) + 509 (S)	3920 (W) + 28539 (S)	42 (W) + 55 (S)
	Finance & Business	293 (W) + 468 (S)	417 (W) + 3413 (S)	8 (W) + 15 (S)
	Human Factors	31 (W) + 43 (S)	362 (W) + 974 (S)	4 (W) + 4 (S)
	Leadership and Strategy	32 (W) + 242 (S)	18 (W) + 500 (S)	1 (W) + 17 (S)
	O&M	557 (W) + 746 (S)	4374 (W) + 24709 (S)	66 (W) + 64 (S)
	Project Management	157 (W) + 965 (S)	20 (W) + 1875 (S)	6 (W) + 53 (S)
	Quality	33 (W) + 200 (S)	61 (W) + 3767 (S)	1 (W) + 2 (S)
	Property Management	57 (W) + 169 (S)	42 (W) + 1087 (S)	7 (W) + 19 (S)
	Risk Management	253 (W) + 267 (S)	1084 (W) + 5319 (S)	17 (W) + 75 (S)
	Sustainability	709 (W) + 794 (S)	7733 (W) + 33413 (S)	59(W) + 106 (S)
	Technology	370 (W) + 1875 (S)	1219 (W) + 30946 (S)	34(W) + 104 (S)
	FM	237 (W) + 236 (S)	71 (W) + 341 (S)	25 (W) + 81 (S)

The query was limited to journals' articles: Table 4 shows how many products were found in each database matching each FM core function with the three tools' categories.

Although the contemporary use of BIM and IoT is relatively recent, some FM core functions like "Sustainability", "O&M", "Communication", and "Technology" have a significant number of publications. On the contrary, perhaps because of the novelty of the two tools' simultaneous use, some functions have a minimal number of publications. Noteworthy, "Sustainability" is the most studied core function even when considering BIM or IoT separately. The number of publications in the several FM core functions is relatively homogeneous if only BIM products are queried.

To narrow down the scope of the review, further analysis was done on the title and abstract of each of the 904 articles dealing with BIM and IoT, discarding articles not directly related to the construction sector and FM (Table 5).

Eventually, duplicates, i.e., articles covering more than one core function or present in both databases, were discarded, and the final list of 99 articles emerged. On these 99 articles published between 2013 and 2021 (Table 6), a bibliometric analysis was carried out using the R package Bibliometrix [46].

Over the period under review, and based on the proposed selection criteria, most of the articles on BIM-IoT integration for FM were published in *Automation in Construction*, with 27 of the total selected articles. Followed by: *Applied Sciences* (6), *Journal of Computing in Civil Engineering* (5), *Sustainability* (5), *Advanced Engineering Informatics* (3), *Building and Environment* (3), *Journal of Construction Engineering and Management* (3) and *Journal of Information Technology in Construction* (3). The remaining journals' publication rates varied between one to two articles during the considered period.

The bibliometric analysis also reveals that a significant number of publications have been conducted in the USA, with 18 publications, followed by China (12), Australia (9), Hong Kong (8), UK (7), Canada (7), Germany (6) and Italy (6). The remaining countries had less than six articles published during the considered period. Furthermore, the top 10 most cited papers are summarised in Table 7.

Table 5. Journals' articles on BIM–IoT integration, for each FM core competence, from Web of Science and Scopus databases.

FM Core Competencies	BIM + IoT Selected Articles	
	Web of Science	Scopus
Communication	20	24
Finance and Business	3	6
Human Factors	5	4
Leadership & Strategy	0	4
O&M	28	28
Project Management	2	14
Quality	0	0
Real Estate and Property Management	4	5
Risk Management	10	23
Sustainability	28	43
Information and Technology Management	9	42
Facility Management	14	37
Total	**68**	**75**

Table 6. Annual publications of journal articles on BIM–IoT integration for FM (* until February 2021).

Year	Articles
2013	5
2014	9
2015	4
2016	4
2017	12
2018	14
2019	19
2020	24
2021 *	8

Table 7. Top 10 of the most cited articles sorted by the number of global citations (until February 2021).

Title	Authors	Year	Source	doi	Cit
A conceptual framework for integrating building information modeling with augmented reality. (IT)	Wang, X.; Love, P.E.D.; Kim, Mi Jeong et al.	2013	Automation in Construction	10.1016/j.autcon.2012.10.012	117
Prefabricated construction enabled by the Internet of Things. (PM)	Zhong, R.Y., Peng, Y., Xue, F. et al.	2017	Automation in Construction	10.1016/j.autcon.2017.01.006	110
A BIM centered indoor localisation algorithm to support building fire emergency response operations (R)	Li, N., Becerik-Gerber, B., Krishnamachari, B., Soibelman, L.	2014	Automation in Construction	10.1016/j.autcon.2014.02.019	106
Case Study of BIM and Cloud-Enabled Real-Time RFID Indoor Localization for Construction Management Applications (PM)	Fang, Y., Cho, Y.K., Zhang, S., Perez, E.	2016	Journal of Construction Engineering and Management	10.1061/(ASCE)CO.1943-7862.0001125	91
Framework of Automated Construction-Safety Monitoring Using Cloud-Enabled BIM and BLE Mobile Tracking Sensors	Park, J., Kim, K., Cho, Y.K.	2017	Journal of Construction Engineering and Management	10.1061/(ASCE)CO.1943-7862.0001223	91
CoSMoS: A BIM and wireless sensor based integrated solution for worker safety in confined spaces	Riaz, Z., Arslan, M., Kiani, A.K., Azhar, S.	2014	Automation in Construction	10.1016/j.autcon.2014.05.010	85

Table 7. *Cont.*

Title	Authors	Year	Source	doi	Cit
An Internet of Things-enabled BIM platform for on-site assembly services in prefabricated construction	Li, C.Z., Xue, F., Li, X., Hong, J., Shen, G.Q.	2018	Automation in Construction	10.1016/j.autcon.2018.01.001	84
Monitoring thermal comfort in subways using building information modeling	Marzouk, M., Abdelaty, A.	2014	Energy and Buildings	10.1016/j.enbuild.2014.08.006	61
A framework for integrating BIM and IoT through open standards	Dave Bhargav, Andrea Buda, Antti Nurminen, Kary Framling	2018	Automation in Construction	10.1016/j.autcon.2018.07.022	60
A review of building information modeling (BIM) and the internet of things (IoT) devices integration: Present status and future trends	Tang, S., Shelden, D.R., Eastman, C.M., Pishdad-Bozorgi, P., Gao, X.	2019	Automation in Construction	10.1016/j.autcon.2019.01.020	59

Although the most cited article overall is the one written by Wang in 2013, the articles with the highest number of citations per year are those of Tang et al. (2019) with 29.5 Citations Per Year (CPY) and Li et al. (2018) with 28 CPY. Other articles most cited per year are Zhong et al. (2017) (27.5 CPY), Park et al. (2017) (22.8 CPY), Dave Bhargav et al. (2018) (20 CPY). It emerges that if we exclude the first, which is a general review on BIM-IoT integration, the most cited articles mainly concern project management and risk management. In Table 8, each article is assigned to a single core function according to the title and abstract analysis made by authors. Accordingly, an FM core competencies content analysis to generate patterns and trends of existing research has been done.

Table 8. Selected articles assignment to a single FM core function according to the title and abstract analysis made by the authors.

FM Core Competencies	References
Finance and Business	Chong et al. (2020) [47]
Human Factors	Marzouk et al. (2014, 2014) [48,49], Natephra et al. (2017) [50], Zhong et al. (2018) [51], Ma et al. (2019) [52], Lin et al. (2020) [53], Zaballos et al. (2020) [54]
Leadership and Strategy	Niu et al. (2016) [55], Chang et al. (2018) [56]
O&M	Rio et al. (2013) [57], Zhang et al. (2015) [58], Delgado et al. (2017) [59], Jeong et al. (2017) [60], Delgado et al. (2018) [61], Boddupalli et al. (2019) [62], Fitz et al. (2019) [63], Mannino et al. (2019) [64], Valinejadshoubi et al. (2019) [65], Cheng JCP et al. (2020) [66], Kameli et al. (2020) [67], Ma et al. (2020) [68], Moretti et al. (2020) [69], O'Shea and Murphy (2020) [70], Xie et al. (2020) [71], Yin et al. (2020) [72]
Project Management	Costin et al. (2015) [73], Fang et al. (2016) [74], Park et al. (2017) [75], Zhong et al. (2017) [76], Chen et al. (2018) [77], Li et al. (2018) [78], Tagliabue et al. (2018) [79], Hamooni et al. (2020) [80], Pan and Zhang (2021) [81]
Quality and Performance	Hernández et al. (2018) [82], Martinez et al. (2019) [83]
Real Estate and Property Management	Atazadeh et al. (2019) [84], Moretti et al. (2020) [85]
Risk Management	Shiau et al. (2013) [86], Arslan et al. (2014) [87], Li et al. (2014) [88], Riaz et al. (2014) [89], Cheng MY (2017) [90], Park and Hong (2017) [91], Park et al. (2017) [92], Riaz et al. (2017) [93], Beata et al. (2018) [94], Chou et al. (2019) [95], Parn et al. (2019) [96], Yu et al. (2019) [97], Asadzadeh et al. (2020) [98], Cheng JCP et al. (2020) [99], Lei et al. (2020) [100], Lin et al. (2021) [101], Tian et al. (2021) [102]

Table 8. *Cont.*

FM Core Competencies	References
Sustainability	Gökçe et al. (2013) [103], Osello et al. (2013) [104], Cheng et al. (2014) [105], Gökçe et al. (2014) [106], Gökçe et al. (2014) [107], Kensek (2014) [108], Ness et al. (2015) [109], Zhao et al. (2015) [110], Habibi (2016) [111], Lee et al. (2016) [112], Habibi (2017) [113], McGlinn et al. (2017) [114], Dave et al. (2018) [115], Francisco et al. (2018) [116], Pasini (2018) [117], Ma et al. (2019) [52], Kang (2020) [118], Mataloto et al. (2020) [119], Hirakraj et al. (2021) [120], Sepasgozar et al. (2021) [121], Tagliabue et al. (2021) [122]
Information and Technology Management	Wang et al. (2013) [123], Alves et al. (2017) [124], Edmondson et al. (2018) [125], Kang et al. (2018) [126], El Ammari et al. (2019) [127], Kazado et al. (2019) [128], Rashid et al. (2019) [129], Rogage et al. (2019) [130], Saar et al. (2019) [131], Tsai et al. (2019) [132], Zhai et al. (2019) [133], Edirisinghe et al. (2020) [134], Pavòn et al. (2020) [135], Quinn et al. (2020) [136], Valinejadshoubi et al. (2020) [137], Zhang et al. (2020) [138], Gao et al. (2021) [139]
Reviews	Wong et al. (2018) [7], Stefanic et al. (2019) [140], Tang et al. (2019) [141], Boje et al. (2020) [142], Panteli et al. (2020) [143], Hou et al. (2021) [144]

5. Discussion

The bibliometric analyses described in this paper identified the main characteristics of the literature in BIM, IoT and the integration of the two aiming to Facilities Management. In this section, an overview of BIM-IoT integration in the several FM core competencies is provided, and areas where further research is required within the scope of each core competence are suggested.

5.1. FM Core Competence: Finance and Business

The Finance and Business core competence concerns economic aspects, and it deals both with significant financial investment and operational expense. The only article concerning this competence [47] proposes a framework in which blockchain technology, smart sensors, smart contract and BIM are integrated. The proposed framework is meant to guide IT developers to design and implement an automated payment system (based on these new technologies) that aims to solve the security of payment problems. This application of multiple advanced technologies simultaneously and its related workflow are new to the current body of knowledge from both technical and managerial perspectives. In the article, smart sensors, at critical points across the entire supply-chain, provide live location and status information automatically onto a BIM model. Furthermore, smart sensor data are also stored on the blockchain network, providing an alternative system that will allow automated payment of fulfilled contractual obligations, resolving late-payment or non-payment-related issues.

Although this research's findings have undeniable advantages, this study is based on a specific blockchain platform. Further studies could adopt other blockchain platforms more suitable in upholding the security of payment. Furthermore, the research does not consider human tampering to commit fraud during the process. Even the authors suggested that subsequent studies should also consider fraud or any other human interventions that may influence systems operation. Hence, additional security layers and/or network security techniques should be investigated.

Finally, a possible main limitation of this framework hindering its adoption in the construction industry is the need for readily available money. The framework, providing automatic payment upon completion of the work, would jam in the case of lack of funds. If the client were temporarily experiencing a shortage of money in the course of the process, the automated payment would be blocked and the entire process would be interrupted.

Smart contracts and blockchain technology will undoubtedly be two essential elements in the future of FM. Future research should focus on these new technologies and challenges presented by them during the facility's whole life cycle. Several blockchain platforms should be investigated to provide the most suitable and secure solutions to the issues addressed.

5.2. FM Core Competence: Human Factor

This core competence focuses on protecting the environment and the people who use the facility, minimising risks and liabilities and positively impacting all stakeholders. All the articles belonging to this competence concern indoor environmental monitoring. Four out of seven articles [48,51–53] are on air quality monitoring. Two articles, [49] and [50], are about thermal comfort. Zaballos et al. (2020) [54] discuss environmental monitoring and emotion detection to provide insights into spaces' comfort level. Almost all articles focus their attention on the BIM–Wireless Sensors Network (WSN) connection, aiming to visualise real-time sensors data on the BIM model. The only exception is the research conducted by Zhong et al. (2018) [51], where the BIM model is used to extract building information stored in a tabular format and converted into ontology instances.

Table 9 shows the type of monitored data in each article.

Table 9. Human factor core competence: type of sensors used in studies on BIM–Wireless Sensors Network (WSN) integration for indoor environmental monitoring.

Reference	Scope	Issue	Sensors
Marzouk et al. (2014)	Indoor Environmental monitoring	Air quality	Temperature Particulate Matter
Zhong et al. (2018)	Indoor Environmental monitoring	Air quality	Temperature Humidity Noise Light Gas (CO, CO_2, Radon, Methane)
Ma et al. (2019)	Indoor Environmental monitoring	Air quality	Temperature Humidity Wind speed around person
Lin et al. (2020)	Indoor Environmental monitoring	Air quality	Carbon Monoxide Temperature Humidity
Marzouk et al. (2014)	Indoor Environmental monitoring	Thermal comfort	Temperature Humidity
Natephra et al. (2017)	Indoor Environmental monitoring	Thermal comfort	Temperature Thermographic camera
Zaballos et al. (2020)	Indoor Environmental monitoring	Air quality Thermal comfort	Temperature Humidity Noise Light CO, CO_2, TVOC

The integration between BIM and WSN offers great advantages to the monitoring systems developed in the various research studies. Through this integration, it is possible to better visualise a multitude of data relating to environmental monitoring and associated with multiple elements and spaces. Following this integration and creating the database containing all environmental data (e.g., temperature, humidity, light, noise, etc.), it is possible not only to monitor thermal/air quality problems to ensure comfort for users but also to detect the need for maintenance of building components.

However, during these processes, interoperability between different information systems and information sharing between various stakeholders remains challenging. Management of these heterogeneous data should be further investigated. Moreover, battery capacity and operation duration could be a significant limitation of a WSN. Therefore, it is necessary to consider adopting high-capacity batteries or a fixed power source for long-term operation.

To conclude, protecting the environment in which people live/work is certainly among the priorities that FM will have to face in the near future. Although the use of new technologies and sensors is widespread and certainly not new, the main challenge for this (but also other) core competencies is data/information interoperability. Finally, a novelty that emerged in this review is the improvement of the comfort level in facilities spaces through users emotion detection. In this direction, more effort should be focused to better fit building spaces to users.

5.3. FM Core Competence: Leadership and Strategy

This core competence focuses on aligning the facility portfolio with the organisation's missions and available resources. According to IFMA [13], sub-competencies in "Leadership and Strategy" include:

1. strategic planning and alignment with the organisation requirements;
2. policies, procedures and compliance;
3. individual and team management;
4. relationship and conflict management;
5. change management;
6. corporate social responsibility;
7. political, social, economic and industry factors affecting facility management.

There are few studies on this competence as for the previous one. After refining the query, only two studies remained (Table 8). Both articles deal with "decision-making" from two different points of view. Niu et al. (2016) [55] discuss several scenarios about using smart construction objects and their augmented capabilities of sensing, processing, computing, networking and reacting to alleviate human beings' incapability in decision-making. The Industry Foundation Classes (IFC) format is adopted to represent these objects in a virtual environment. With their innovative properties, smart construction objects can contribute to data collection and information processing and make autonomous decisions, eliminating human errors in the process and saving time. Although smart construction objects have undeniable advantages, there are still several limitations and challenges to fully exploit their potential, particularly cultural changes, new costs, Artificial Intelligence (AI) acceptance and organisation readiness.

On the other hand, Chang et al. (2018) [56] try to support complex decisions requiring interdisciplinary information using sensor data and the BIM model. Their research also deals with the design of a common platform allowing communication among sensors with different protocols and how visualisation may help make energy-saving management decisions. This visualisation allows us to see different values distribution in different contexts and make appropriate adjustments in each context. In addition, in this core competence, the key point is data/information integration from different sources. In the near future, this is undoubtedly the main problem that researchers will have to face due to the multiple and varied data sources and platforms.

5.4. FM Core Competence: Operations and Maintenance (O&M)

An important role in FM is to manage operations and maintenance of the facility. To do this, a good knowledge of building systems, structure, interiors and exteriors is required to ensure that systems operate efficiently, reliably, safely and in compliance with standards and regulations.

This core competence is one of the most investigated, and one of the earliest studies was conducted by Rio et al. (2013) [57]. The review showed a high growth rate in this category: ten out of fifteen articles were published in the past two years.

Among the sixteen reviewed articles in this category, nine [57–63,65,70] are about Structural Health Monitoring (SHM). BIM–sensor integration for SHM has been addressed since 2013, and interest in this topic has remained constant over the years. Through data-driven SHM techniques, it is possible to improve information management on structures health, safety and hazard mitigation. However, traditional approaches are insufficient to manage a large amount of data and information to conduct systematic decision-making for future maintenance.

The first attempt to create a connection between BIM and real-time data was made in 2013 by Rio et al. [57]. In their research, sensors data are stored within the BIM model. This strategy, however, could prove counterproductive as too many data from different types of sensors could weigh down the model.

Subsequent studies [58–63,65] propose an information modelling framework for supporting SHM, which includes an external database to facilitate storage, sharing and utilisation of gathered data. Authors, in their studies, propose approaches that support dynamic visualisation (within the BIM model) of some key structural performance parameters and enable continuous updating and long-term data management, generating models compliant with the IFC standard.

Such tools aim to facilitate decision-making on maintenance and risk management, avoiding manual errors resulting from visual inspection of the structures.

Furthermore, in their study, Fitz, Theiler and Smarsly [63] introduce the concept of the Cyber-Physical System (CPS) and present a metamodel for describing it. In their paper, communication-related properties and behaviour of CPS applied for SHM are described. Moreover, system components relevant to communication are specified. Then, the metamodel to formally define a CPS is proposed and mapped into the IFC schema.

On the other hand, the remaining articles that do not deal with SHM address equipment maintenance [66–68] and space management [64,69,71,72]. Here too, collected sensors data are stored in an external database.

The most relevant work in this area is, probably, the research of Cheng et al. [66]. They developed a data-driven predictive maintenance framework based on BIM, IoT and machine learning algorithms. Both Artificial Neural Network (ANN) and Support Vector Machine (SVM) are used to predict Mechanical, Electrical and Plumbing (MEP) components' future conditions with reasonably accurate results. Even if other prediction methods are taken into consideration, the proposed framework has significant implications: (a) fault alarming in an early stage avoiding failures; (b) future condition prediction (knowing in advance the failure timing); (c) minimising or avoiding overtime costs by preparing maintenance materials and tools ahead of time.

Articles that deal with management and maintenance services [64,69] focus primarily on occupancy control even if these systems are not always reliable due to their difficulty in counting people in crowded spaces.

In the O&M context, a first conclusion may be made: reviewed articles suggest that future research should focus on facing challenges presented by managing and visualising data acquired during the whole life cycle of the facility, not only during a single phase. Data-rich BIM models will be necessary to support facilities monitoring and applications fully.

Furthermore, many proprietary file formats are used in most articles. To streamline workflows and improve interoperability, it may be appropriate to increase the use of open formats.

Finally, further studies are required to automatically identify critical locations in which sensors are needed, types of sensors required to monitor critical elements and sensors data integration to improve O&M management.

5.5. FM Core Competence: Project Management

Another essential core skill in FM is Project Management (PM). Projects can vary in scope, complexity, duration and financial risk. According to IFMA, sub-competencies of PM include planning and design, execution and delivery, and evaluation.

Most of the articles concerning the PM [73–79] deal with the topics of real-time tracking of personnel, materials and equipment to enhance the security, safety, quality control, logistics and productivity monitoring. To do this, BIM and Radio Frequency IDentification (RFID) are the most used technologies [73,74,76–78] to implement localisation of people and objects. The proposed systems have a reliable accuracy rate, and RFID localisation systems have great potential in practical applications and could improve resource allocation efficiency and decrease human errors. Instead, Park et al. (2017) [75] developed a tracking system based on the integration of Bluetooth Low Energy (BLE) technology, motion sensors and BIM. This integration aims to achieve more accurate tracking that reduces and compensates for the sensors' errors.

On the other hand, Hamooni et al. (2020) [80] proposed a method that uses BIM interoperability and wireless sensors to monitor concrete maturity and control the concrete formwork process. BIM allows for the calculation of formwork removal time based on the maturity and strength data collected from sensors inside the concrete. This system will allow the concrete placement process to be continuously monitored and controlled and the curing time before formwork removal to be reduced, thus affecting construction management and project controls by (a) reducing the time required to complete the work, (b) avoiding project delays and (c) lowering unnecessary formwork rental expenses.

In conclusion, the main technological challenges found are related to the location and coverage of the sensors network and the signal strength of the router/hotspot. A significant problem that could arise is the stability of networks for communicating information.

Future research may include improvement of these systems and platforms by incorporating more functions related to the PM sub-competencies and productive analysis (e.g., future workforce estimation or a deep investigation of impacts on the total cost and time of a construction project resulting from the use of BIM–WSN integration).

5.6. FM Core Competence: Quality

Quality is one of the less investigated core competencies. It concerns needs and expectations on the facility and facility's services, aiming to improve facility organisations' and service providers' performance.

Both the articles on this topic [82,83] concern the quality of building components/ construction work to ensure that specifications are implemented according to the project. Digitising information allows detecting design errors or poor performance. Both research studies integrate BIM validation tools to assure BIM quality.

5.7. FM Core Competence: Real Estate and Property Management

"Real Estate and property management" core competence is about the management of physical assets to enhance users' experience to meet asset owners' strategic objectives and to optimise real estate value. It is one of the least investigated core competencies. Among the reviewed articles, only two of these belongs to real estate and property management competence. Notable is the article of Atazadeh et al. [84], which discusses the use of BIM for defining the legal ownership of IoT-generated data, which are part of the asset value. There are no specific regulations or laws that define the retrieval and use of IoT data considering the appropriate legal rights and responsibilities. Rights, restrictions and responsibilities related to the use of IoT data in multi-owned buildings could be better defined using the BIM environment.

To conclude, as also highlighted by Moretti et al. [85] in their article, future developments in FM aiming at Real Estate and property management should focus attention again on interoperability and openBIM methodology to support dynamic assets management applications. The main issue in this context is the scarce as-built information. Supporting data integration, open formats and interoperability makes it possible to achieve better solutions for building management.

5.8. FM Core Competence: Risk Management

Risk Management plays a central role in FM and, unsurprisingly, is among the core competencies most investigated by researchers worldwide.

The articles belonging to this core competence address various issues related to risk management, Table 10 groups them by topic. Most of the articles, twelve out of seventeen, are fairly distributed between fire risk issues and safety in the workplace.

All the research agrees that the integration of data between BIM and WSN will provide an invaluable result for future applications in managing users and workers' health and safety. Fire risk studies focus mainly on (a) defining the fire conditions as well as the location and types of relevant fire-extinguishing tools needed; (b) localisation of trapped occupants in a fire emergency scene; and (c) evacuation/rescue paths optimisation. Proposed workflows and algorithms are BIM-centred, where BIM is integrated to provide geometric information and a graphical interface for user interaction.

Table 10. Risk management core competence: articles grouped by topic addressed.

References	Topic
[86,88,90,94,95,99]	Fire risk
[87,89,92,93,96,98]	Workers/Users Health and Safety (H&S)
[91,100]	Multi disaster (earthquake, flooding, and fire) countermeasure
[97]	Emergency response of utility tunnels
[101,102]	Excavation risks

Relevant studies in this field have been conducted by Cheng et al. [90] and Chou et al. [95]. Their studies are quite similar and propose a system based on a Bluetooth sensor network that can be used (a) to early detect a fire (b) to plan evacuation/rescue routes and to guide building users in emergencies, (c) for dynamic 3D visualisation of fire events and (d) for bidirectional human-machine interactions to optimise evacuation/rescue efforts. The proposed systems could reduce the number of casualties, support the rescue process and emergency evacuation, and mitigate the panic among people in cases of fire.

Another interesting research study in "fire risk management" was conducted by Cheng et al. [99]. Their study proposes an approach for adaptive path planning against the rapid environmental changes in fires. To detect the number of people in a building space, the network uses real-time videos from Closed-Circuit Television (CCTV) cameras and deep learning algorithms. In addition, an IoT sensor network (detect temperature, carbon dioxide and carbon monoxide) is used to detect hazardous areas. The BIM model provides floor plan information, sensors location and a simplified visualisation model during evacuation. Eventually, research suggests that it is possible to evacuate people through AR devices along the shortest path while avoiding congested and hazardous areas.

Research studies concerning workers' and building users' health and safety have investigated the integration of BIM with several types of wireless sensors to provide a centralised database with updated real-time data throughout the building lifecycle, starting from the construction phase. Currently, safety monitoring practices primarily rely on "manual" observation, which is labour-intensive and error-prone [92]. Therefore, the impact of sensor-based safety management systems, coupled with the BIM environment, could improve health, safety and emergency management.

In conclusion, systems coupling BIM models and sensors can continuously monitor the built environment in an automated way. They can be used for various purposes: (a) to prevent accidents by notifying workers of incoming hazards; (b) to notify safety managers or site supervisors about unidentified or newly appearing threats; (c) to monitor the environmental conditions of a confined space; (d) to better manage emergencies (e.g., fires) by providing optimised escape (or rescue) routes.

However, more research needs to be conducted to make these systems interoperable with existing sensor systems and minimise computational times to avoid any delay in emergency response operations. Furthermore, these systems could automatically detect and document near misses to prevent better accidents and, thus, to improve safety further.

One of the major limitations is the sensors reliability and transmission over long distances, which can cause false alarms. Other important limitations are related to the accuracy of deployed devices, which may be reduced due to water presence, to the presence of fire and/or smoke, which could impact accuracy and signal propagation. Moreover, energy demands may limit continuously monitoring sensors if not wired. Eventually, these BIM-based systems' performances rely on the accuracy of the BIM model. Therefore, having an updated BIM model is essential.

5.9. FM Core Competence: Sustainability

Sustainability, which is a legal obligation in some countries, could also be considered as a social responsibility that often turns into an economic advantage for asset owners. Facility managers are expected to act in order to protect the environment and the people who use their facilities while supporting organisational effectiveness and minimising risks and liabilities. Subcompetencies of sustainability include energy management, water management, materials and consumables management, waste management and workplace and site management.

Sustainability is among the most investigated FM core competencies. Most of the articles concerning sustainability deal with how buildings' energy demand can be reduced through Information and Communication Technologies (ICT). Furthermore, the ICT application in several processes (e.g., BIM) and scenarios have been investigated. In most cases [52,103,104,106–108,110–117,119] BIM is used to process or display buildings' geometric data, FM data and energy data collected through sensors. Among these articles, References [104,111,113,115–117,119] focus on users' behaviours. These researches aim to raise users' awareness of energy efficiency and consider building users as a primary factor to improve energy efficiency and IEQ. Results put forth the use of real-time monitoring systems and suggest a controlled interaction among users and heating systems to improve energy performances and comfort.

An interesting approach to interact with users has been made by Francisco et al. [116]. They propose a method of combining data and graphic representation through spatial and colour coding techniques in BIM. Through this type of information representation, users can access complex information through a simple interface. In this way, it is possible to improve the interpretation of energy data and increase the user's involvement in the building's management, consequently improving building consumption, comfort and health. Furthermore, the proposed method could be applied to other factors such as water consumption, room temperature or indoor air quality.

Only two studies [52,114] deal with/involve artificial neural networks (ANNs) in sustainability, probably because it is a relatively new topic in the AECO sector. The study of McGlinn et al. is relevant. [114]. They propose an intelligent monitoring and control interface for efficient energy management using BIM and Semantic Web to integrate smart buildings, sensors and software components like artificial neural network (ANN) and genetic algorithms (GA). This interface provides suggestions based on the building's sensor measurements and proposes these suggestions to the Facility Manager. However, there are still issues related to interface usability for non-technical users.

The other two studies dealing with sustainability discuss: (a) a framework integrating the information necessary for green buildings design and their automated evaluation process [105] and (b) a BIM–RFID-based approach with the potential to improve resource reuse and efficiency [109]. Although through the two approaches presented in the articles mentioned above, it is possible to gain advantages in the design processes of Green Buildings, it is evident that research in these fields is still at an early stage.

When discussing sustainable approaches aimed at controlling energy consumption in a building, it is impossible not to mention the Building Energy Management Systems (BEMSs). A BEMS can fully monitor and control the building's energy needs through building energy data collection, performance analysis and equipment control [139]. Through better energy management in a building, a BEM not only reduces energy consumption and costs but also improves occupants' comfort [139]. Conceptually, a BEMS architecture has different layers: a sensor/actuation layer, a computational layer, an application layer and, in some cases, also a user interaction layer [114].

Hence, on the one hand, the BIM model provides a series of static data relating to the building (not only geometric and spatial data but also other information according to the BIM's several dimensions). On the other hand, the BEM system takes the role of collecting data from sensors in the building on-site. The synergy between these two environments (BIM–BEM) can positively affect building energy management, especially by users' involvement. Among selected articles, the first attempt at BIM-BEM integration dates back to 2013 with the research carried out by Osello et al. [104]. They developed an ICT infrastructure made of heterogeneous monitoring and actuation devices to reduce energy consumption. Finally, they used BIM and interoperability to process and visualise all data essential for energy simulations and for FM. Other studies, from Lee et al. [112] and McGlinn et al. [114], show that BIM is a useful approach for the visual representation, management and exchange of information on all aspects of a building. In particular, in Lee's research, BIM was used to develop an energy management platform. In their study, BIM was used to visualise gathered environmental and energy consumption data. In this way, facility managers could better manage buildings energy consumption and control buildings' equipment.

Another significant attempt at BIM–BEM integration has been made by Kang (2020) [118]. Kang, in his research, proposes a BIM-based Human Machine Interface (HMI) framework for space-based energy management. The proposed framework links data between BIM and BEMS, which are heterogeneous systems, aiming at space-based real-time energy monitoring. Furthermore, as it is challenging to use a BIM data structure if it does not fit into the energy management system, this researcher also defines requirements for developing a BIM database suitable for the proposed framework.

In conclusion, although there is much to be done in built environment sustainability challenges, four major steps should be accomplished: (a) fine-tuning of the interaction between environmental sensors data and Artificial Intelligence (AI) or optimisation algorithms; (b) developing sustainable and innovative user interaction strategies; (c) focusing attention on other sustainability sub-competencies (e.g., water management, materials and consumables management, waste management); (d) aiming at BIM–BEM integration, overcoming problems due to the differentiation of communication protocols.

5.10. FM Core Competence: Information and Technology Management

Although the whole topic of BIM–IoT integration could be discussed in this section, only articles concerning sub-competencies such as technology needs assessment and implementation, maintenance and upgrade of technology systems or protection and cybersecurity are addressed here. Some of the analysed articles may be associated with other FM competencies, but they are discussed in this FM core competence if:

a. the research on BIM–IoT integration encompasses more than one FM competence;
b. the article discusses the integration of technologies such as Augmented Reality (AR)/Mixed Reality (MR)/Virtual Reality (VR).

Many of the reviewed articles suggest no generic approach to assist in creating software services and applications combining sensor data with BIM models. Articles address the engineering complexity associated with integrating sensor data with BIM to facilitate real-time operational performance information management and permit proactive operational and maintenance decisions in many ways. Only three articles [123,127,131] discuss the development of collaborative BIM-based AR/MR/VR approaches. Interesting research has been carried out by El Ammari et al. [127], who developed a Mixed-Reality framework for facilities management with two modules: a field AR module and an office Immersive Augmented Virtuality (IAV) module. These modules can be used independently or combined using interactive visual collaboration, with an improvement in field task efficiency.

Another noteworthy research study was carried out by Kazado et al. (2019) [128], who presented three approaches for BIM-sensors integration to enable visualisation and analysis of real-time and historical data. Despite being probably the first work using Autodesk Navisworks software to implement a user-friendly interface that integrates the existing building sensor technology and BIM process, the use of a closed data format (Autodesk's files format) instead of an open one could be a limitation.

Most of the studies highlight the risk of losing competitiveness both on the local and international markets if stakeholders in the construction sector slow down the adoption of new technology. The construction industry appears to be already outdated when compared to other industrial sectors. Nevertheless, according to the articles, stakeholders seem reluctant to invest, especially in costly innovative technological devices. Therefore, many researchers aim to reduce the initial investment costs while proposing innovative solutions that bring added value to the FM processes.

6. Conclusions

Although still in its infancy, the construction sector's digitisation process is underway, aiming to create an ever-larger network of cyber-physical connections, supported by the abundance of sensorized and networked elements. The analysis of data generated by sensorized building components and systems will allow using connected digital models to improve future design and increase the environmental, safety and financial performance of the digitally built environment.

This document provides an overview of BIM and IoT integration in FM. From a query on Scopus and Web of Science with more than 900 results, 99 articles were identified and reviewed as the most relevant references. Existing gaps and future research directions were outlined.

BIM now supports many technological advances that the industry is witnessing, albeit with some limitations. Although BIM is widely used in the building design phase, there is still much to do for its use in Facility Management. Nevertheless, BIM can be considered a natural interface for IoT/real-time data implementation. Several researchers have begun to explore the potential synergy between these two environments.

From the literature review, it emerges that the BIM and IoT integration research is in an early phase. Most research works are still in the conceptual stage, even though some studies are quite thorough and propose solutions tested in real-world applications. The main obstacles preventing the uptake of these new technologies include (1) in most cases, the lack of a BIM approach that meets the information requirements and fully exploits the potential of the digital model; (2) the fragmented nature of the AEC sector; and (3) shortage of real-life use cases demonstrating potential benefits.

General remarks found on BIM for FM are related to the need to

1. enhance the interoperability of data from as-designed to as-built for FM purposes;
2. review and improve IFC open standards and data specifications to satisfy data and information required for FM.

Furthermore, one of the main challenges in BIM–IoT integration is coupling dynamic real-time data to the model database. In this context, future studies are needed to

1. find a standardised way to integrate and manage data;
2. enhance further exploration in cloud computing;
3. improve other digital technologies such as Augmented, Virtual and Mixed Reality and their application in the AEC sector, integrating them with BIM

BIM–IoT data integration has a new added value in the market: the physical object is a product that carries information throughout its life cycle. This will significantly help the construction industry, which, mimicking more industrialised sectors, has just begun its journey from being product-oriented to service-oriented. However, to take advantage of this transformation, the integration of data into BIM models needs to be managed in the best possible way.

BIM and IoT studies are often based on proprietary files and closed ecosystems, where information is not yet shared openly among stakeholders. Hence, subsequent studies within the BIM-IoT integration domain should focus their attention on open data and open communication standards.

On the other hand, WSN could be considered the IoT solution for monitoring and recording the physical condition of buildings and environmental monitoring management. Research hass proved that both high costs and ineffectiveness of WSN devices can be avoided if information requirements (data types, data frequencies, WSN devices' location, etc.) are appropriately set at the very beginning of the asset lifecycle.

The major problems encountered in the use of WSN concern:

1. signal transmission: the number of devices necessary for the correct transmission of data and information can influence costs or be limited by the type of construction;
2. powering devices: batteries may not be efficient enough in continuously monitored devices, and the main power supply cannot always reach all devices.

In conclusion, this review highlights four key areas to be further studied:

1. Learning from errors. Future asset design should be influenced by lessons learned from existing ones. Architects and engineers rarely study buildings' performances or act on occupant feedback (post-occupancy evaluations) once they are handed over to the client. Therefore, it is very probable that errors made are often replicated in other future building design developments. Data coming from the use phase of the assets allows essential information to be collected to better guide the design phase, aiming for improvements in FM activities. However, few researchers have addressed the problem of handing back the information collected in the life cycle to the design team. There is no evidence in the literature about what information is needed or how best to collect and present it.
2. Exploiting new technologies. Advancement in Information and Communications Technology is giving the construction sectors hardware and software tools with unprecedented capabilities and with ever lower costs. First exploratory examples of the use of these technologies for district and cities have appeared, giving a new vision of the idea of smart cities. In this context, new research should focus on both adopting new technologies and finding the best way to satisfy the need for specific training of FM managers and workers. Additionally, although interoperability frameworks have already been investigated and developed, future research should address the potential of BIM to handle a dynamic environment and overcome the differentiation of communication protocols.
3. Application of machine learning algorithms to gather data to develop self-learning and self-improving algorithms to enhance building management. Edge computing and edge analytics seem to allow for an improved system, where data are used where they are produced.
4. User awareness/involvement through simplified interfaces/interactions. BIM remains a tool for experts, and hardly any users or FM managers are able to use the information it contains. Although research has proved that users' involvement is vital for more sustainable buildings, few studies collected real-time data involving asset users in buildings use and management. First research studies have proved that AR can improve building operations and maintenance, but there may still be ground for improvement in this field.

A deep review of 99 articles related to the eleven IFMA FM core competencies highlights four main knowledge gaps in the emerging sector of BIM and IoT integration for FM. These are related to the back-propagation of information from the use stage to the design one, to new technologies exploitation and final users' involvement in improving buildings sustainability. This may help further research advancement for studies to improve built environment management.

Author Contributions: Conceptualisation, A.M., M.C.D. and F.R.C.; methodology, A.M., M.C.D. and F.R.C.; software, A.M.; data curation, A.M.; writing—original draft preparation, A.M.; writing—review and editing, A.M., M.C.D. and F.R.C.; supervision, M.C.D. and F.R.C. All authors have read and agreed to the published version of the manuscript.

Funding: This research received no external funding.

Conflicts of Interest: The authors declare no conflict of interest.

References

1. Haghsheno, S.; Deubel, M.; Spenner, L. Identification and description of relevant digital technologies for the construction industry. *Bauingenieur* **2019**, *94*, 45–55. [CrossRef]
2. Hossain, M.A.; Nadeem, A. Towards digitising the construction industry: State of the art of construction 4.0. In Proceedings of the ISEC 10: Interdependence between Structural Engineering and Construction Management, Chicago, IL, USA, 20–25 May 2019; pp. 1–6.
3. Oesterreich, T.D.; Teuteberg, F. Understanding the implications of digitisation and automation in the context of Industry 4.0: A triangulation approach and elements of a research agenda for the construction industry. *Comput. Ind.* **2016**, *83*, 121–139. [CrossRef]
4. Blanco, J.; Mullin, A.; Pandya, K.; Sridhar, M. The new age of engineering and construction technology. *McKinsey Q.* **2017**, 1–16. [CrossRef]
5. McKinsey Global Institute. *Reinventing Construction: A Route To Higher Productivity*; McKinsey Global Institute: Chicago, IL, USA, 2017. [CrossRef]
6. Chung, S.W.; Kwon, S.W.; Moon, D.Y.; Ko, T.K. Smart facility management systems utilising open BIM and augmented/virtual reality. In Proceedings of the 35th International Symposium on Automation and Robotics in Construction (ISARC2018), Berlin, Germany, 20–25 July 2018.
7. Wong, J.K.W.; Ge, J.; He, S.X. Digitisation in facilities management: A literature review and future research directions. *Autom. Constr.* **2018**, *92*, 312–326. [CrossRef]
8. Cheng, J.C.P.; Chen, K.; Chen, W.; Li, C.T. A Bim-Based Location Aware Ar Collaborative Framework for Facility Maintenance Management. *J. Inf. Technol. Constr.* **2019**, *24*, 30–39. Available online: http://www.itcon.org/2019/19 (accessed on 20 February 2021).
9. Glaessgen, E.; Stargel, D. The Digital Twin Paradigm for Future NASA and US Air Force Vehicles. In Proceedings of the 53rd AIAA/ASME/ASCE/AHS/ASC Structures, Structural Dynamics and Materials Conference 20th AIAA/ASME/AHS Adaptive Structures Conference 14th AIAA, Honolulu, HI, USA, 23–26 April 2012. [CrossRef]
10. Bonci, A.; Carbonari, A.; Cucchiarelli, A.; Messi, L.; Pirani, M.; Vaccarini, M. A cyber-physical system approach for building efficiency monitoring. *Autom. Constr.* **2019**, *102*, 68–85. [CrossRef]
11. Love, P.E.D.; Matthews, J. The 'how' of benefits management for digital technology: From engineering to asset management. *Autom. Constr.* **2019**, *107*, 102930. [CrossRef]
12. *ISO 41011:2017. International Standard Facility Management—Vocabulary*; ISO: Geneva, Switzerland, 2017.
13. International Facility Management Association (IFMA). Certified Facility Manager®(CFM®)—Competency Guide. 2018. Available online: https://www.ifma.org/about/what-is-facility-management (accessed on 20 February 2021).
14. Omar, N.S.; Najy, H.I.; Ameer, H.W. Developing of Building Maintenance Management by Using BIM. *Int. J. Civ. Eng. Technol.* **2018**, *9*, 1371–1383. Available online: https://www.iaeme.com/MasterAdmin/uploadfolder/IJCIET-09-11-132/IJCIET-09-11-132.pdf (accessed on 20 February 2021).
15. Vanier, D.J. Advanced asset management: Tools and techniques. In Proceedings of the APWA International Public Works Congress NRCC/CPWA Seminar Series "Innovations in Urban Infrastructure", Louisville, KY, USA, 9 September 2000.
16. Cheng, J.C.P.; Chen, W.; Tan, Y.; Wang, M. A BIM-Based Decision Support System Framework for Predictive Maintenance Management of Building Facilities. In Proceedings of the 16th International Conference on Computing in Civil and Building Engineering (ICCCBE2016), Osaka, Japan, 6–8 July 2016; pp. 711–718.
17. Xiao, Y.-Q.; Li, S.-W.; Hu, Z.-Z. Automatically Generating a MEP Logic Chain from Building Information Models with Identification Rules. *Appl. Sci.* **2019**, *9*, 2204. [CrossRef]
18. Wanigarathna, N.; Jones, K.; Bell, A.; Kapogiannis, G. Building information modelling to support maintenance management of healthcare built assets. *Facilities* **2019**, *37*, 415–434. [CrossRef]
19. Rajput, S.; Singh, S.P. Industry 4.0—Challenges to implement circular economy. *Benchmarking* **2019**. [CrossRef]
20. De Sousa Jabbour, A.B.L.; Jabbour, C.J.C.; Filho, M.G.; Roubaud, D. Industry 4.0 and the circular economy: A proposed research agenda and original roadmap for sustainable operations. *Ann. Oper. Res.* **2018**, *270*, 273–286. [CrossRef]
21. Ashworth, S.; Tucker, M.; Druhmann, C.K. Critical success factors for facility management employer's information requirements (EIR) for BIM. *Facilities* **2019**, *37*, 103–118. [CrossRef]
22. NIBS. National BIM Guide for Owners. NIBS, 2017; 46p, Available online: https://www.nibs.org/files/pdfs/NIBS_BIMC_NationalBIMGuide.pdf (accessed on 20 February 2021).
23. Hilal, M.; Maqsood, T.; Abdekhodaee, A. A hybrid conceptual model for BIM in FM. *Constr. Innov.* **2019**, *19*, 531–549. [CrossRef]
24. *ISO 19650-2:2018. Organization and Digitisation of Information about Buildings and Civil Engineering Works, Including Building Information Modelling (BIM)—Information Management Using Building Information Modelling*; ISO: Geneva, Switzerland, 2018.
25. BIM Forum. Level of Development (LOD) Specification Part I & Commentary, Bim-Bep. 2019; 254p, Available online: https://bimforum.org/resources/Documents/BIMForum_LOD_2019_reprint.pdf (accessed on 20 February 2021).
26. Love, P.E.D.; Simpson, I.; Hill, A.; Standing, C. From justification to evaluation: Building information modeling for asset owners. *Autom. Constr.* **2013**, *35*, 208–216. [CrossRef]
27. Becerik-Gerber, B.; Jazizadeh, F.; Li, N.; Calis, G. Application Areas and Data Requirements for BIM-Enabled Facilities Management. *J. Constr. Eng. Manag.* **2012**, *138*, 431–442. [CrossRef]

28. Bortoluzzi, B.; Efremov, I.; Medina, C.; Sobieraj, D.; McArthur, J.J. Automating the creation of building information models for existing buildings. *Autom. Constr.* **2019**, *105*, 102838. [CrossRef]
29. Matarneh, S.T.; Danso-Amoako, M.; Al-Bizri, S.; Gaterell, M.; Matarneh, R.T. BIM for FM: Developing information requirements to support facilities management systems. *Facilities* **2019**, *38*, 378–394. [CrossRef]
30. Wang, Y.; Wang, X.; Wang, J.; Yung, P.; Jun, G. Engagement of facilities management in design stage through BIM: Framework and a case study. *Adv. Civ. Eng.* **2013**, *2013*, 189105. [CrossRef]
31. Ismail, Z.A. An Integrated Computerised Maintenance Management System (I-CMMS) for IBS building maintenance. *Int. J. Build. Pathol. Adapt.* **2019**, *37*, 326–343. [CrossRef]
32. Chen, W.; Chen, K.; Cheng, J.C.P.; Wang, Q.; Gan, V.J.L. BIM-based framework for automatic scheduling of facility maintenance work orders. *Autom. Constr.* **2018**, *91*, 15–30. [CrossRef]
33. Golabchi, A.; Akula, M.; Kamat, V. Automated building information modeling for fault detection and diagnostics in commercial HVAC systems. *Facilities* **2016**, *34*, 233–246. [CrossRef]
34. Munir, M.; Kiviniemi, A.; Jones, S.W. Business value of integrated BIM-based asset management. *Eng. Constr. Arch. Manag.* **2019**, *26*, 1171–1191. [CrossRef]
35. Dixit, M.K.; Venkatraj, V.; Ostadalimakhmalbaf, M.; Pariafsai, F.; Lavy, S. Integration of facility management and building information modeling (BIM). *Facilities* **2019**, *37*, 455–483. [CrossRef]
36. Pärn, E.A.; Edwards, D.J.; Sing, M.C.P. The building information modelling trajectory in facilities management: A review. *Autom. Constr.* **2017**, *75*, 45–55. [CrossRef]
37. Patacas, J.; Dawood, N.; Vukovic, V.; Kassem, M. BIM for facilities management: Evaluating BIM standards in asset register creation and service life planning. *J. Inf. Technol. Constr.* **2015**, *20*, 313–331.
38. Asghari, P.; Rahmani, A.M.; Javadi, H.H.S. Internet of Things applications: A Systematic Review. *Comput. Netw.* **2019**, *148*, 241–261. [CrossRef]
39. Gubbi, J.; Buyya, R.; Marusic, S.; Palaniswami, M. Internet of Things (IoT): A vision, architectural elements, and future directions. *Futur. Gener. Comput. Syst.* **2013**, *29*, 1645–1660. [CrossRef]
40. Pradeep, S.; Kousalya, T.; Suresh, K.M.A.; Edwin, J. IoT and its connectivity challenges in smart home. *Int. Res. J. Eng. Technol.* **2016**, *3*, 1040–1043. Available online: http://www.indjst.org/index.php/indjst/article/view/109387 (accessed on 20 February 2021).
41. Rizal, R.; Hikmatyar, M. Investigation Internet of Things (IoT) Device using Integrated Digital Forensics Investigation Framework (IDFIF). *J. Phys. Conf. Ser.* **2019**, *1179*, 012140. [CrossRef]
42. Rondeau, E.P.; Brown, R.K.; Lapides, P.D. *Facility Management*; John Wiley & Sons: Hoboken, NJ, USA, 2012.
43. Jeon, Y.; Cho, C.; Seo, J.; Kwon, K.; Park, H.; Oh, S.; Chung, I.J. IoT-based occupancy detection system in indoor residential environments. *Build. Environ.* **2018**, *132*, 181–204. [CrossRef]
44. Rafsanjani, H.N.; Ghahramani, A. Towards utilising internet of things (IoT) devices for understanding individual occupants' energy usage of personal and shared appliances in office buildings. *J. Build. Eng.* **2020**, *27*, 100948. [CrossRef]
45. Madrid, N.; Boulton, R.; Knoesen, A. Remote monitoring of winery and creamery environments with a wireless sensor system. *Build. Environ.* **2017**, *119*, 128–139. [CrossRef]
46. Aria, M.; Cuccurullo, C. Bibliometrix: An R-tool for comprehensive science mapping analysis. *J. Inf.* **2017**, *11*, 959–975. [CrossRef]
47. Chong, H.Y.; Diamantopoulos, A. Integrating advanced technologies to uphold security of payment: Data flow diagram. *Autom. Constr.* **2020**, *114*, 103158. [CrossRef]
48. Marzouk, M.; Abdelaty, A. BIM-based framework for managing performance of subway stations. *Autom. Constr.* **2014**, *41*, 70–77. [CrossRef]
49. Marzouk, M.; Abdelaty, A. Monitoring thermal comfort in subways using building information modelling. *Energy Build.* **2014**, *84*, 252–257. [CrossRef]
50. Natephra, W.; Motamedi, A.; Yabuki, N.; Fukuda, T. Integrating 4D thermal information with BIM for building envelope thermal performance analysis and thermal comfort evaluation in naturally ventilated environments. *Build. Environ.* **2017**, *124*, 194–208. [CrossRef]
51. Zhong, B.; Gan, C.; Luo, H.; Xing, X. Ontology-based framework for building environmental monitoring and compliance checking under BIM environment. *Build. Environ.* **2018**, *141*, 127–142. [CrossRef]
52. Ma, G.; Liu, Y.; Shang, S. A building information model (BIM) and artificial neural network (ANN) based system for personal thermal comfort evaluation and energy efficient design of interior space. *Sustainability* **2019**, *11*, 4972. [CrossRef]
53. Lin, Y.-C.; Cheung, W.-F. Developing WSN/BIM-Based Environmental Monitoring Management System for Parking Garages in Smart Cities. *J. Manag. Eng.* **2020**, *36*. [CrossRef]
54. Zaballos, A.; Briones, A.; Massa, A.; Centelles, P.; Caballero, V. A smart campus' digital twin for sustainable comfort monitoring. *Sustainability* **2020**, *12*, 9196. [CrossRef]
55. Niu, Y.; Lu, W.; Chen, K.; Huang, G.G.; Anumba, C. Smart construction objects. *J. Comput. Civ. Eng.* **2016**, *30*. [CrossRef]
56. Chang, K.-M.; Dzeng, R.-J.; Wu, Y.-J. An Automated IoT Visualization BIM Platform for Decision Support in Facilities Management. *Appl. Sci.* **2018**, *8*, 1086. [CrossRef]
57. Rio, J.; Ferreira, B.; Poças-Martins, J. Expansion of IFC model with structural sensors. *Inf. La Constr.* **2013**, *65*, 219–228. [CrossRef]

58. Zhang, Y.; Bai, L. Rapid structural condition assessment using radio frequency identification (RFID) based wireless strain sensor. *Autom. Constr.* **2015**, *54*, 1–11. [CrossRef]
59. Delgado, J.M.D.; Butler, L.J.; Gibbons, N.; Brilakis, I.; Elshafie, M.Z.E.B.; Middleton, C. Management of structural monitoring data of bridges using BIM. *Proc. Inst. Civ. Eng. Eng.* **2017**, *170*, 204–218. [CrossRef]
60. Jeong, S.; Hou, R.; Lynch, J.P.; Sohn, H.; Law, K.H. An information modeling framework for bridge monitoring. *Adv. Eng. Softw.* **2017**, *114*, 11–31. [CrossRef]
61. Delgado, J.M.D.; Butler, L.J.; Brilakis, I.; Elshafie, M.Z.E.B.; Middleton, C.R. Structural Performance Monitoring Using a Dynamic Data-Driven BIM Environment. *J. Comput. Civ. Eng.* **2018**, *32*. [CrossRef]
62. Boddupalli, C.; Sadhu, A.; Azar, E.R.; Pattyson, S. Improved visualisation of infrastructure monitoring data using building information modelling. *Struct. Infrastruct. Eng.* **2019**, *15*, 1247–1263. [CrossRef]
63. Fitz, T.; Theiler, M.; Smarsly, K. A metamodel for cyber-physical systems. *Adv. Eng. Inform.* **2019**, *41*, 100930. [CrossRef]
64. Mannino, A.; Moretti, N.; Dejaco, M.C.; Baresi, L.; Cecconi, F.R. Office building occupancy monitoring through image recognition sensors. *Int. J. Saf. Secur. Eng.* **2019**, *9*, 371–380. [CrossRef]
65. Valinejadshoubi, M.; Bagchi, A.; Moselhi, O. Development of a BIM-Based Data Management System for Structural Health Monitoring with Application to Modular Buildings: Case Study. *J. Comput. Civ. Eng.* **2019**, *33*, 1–16. [CrossRef]
66. Cheng, J.C.P.; Chen, W.; Chen, K.; Wang, Q. Data-driven predictive maintenance planning framework for MEP components based on BIM and IoT using machine learning algorithms. *Autom. Constr.* **2020**, *112*, 103087. [CrossRef]
67. Kameli, M.; Sardroud, J.M.; Hosseinalipour, M.; Behruyan, M.; Ahmed, S.M. An application framework for development of a maintenance management system based on building information modeling and radio-frequency identification: Case study of a stadium building. *Can. J. Civ. Eng.* **2020**, *47*, 736–748. [CrossRef]
68. Ma, Z.; Ren, Y.; Xiang, X.; Turk, Z. Data-driven decision-making for equipment maintenance. *Autom. Constr.* **2020**, *112*, 103103. [CrossRef]
69. Moretti, N.; Cadena, J.D.B.; Mannino, A.; Poli, T.; Cecconi, F.R. Maintenance service optimisation in smart buildings through ultrasonic sensors network. *Intell. Build. Int.* **2021**, *13*, 4–16. [CrossRef]
70. O'Shea, M.; Murphy, J. Design of a BIM integrated structural health monitoring system for a historic offshore lighthouse. *Buildings* **2020**, *10*, 131. [CrossRef]
71. Xie, X.; Lu, Q.; Rodenas-Herraiz, D.; Parlikad, A.K.; Schooling, J.M. Visualised inspection system for monitoring environmental anomalies during daily operation and maintenance. *Eng. Constr. Archit. Manag.* **2020**, *27*, 1835–1852. [CrossRef]
72. Yin, X.; Liu, H.; Chen, Y.; Wang, Y.; Al-Hussein, M. A BIM-based framework for operation and maintenance of utility tunnels. *Tunn. Undergr. Space Technol.* **2020**, *97*, 103252. [CrossRef]
73. Costin, A.M.; Teizer, J.; Schoner, B. RFID and bim-enabled worker location tracking to support real-time building protocol control and data visualisation. *J. Inf. Technol. Constr.* **2015**, *20*, 495–517.
74. Fang, Y.; Cho, Y.K.; Zhang, S.; Perez, E. Case Study of BIM and Cloud-Enabled Real-Time RFID Indoor Localization for Construction Management Applications. *J. Constr. Eng. Manag.* **2016**, *142*. [CrossRef]
75. Park, J.W.; Chen, J.; Cho, Y.K. Self-corrective knowledge-based hybrid tracking system using BIM and multimodal sensors. *Adv. Eng. Inform.* **2017**, *32*, 126–138. [CrossRef]
76. Zhong, R.Y.; Peng, Y.; Xue, F.; Fang, J.; Zou, W.; Luo, H.; Ng, S.T.; Lu, W.; Shen, G.Q.P.; Huang, G.Q. Prefabricated construction enabled by the Internet-of-Things. *Autom. Constr.* **2017**, *76*, 59–70. [CrossRef]
77. Chen, K.; Xu, G.; Xue, F.; Zhong, R.Y.; Liu, D.; Lu, W. A Physical Internet-enabled Building Information Modelling System for prefabricated construction. *Int. J. Comput. Integr. Manuf.* **2018**, *31*, 349–361. [CrossRef]
78. Li, C.Z.; Xue, F.; Li, X.; Hong, J.; Shen, G.Q. An Internet of Things-enabled BIM platform for on-site assembly services in prefabricated construction. *Autom. Constr.* **2018**, *89*, 146–161. [CrossRef]
79. Tagliabue, L.C.; Ciribini, A.L.C. A BIM Based IoT Approach to the Construction Site Management. *Bo-Ricerche E Progett. Per Territ. La Citta E L Archit.* **2018**, *9*, 136–145.
80. Hamooni, M.; Maghrebi, M.; Sardroud, J.M.; Kim, S. Extending BIM Interoperability for Real-Time Concrete Formwork Process Monitoring. *Appl. Sci.* **2020**, *10*, 1085. [CrossRef]
81. Pan, Y.; Zhang, L. A BIM-data mining integrated digital twin framework for advanced project management. *Autom. Constr.* **2021**, *124*, 103564. [CrossRef]
82. Hernández, J.L.; Lerones, P.M.; Bonsma, P.; van Delft, A.; Deighton, R.; Braun, J.D. An IFC interoperability framework for self-inspection process in buildings. *Buildings* **2018**, *8*, 32. [CrossRef]
83. Martinez, P.; Ahmad, R.; Al-Hussein, M. A vision-based system for pre-inspection of steel frame manufacturing. *Autom. Constr.* **2019**, *97*, 151–163. [CrossRef]
84. Atazadeh, B.; Olfat, H.; Rismanchi, B.; Shojaei, D.; Rajabifard, A. Utilising a building information modelling environment to communicate the legal ownership of internet of things-generated data in multi-owned buildings. *Electronics* **2019**, *8*, 1258. [CrossRef]
85. Moretti, N.; Xie, X.; Merino, J.; Brazauskas, J.; Parlikad, A.K. An openbim approach to iot integration with incomplete as-built data. *Appl. Sci.* **2020**, *10*, 8287. [CrossRef]
86. Shiau, Y.-C.; Tsai, Y.-Y.; Hsiao, J.-Y.; Chang, C.-T. Development of building fire control and management system in BIM environment. *Stud. Inform. Control.* **2013**, *22*, 15–24. [CrossRef]

87. Arslan, M.; Riaz, Z.; Kiani, A.K.; Azhar, S. Real-Time Environmental Monitoring, Visualisation And Notification System For Construction H&S Management. *J. Inf. Technol. Constr.* **2014**, *19*, 72–91. Available online: http://www.itcon.org/2014/4 (accessed on 20 February 2021).
88. Li, N.; Becerik-Gerber, B.; Krishnamachari, B.; Soibelman, L. A BIM centered indoor localisation algorithm to support building fire emergency response operations. *Autom. Constr.* **2014**, *42*, 78–89. [CrossRef]
89. Riaz, Z.; Arslan, M.; Kiani, A.K.; Azhar, S. CoSMoS: A BIM and wireless sensor based integrated solution for worker safety in confined spaces. *Autom. Constr.* **2014**, *45*, 96–106. [CrossRef]
90. Cheng, M.Y.; Chiu, K.C.; Hsieh, Y.M.; Yang, I.T.; Chou, J.S.; Wu, Y.W. BIM integrated smart monitoring technique for building fire prevention and disaster relief. *Autom. Constr.* **2017**, *84*, 14–30. [CrossRef]
91. Park, S.; Hong, C. Roles and scope of system interface in integrated control system for multi disaster countermeasure. *Int. J. Saf. Secur. Eng.* **2017**, *7*, 361–366. [CrossRef]
92. Park, J.; Kim, K.; Cho, Y.K. Framework of Automated Construction-Safety Monitoring Using Cloud-Enabled BIM and BLE Mobile Tracking Sensors. *J. Constr. Eng. Manag.* **2017**, *143*. [CrossRef]
93. Riaz, Z.; Parn, E.A.; Edwards, D.J.; Arslan, M.; Shen, C.; Pena-Mora, F. BIM and sensor-based data management system for construction safety monitoring. *J. Eng. Des. Technol.* **2017**, *15*, 738–753. [CrossRef]
94. Beata, P.A.; Jeffers, A.E.; Kamat, V.R. Real-Time Fire Monitoring and Visualization for the Post-Ignition Fire State in a Building. *Fire Technol.* **2018**, *54*, 995–1027. [CrossRef]
95. Chou, J.S.; Cheng, M.Y.; Hsieh, Y.M.; Yang, I.T.; Hsu, H.T. Optimal path planning in real time for dynamic building fire rescue operations using wireless sensors and visual guidance. *Autom. Constr.* **2019**, *99*, 1–17. [CrossRef]
96. Parn, E.A.; Edwards, D.; Riaz, Z.; Mehmood, F.; Lai, J. Engineering-out hazards: Digitising the management working safety in confined spaces. *Facilities* **2019**, *37*, 196–215. [CrossRef]
97. Yu, G.; Mao, Z.; Hu, M.; Li, Z.; Sugumaran, V. BIM+ Topology Diagram-Driven Multiutility Tunnel Emergency Response Method. *J. Comput. Civ. Eng.* **2019**, *33*, 1–19. [CrossRef]
98. Asadzadeh, A.; Arashpour, M.; Li, H.; Ngo, T.; Bab-Hadiashar, A.; Rashidi, A. Sensor-based safety management. *Autom. Constr.* **2020**, *113*, 103128. [CrossRef]
99. Cheng, J.C.P.; Chen, K.; Wong, P.K.-Y.; Chen, W.; Li, C.T. Graph-based network generation and CCTV processing techniques for fire evacuation. *Build. Res. Inf.* **2021**, *49*, 179–196. [CrossRef]
100. Lei, Y.; Rao, Y.; Wu, J.; Lin, C.H. BIM based cyber-physical systems for intelligent disaster prevention. *J. Ind. Inf. Integr.* **2020**, *20*, 100171. [CrossRef]
101. Lin, S.S.; Shen, S.L.; Zhou, A.; Xu, Y.S. Risk assessment and management of excavation system based on fuzzy set theory and machine learning methods. *Autom. Constr.* **2021**, *122*, 103490. [CrossRef]
102. Tian, W.; Meng, J.; Zhong, X.J.; Tan, X. Intelligent Early Warning System for Construction Safety of Excavations Adjacent to Existing Metro Tunnels. *Adv. Civ. Eng.* **2021**, *2021*, 8833473. [CrossRef]
103. Goekce, H.U.; Goekce, K.U. Holistic system architecture for energy efficient building operation. *Sustain. CITIES Soc.* **2013**, *6*, 77–84. [CrossRef]
104. Osello, A.; Acquaviva, A.; Aghemo, C.; Blaso, L.; Dalmasso, D.; Erba, D.; Fracastoro, G.; Gondre, D.; Jahn, M.; Macii, E.; et al. Energy saving in existing buildings by an intelligent use of interoperable ICTs. *Energy Effic.* **2013**, *6*, 707–723. [CrossRef]
105. Cheng, J.C.P.; Das, M. A bim-based web service framework for green building energy simulation and code checking. *J. Inf. Technol. Constr.* **2014**, *19*, 150–168. Available online: https://www.scopus.com/inward/record.uri?eid=2-s2.0-84907376890&partnerID=40&md5=2f62f7e068a13717acf819b1e0948d62 (accessed on 20 February 2021).
106. Gökçe, H.U.; Gökçe, K.U. Integrated system platform for energy efficient building operations. *J. Comput. Civ. Eng.* **2014**, *28*, 1–11. [CrossRef]
107. Goekce, H.U.; Goekce, K.U. Multi dimensional energy monitoring, analysis and optimisation system for energy efficient building operations. *Sustain. CITIES Soc.* **2014**, *10*, 161–173. [CrossRef]
108. Kensek, K.M. Integration of Environmental Sensors with BIM: Case studies using Arduino. *Dynamo Revit API Inf. La Constr.* **2014**, *66*. [CrossRef]
109. Ness, D.; Swift, J.; Ranasinghe, D.C.; Xing, K.; Soebarto, V. Smart steel: New paradigms for the reuse of steel enabled by digital tracking and modelling. *J. Clean. Prod.* **2015**, *98*, 292–303. [CrossRef]
110. Zhao, W.; Liang, Y. Energy-Efficient and Robust In-Network Inference in Wireless Sensor Networks. *IEEE Trans. Cybern.* **2015**, *45*, 2105–2118. [CrossRef] [PubMed]
111. Habibi, S. Smart innovation systems for indoor environmental quality (IEQ). *J. Build. Eng.* **2016**, *8*, 1–13. [CrossRef]
112. Lee, D.; Cha, G.; Park, S. A study on data visualisation of embedded sensors for building energy monitoring using BIM. *Int. J. Precis. Eng. Manuf.* **2016**, *17*, 807–814. [CrossRef]
113. Habibi, S. Micro-climatization and real-time digitalisation effects on energy efficiency based on user behaviour. *Build. Environ.* **2017**, *114*, 410–428. [CrossRef]
114. McGlinn, K.; Yuce, B.; Wicaksono, H.; Howell, S.; Rezgui, Y. Usability evaluation of a web-based tool for supporting holistic building energy management. *Autom. Constr.* **2017**, *84*, 154–165. [CrossRef]
115. Dave, B.; Buda, A.; Nurminen, A.; Främling, K. A framework for integrating BIM and IoT through open standards. *Autom. Constr.* **2018**, *95*, 35–45. [CrossRef]

116. Francisco, A.; Truong, H.; Khosrowpour, A.; Taylor, J.E.; Mohammadi, N. Occupant perceptions of building information model-based energy visualisations in eco-feedback systems. *Appl. Energy.* **2018**, *221*, 220–228. [CrossRef]
117. Pasini, D. Connecting BIM and IoT for addressing user awareness toward energy savings. *J. Struct. Integr. Maint.* **2018**, *3*, 243–253. [CrossRef]
118. Kang, T. Bim-based human machine interface (Hmi) framework for energy management. *Sustainability* **2020**, *12*, 8861. [CrossRef]
119. Mataloto, B.; Mendes, H.; Ferreira, J.C. Things2people interaction toward energy savings in shared spaces using BIM. *Appl. Sci.* **2020**, *10*, 5709. [CrossRef]
120. Hirakraj, B.; Debasis, S.; Rajesh, G. Application of integrated fuzzy FCM-BIM-IoT for sustainable material selection and energy management of metro rail station box project in western India. *Innov. Infrastruct. Solut.* **2021**, *6*, 73. [CrossRef]
121. Sepasgozar, S.M.E.; Hui, F.K.P.; Shirowzhan, S.; Foroozanfar, M.; Yang, L.; Aye, L. Lean practices using building information modeling (Bim) and digital twinning for sustainable construction. *Sustainability* **2021**, *13*, 161. [CrossRef]
122. Tagliabue, L.C.; Cecconi, F.R.; Maltese, S.; Rinaldi, S.; Ciribini, A.L.C.; Flammini, A. Leveraging digital twin for sustainability assessment of an educational building. *Sustainability* **2021**, *13*, 480. [CrossRef]
123. Wang, X.; Love, P.E.D.; Kim, M.J.; Park, C.S.; Sing, C.P.; Hou, L. A conceptual framework for integrating building information modeling with augmented reality. *Autom. Constr.* **2013**, *34*, 37–44. [CrossRef]
124. Alves, M.; Carreira, P.; Costa, A.A. BIMSL: A generic approach to the integration of building information models with real-time sensor data. *Autom. Constr.* **2017**, *84*, 304–314. [CrossRef]
125. Edmondson, V.; Cerny, M.; Lim, M.; Gledson, B.; Lockley, S.; Woodward, J. A smart sewer asset information model to enable an 'Internet of Things' for operational wastewater management. *Autom. Constr.* **2018**, *91*, 193–205. [CrossRef]
126. Kang, K.; Lin, J.; Zhang, J. BIM- and IoT-based monitoring framework for building performance management. *J. Struct. Integr. Maint.* **2018**, *3*, 254–261. [CrossRef]
127. El Ammari, K.; Hammad, A. Remote interactive collaboration in facilities management using BIM-based mixed reality. *Autom. Constr.* **2019**, *107*, 102940. [CrossRef]
128. Kazado, D.; Kavgic, M.; Eskicioglu, R. Integrating Building Information Modeling (Bim) And Sensor Technology For Facility Management. *J. Inf. Technol. Constr.* **2019**, *24*, 441. Available online: http://www.itcon.org/2019/23 (accessed on 20 February 2021).
129. Rashid, K.M.; Louis, J.; Fiawoyife, K.K. Wireless electric appliance control for smart buildings using indoor location tracking and BIM-based virtual environments. *Autom. Constr.* **2019**, *101*, 48–58. [CrossRef]
130. Rogage, K.; Clear, A.; Alwan, Z.; Lawrence, T.; Kelly, G. Assessing building performance in residential buildings using BIM and sensor data. *Int. J. Build. Pathol. Adapt.* **2019**, *38*, 176–191. [CrossRef]
131. Saar, C.C.; Klufallah, M.; Kuppusamy, S.; Yusof, A.; Shien, L.C.; Han, W.S. Bim integration in augmented reality model. *Int. J. Technol.* **2019**, *10*, 1266–1275. [CrossRef]
132. Tsai, Y.-H.; Wang, J.; Chien, W.-T.; Wei, C.-Y.; Wang, X.; Hsieh, S.-H. A BIM-based approach for predicting corrosion under insulation. *Autom. Constr.* **2019**, *107*, 102923. [CrossRef]
133. Zhai, Y.; Chen, K.; Zhou, J.X.; Cao, J.; Lyu, Z.; Jin, X.; Shen, G.Q.P.; Lu, W.; Huang, G.Q. An Internet of Things-enabled BIM platform for modular integrated construction: A case study in Hong Kong. *Adv. Eng. Inform.* **2019**, *42*, 100997. [CrossRef]
134. Edirisinghe, R.; Woo, J. BIM-based performance monitoring for smart building management. *Facilities* **2020**, *39*, 19–35. [CrossRef]
135. Pavón, R.M.; Alvarez, A.A.A.; Alberti, M.G. BIM-based educational and facility management of large university venues. *Appl. Sci.* **2020**, *10*, 7976. [CrossRef]
136. Quinn, C.; Shabestari, A.Z.; Misic, T.; Gilani, S.; Litoiu, M.; McArthur, J.J. Building automation system—BIM integration using a linked data structure. *Autom. Constr.* **2020**, *118*, 103257. [CrossRef]
137. Valinejadshoubi, M.; Moselhi, O.; Bagchi, A. Integrating BIM into sensor-based facilities management operations. *J. Facil. Manag.* **2021**. Ahead of Print. [CrossRef]
138. Zhang, Y.Y.; Kang, K.; Lin, J.R.; Zhang, J.P.; Zhang, Y. Building information modeling–based cyber-physical platform for building performance monitoring. *Int. J. Distrib. Sens. Netw.* **2020**, *16*, 2020. [CrossRef]
139. Gao, X.; Pishdad-Bozorgi, P.; Shelden, D.R.; Tang, S. Internet of Things Enabled Data Acquisition Framework for Smart Building Applications. *J. Constr. Eng. Manag.* **2021**, *147*, 04020169. [CrossRef]
140. Štefanič, M.; Stankovski, V. A review of technologies and applications for smart construction. *Proc. Inst. Civ. Eng. Civ. Eng.* **2018**, *172*, 83–87. [CrossRef]
141. Tang, S.; Shelden, D.R.; Eastman, C.M.; Pishdad-Bozorgi, P.; Gao, X. A review of building information modeling (BIM) and the internet of things (IoT) devices integration: Present status and future trends. *Autom. Constr.* **2019**, *101*, 127–139. [CrossRef]
142. Boje, C.; Guerriero, A.; Kubicki, S.; Rezgui, Y. Towards a semantic Construction Digital Twin: Directions for future research. *Autom. Constr.* **2020**, *114*, 103179. [CrossRef]
143. Panteli, C.; Kylili, A.; Fokaides, P.A. Building information modelling applications in smart buildings: From design to commissioning and beyond A critical review. *J. Clean. Prod.* **2020**, *265*. [CrossRef]
144. Hou, L.; Wu, S.; Zhang, G.K.; Tan, Y.; Wang, X. Literature review of digital twins applications in constructionworkforce safety. *Appl. Sci.* **2021**, *11*, 339. [CrossRef]

applied sciences

MDPI

Article

Building Information Modelling and Energy Simulation for Architecture Design

Marina Bonomolo [1,*], Simone Di Lisi [2] and Giuliana Leone [1]

[1] Department of Engineering, University of Palermo, 90128 Palermo, Italy; giuliana.leone@unipa.it
[2] Via Del Mazziere, 73, 90018 Termini Imerese, Italy; simone_di_lisi_95@libero.it
* Correspondence: marina.bonomolo@unipa.it

Abstract: Over the years, building information modelling (BIM) has undergone a significant increase, both in terms of functions and use. This tool can almost completely manage the entire process of design, construction, and management of a building internally. However, it is not able to fully integrate the functions and especially the information needed to conduct a complex energy analysis. Indeed, even if the energy analysis has been integrated into the BIM environment, it still fails to make the most of all the potential offered by building information modelling. The main goals of this study are the analysis of the interaction between BIM and energy simulation, through a review of the main existing commercial tools (available and user-friendly), and the identification and the application of a methodology in a BIM environment by using Graphisoft's BIM software Archicad and the plug-in for dynamic energy simulation EcoDesigner STAR. The application on a case study gave the possibility to explore the advantages and the limits of these commercial tools and, consequently, to provide some possible improvements. The results of the analysis, satisfactory from a quantitative and qualitative point of view, validated the methodology proposed in this study and highlighted some limitations of the tools used, in particular for the aspects concerning the personalization of heating systems.

Keywords: BIM; BEM; simulation modelling; dynamic simulation

Citation: Bonomolo, M.; Di Lisi, S.; Leone, G. Building Information Modelling and Energy Simulation for Architecture Design. *Appl. Sci.* **2021**, *11*, 2252. https://doi.org/10.3390/app11052252

Academic Editors: Jürgen Reichardt and Igal Shohet

Received: 27 January 2021
Accepted: 1 March 2021
Published: 4 March 2021

1. Introduction

The large-scale diffusion of building information modelling (BIM) tools for architecture has led to an enormous evolution of these digital means [1]. Today, it includes multiple functions capable of carrying out numerous analyses (e.g., structural, energy, metric-estimative, etc.) on a single virtual model of the building. In particular, some aspects, e.g., energy ones can be implemented by using a monitoring system connected to the BIM. Jen-tu and Vernatha [2] proposed an application of Building Information Modelling in establishing the 'BIM based Energy Management Support System' (BIM-EMSS) to assist individual departments within universities in their energy management tasks. They installed sensors for occupants and other equipment such as electricity sub-meters that constantly logging consumption, and developing BIM models of all rooms within individual departments' facilities, data warehouse, building energy management system that provides energy managers with various energy management functions, and energy simulation tools (such as eQuest). In addition, Alahmad et al. [3] integrated BIM with a Real Time Power Monitoring (RTPM) System, and Jeong-Han Woo et al. [4] presented a prototype of BIM-based Baseline Building Model (B3M) for ageing commercial buildings. When the aim is not the real-time monitoring and there is not an existing building or the possibility to install sensors and other devices, it is necessary to use energy simulation tools and connect them with the BIM. Although BIM software is technologically advanced and able to best meet the needs of professionals in various applications, the energy simulation function needs many improvements. Some researches started to investigate the topic of the "green dimension" [5]. Indeed, to optimize the design in terms of energy efficiency, it could be useful that the energy simulation phase be carried out in meantime with the development of the project [6] and to make choices based on intermediate calculations, and therefore constantly modify the project until the objectives are achieved [7]. To do this, it is often necessary to build another digital model, called BEM (Building Energy Model), or implement the BIM model with all the information needed for energy simulation (occupation of the rooms, calculation of thermal bridges, analysis based on hourly climate data, etc.) [8]. The main issue is the lack of bidirectional interaction between the two models. Furthermore, one of the main issues is data transmission. For this reason, BIM for energy simulation is often used only for early design step [9]. The aim should be to work with a unique model [10]. In last decade, many researchers presented studies on BIM's application in energy analysis and simulation [11] and proposed solutions for the interoperability between BIM and energy simulation tools [12]. Some of them developed tools to integrate BIMs. Bratch et al. [13] evaluated the possibility of integrating a thermal load prediction metamodel to building information models to facilitate the data exchange process. To do this, they developed a tool to validate the viability of this integration using gbXML, and it was submitted to validation tests. Ramaji et al. developed an extension for OpenStudio able to transform building information models represented in Industry Foundation Classes (IFC) files into building energy analysis models in the OpenStudio data format. Kamel et al. [14] developed and presented a new tool called ABEMAT (Automated Building Energy Modelling and Assessment Tool), able to make the building energy simulation using BIM automatic and to give fine-grained outputs. It receives a gbXML file and provides users with the amount of heat transfer through each building envelope element.

Kim et al. [15] developed and validated a library for BIM for the building energy simulation (ModelicaBIM library) using an Object-Oriented Physical Modeling (OOPM) in the scope of the building envelope.

A design-decision-supporting tool for the conceptual phases of design and throughout the design process based on a BIM template has been designed by El Sayary and Omar [16]. In particular, the aim was to develop a simple tool to calculate how many photovoltaic solar panels can be installed to reach a zero-energy building by substituting all electric devices. Xu et al. [17] investigated the application of BIM for addressing the building energy performance gap. The authors provided a clear set of guidelines for how BIM could be used, by each function, to overcome the BEPG to reduce global emissions driven from the building and construction sector. These cited studies provided good outcomes and results. Nevertheless, a goal of this study is to test a commercial existing tool, easy to be found and to be used. Other authors applied existing commercial tools.

Additionally, Utkucun and Sözer [18] proposed a method to determine the interoperability of the utilized programs for evaluating a building's energy performance and indoor comfort through the BIM approach. To do this, they built three main analysis models: An architecture of the building (with the 3D building model), indoor comfort conditions (with the computational fluid dynamics for natural ventilation simulation), and energy performance (with a building energy model specified by the building architecture and its systems). Then, they integrated them through a BIM platform. In order to investigate the potential and limitations of applying BIM to energy management and simulation in the operation lifecycle phase of a service building, Rodrigues et al. [19] developed a service building BIM model in Autodesk Revit and used Energy Analysis for Autodesk Revit that automatically generated the Building Energy Model (BEM) from the BIM model and performed a cloud-based simulation in Autodesk Green Building Studio (GBS). They found input limitations of GBS, mainly in HVAC systems customization, compromise the representation and energy performance evaluation of the building under actual operating conditions. For this reason, they affirmed that GBS is more adequate for early buildings' lifecycle stages where energy simulation results may support decisions that aim to improve the buildings' energy performance during the operation phase.

Tushar et al. [20] used the software Autodesk Revit together with the energy rating tool (FirstRate5), and BIM-enabled life cycle assessment (LCA-Tally) to quantify, compare, and improve the building design options to reduce carbon footprint and energy consumptions in residential dwellings. Alam and Ham [21] compared FirstRate5 with Archicad EcoDesigner developing three building types. They found significant differences between simulations, being, measured areas, thermal loads, and potentially serious shortcomings within FirstRate5, that were discussed along with the future potential of a fully BIM-integrated model for energy rating certification in Victoria. In Farzad Jalaei and Ahmad Jrade's study [22], an integrated method that links BIM, energy analysis, and cost estimating tools with a green building certification system was presented. The aim was to calculate the potential gain or loss of energy for the building and to evaluate its sustainability based on the US and/or Canadian Green Building Council.

Right now, the data transmission between BIM and BEM is possible through two types of file format: IFC (Industry Foundation Classes) and gbXML (Green Building Extensible Markup Language). Both have important advantages. The IFC format is the standard format for information exchange in BIM modelling and is the only format to have a certification. It is possible to load most of the information concerning a building except for the energy analysis data, such as occupancy profiles, data relating to external and internal temperatures, systems, etc. The gbXML is a format based on IFC, but containing all the information of an energetic nature. It was developed to operate in this field, and it is the most popular among energy analysis software. Indeed, some software have been imported only in this format to the detriment of the IFC (IES-VE [23], EnergyPlus [24], eQUEST [25]). However, in a review paper, Gao et al. [26] investigated the data transfer between BIM to building energy modelling and they found that the development of BIM-based building energy modelling is still at the initial stage and few methods can be guaranteed to generate reliable building energy models from building information models without errors.

In this light, it has to be remembered that the choice of the tool is very important because, based on the software used, working times can increase or be reduced. On the other hand, it is possible that once the BIM model of a building is completed, it is exported to software for energy analysis, and this does not interpret the geometries well or even fails to import a lot of information previously entered. It makes the work useless. In this case, the technician has to manually enter each missing item or in the worst case, he/she has to build a model specifically for the energy analysis. This procedure can be more or less long depending on the building under consideration. It certainly differs from the BIM aim that provides a faster and interoperable workflow by means of a single model.

The most common choice in the architectural-energy field is to proceed with the architectural design and postpone the energy analysis at the end. It provides a separate and specific calculation, and this process often leads to a meaningless analysis as this is ascertained.

The process to be adopted must be the opposite: Support the design with energy analysis from the early stages (which by its nature will be a summary, as you will not yet have all the parameters to be able to make a more complex analysis) and direct the project towards a more eco-sustainable way. In this light, obviously, a real transition, from BIM to BEM is required. It has to be managed by single software that is based on the digital model that is being built, and which automatically updates the BEM on this basis, allowing for a reduction in work-time and a more energy-saving design.

In this study, a methodology based on the full interaction between Graphisoft's BIM software Archicad and the plug-in for dynamic energy simulation EcoDesigner STAR was tested. They have been selected after a study of the rules on energy analysis, an examination of the operational potential of different software on the market, and a research conducted by a wide scientific community interested in various capacities in issues related to the interaction between architecture and energy analysis. In particular, the choice has been based on some main criteria such as versatility, i.e., the presence of integrated functions that allow BIM and BEM modelling, compliance with standards (the software or plug-in for energy analysis must meet the requirements set by the most advanced energy diagnosis standards, such as UNI EN ISO 52016-1 for the calculation in dynamic hourly regime, ASHRAE 140-2017 and UNI/TS 11300 on the monthly average stationary calculation and certification required by buildingSMART for IFC certification; and versatility thanks to the presence of integrated functions that allow BIM modelling and BEM modelling.

The main goals are the analysis of the interaction between BIM and energy simulation and the identification of a methodology that allows overcoming the above-mentioned limits. With this purpose, Graphisoft's BIM software Archicad and the plug-in for dynamic energy simulation, EcoDesigner STAR, were used. They have been selected after a careful study of the rules on energy analysis, an examination of the operational potential of different software on the market, and research conducted by a wide scientific community interested in various capacities in issues related to the interaction between architecture and energy analysis. Thanks to the state-of-art analysis, the advantages and the disadvantages of the existing tools were highlighted and compared. The application of the selected tool on an existing case study gave the possibility to further study and to test the methodology. Finally, the analysis of the lacks suggested some improvements that can be done.

2. Methodology of the Study

This study *starts* from the idea that BEM can be useful for energy analysis building, but there is some issue due to the lack of bidirectional interaction between BIM and BEM, as *observed* in literature.

This paper aims to study a method to carry out energy analysis using these tools. To do this, it was *hypothesized* to test a combination of two tools (selected after an investigation of some existing tools doing *background research*). They were used to perform the *experiment* (test) that provides information and data. These latter were analyzed and interpreted to formulate the *results* by including advantages and disadvantages to performing an energy analysis by using the selected tools. In *conclusion*, the study shows that it is possible to perform energy analysis through the use of BIM, but it needs some improvements.

The research was conducted in several steps. In the first phase, some of the most common new design tools and the most recent standards for energy analysis were studied. From this first analysis, it was possible to list the disadvantages and the advantages. Once a tool was selected that matched the required characteristics, it was validated using a simple building. Then, an existing building was chosen as a case study to apply the energy analysis methodology in a BIM environment. It was defined through the application of a test model. Figure 1 shows the workflow diagram.

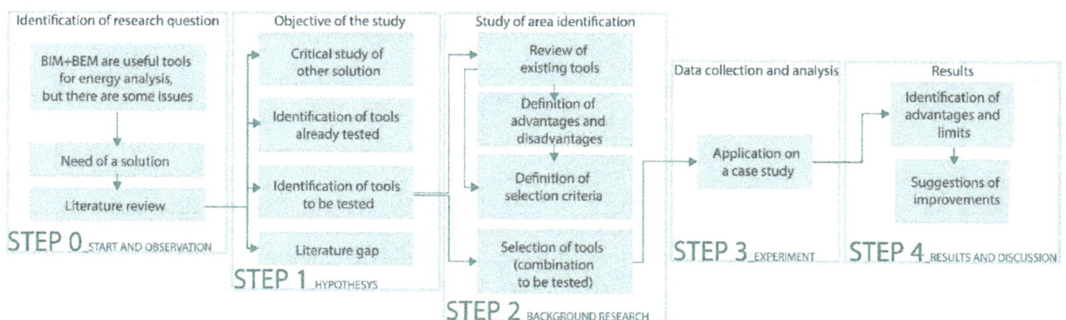

Figure 1. Workflow diagram with research framework.

The studies conducted in the first phase highlighted one of the main problems that hindered the diffusion and use of energy analysis plug-ins in the BIM environment, namely the lack of integration between the parametric architectural model with information content (BIM, Building Information Model) and the energy model (BEM, Building Energy Model). The problem is the possible loss of information during the conversion from BIM to BEM, or worse, the lack of interpretation of the entire model. It is clear that this problem diverges the actual workflow from the characteristic one of BIM, based on the principle of interoperability, i.e., the possibility that allows different professionals to work and exchange information on a single digital model.

As said, the lack of integration between BIM and BEM means that, during the energy analysis phase, it is necessary to build a new model or implement the Building Information Model with new information not managed by it.

Once the objective of the study was defined, we moved on to research and the choice of software suitable for supporting a working methodology that allows full integration between BIM and BEM already in the design phase. Various digital tools for energy analysis were evaluated, and the choice of the software used in this study was determined by the evaluation of some of its specific characteristics, such as the validity of the calculations (with reference to current legislation) and certifications in the BIM field. To verify the validity of the calculations, a sample building with characteristics suitable for evaluating the effectiveness of the software was simulated.

The last phase of the study was dedicated to applying the digital tool and the methodology defined for the sample model to an existing building. The choice was determined by the value and also by the simplicity of the building, which allows you to control the result of the simulation process more precisely. The simplicity to which reference is made does not concern the spatial and architectural quality of the building or housing, but rather the relationships with the ground and the characteristics of use that allow a more effective control of the results of the BEM energy analysis. The chosen building, in fact, has simple thermal zones, with no particularities (of use or construction) that could alter the calculation of the energy simulation. Finally, based on the detection of the limits and the lacks, some further possible improvements of the tool have been suggested.

3. From BIM to BEM: Advantages and Limitations

The study and the implementation of methods to transform a "BIM" to a "BEM" are a topic always more common in the scientific community. Indeed, BEM has a large number of applications in the most varied cases of energy analysis. Moreover, as said, the available tools have limitations (sometimes highly restrictive). For this reason, technicians are not encouraged to use them. The main identified problem is the transmission of data between the BIM modelling tools and the energy simulation tools. It limits the possibility to operate with the least possible number of digital models.

The main formats are:

- The IFC is the standard format for exchanging information in BIM modelling. It is the only format to have a certification. By using it, it is possible to upload most of the information relating to a building with the exception of data relating to energy analysis (such as occupancy profiles, data relating to external and internal temperatures, systems, etc.);
- the gbXML is a format based on the IFC; it contains all the energy information. It was developed specifically to operate in this field and it is the most popular among energy analysis software. Indeed, some software only allows the import of this format to the detriment of IFC (IES-VE, EnergyPlus, eQUEST).

It is therefore understood that, based on the software used, working times can increase or be reduced to a minimum. It is possible that, once the BIM model of a building is completed, it is exported to software for energy analysis. Sometimes, it happens that it does not interpret the geometries well or even fails to import a lot of information previously entered, effectively making the work already done useless. In this case, the technician has to manually enter each missing item or, in the worst case, he/she has to build a model specifically for the energy analysis. This procedure can be more or less long depending on the building under consideration. This certainly differs from the BIM methodology aim. Indeed, it provides for a faster and interoperable workflow by means of a single model. The most commonly used choice in the architectural-energy field is to proceed with the architectural design and postpone the energy analysis at the end with a separate and specific calculation. Often this process leads to a meaningless analysis as this is ascertained, which was designed with no room for improvement [26]. Obviously, the process to be adopted should be the opposite: Support the design with energy analysis from the early stages (which by its nature will be a summary, as you will not yet have all the parameters to be able to make a more complex analysis) and direct the project towards a more eco-sustainable way. With this in mind, obviously, a real transition from BIM to BEM is required, instantaneous and managed by a single program. It is based on the digital model that is being built. Consequently, the BEM should be automatically updated. This procedure should reduce the time, the energy spent to work, and should allow a more energy-saving design.

4. Short Review on Existing Tools

The quality of the product, the possibilities offered, and the use the user must do influence the diffusion of a certain program. The most famous and common software for BIM are Revit (Autodesk, San Rafael, United States) [27] and ArchiCAD (Graphisoft, Budapest, Ungheria) [28] and then, Allplan (Nemetschek, Munich, Germany) [29], and Edificius [30] (Acca Software, Avellino, Italy) [31]. The tools included in these software are quite similar. They are equivalent to many functions and just a few advantages are different. In all cases, as regards energy diagnosis, the use of one of these tools often is not suitable.

The choices on the market are different, each with its peculiarities, therefore the main characteristics of each software are examined below, giving precedence to the most common BIM software and related plug-ins and then to independent programs.

4.1. Revit

Revit [27] is one of the most popular BIM software, both for its performance and for its compatibility with other programs widely used in the construction sector (also produced by Autodesk). It must be specified in this regard that Revit has not developed great connectivity with software that is not part of the Autodesk suite. It is possible to find compatibility problems even if almost all manufacturers try to interface as much as possible with Revit. Energy diagnosis is allowed through an additional module, Energy Analysis, which integrates the design features of Revit with the analysis features of Autodesk Green Building Studio, an independent cloud service for energy diagnosis based on the DOE-2.2 simulation engine [25] (which complies with the ASHRAE 140-2007 [32] standard).

4.2. ArchiCAD

ArchiCAD [28], developed by the US company Graphisoft, is one of the two most popular software for BIM design and has the IFC certification of buildingSMART. The energy diagnosis can be carried out both by functions integrated into the program and by a plug-in: EcoDesigner STAR. This latter is an integral part of the program itself. Its calculation engine (VIP-Core by StruSoft) operates in compliance with the ASHRAE 140-2007 and ASHRAE 90.1-2007 (LEED Energy) standards. Therefore, it operates in a dynamic regime. The main novelty of the plug-in is the integration of the missing tools in the package of standard tools (such as the calculation of thermal bridges or renewable energy) and the ability to export files in .gbXML and .PHPP format, for easier collaboration between professionals and technicians.

4.3. Allplan

Allplan [29], developed by Nemetschek, is the leading BIM-based software used in Germany. It is among the programs certified by buildingSMART. Regarding the energy functions, it does not have sufficient tools to conduct a correct simulation. So, in 2009, it was implemented with a new module: AX-Energy. This module integrates the software tools allowing it to carry out energy analyses according to Decrees 311/2006 [33] and 115/2008 [34] and UNI/TS 11300-1 [35] and 2 [36] standards, thus relying on an almost stationary, rather than the dynamic, regime.

4.4. Edificius

Edificius [31] is a software produced by ACCA Software. It is the only Italian program to have received the buildingSMART IFC certification. The construction of a BIM model is accompanied, through an external program by the same company, by the construction of a BEM model for the energy analysis of the architectural building. TerMus used the EnergyPlus energy simulation engine based on the ASHRAE 140-2007 standards allowing analysis in a dynamic regime. Unluckily, it is necessary to install a series of modules, each with its own specific function (for example TerMus-PT calculates thermal bridges and mold risk, TerMus-DIM deals with energy diagnosis and improvement interventions, TerMus-PLUS for dynamic calculation and so on).

4.5. Design Builder

Design Builder [37] is an independent program based on the EnergyPlus simulation engine (it is its graphic interface), capable of analyzing a building under dynamic conditions from the energy point of view. The software has 3D modelling tools, but it is still possible to import into it a model built with an external program compressed in .gbXML format. Since it is not a software used purely for parametric modelling, it does not have the buildingSMART IFC certification.

4.6. Open Studio

Open Studio [38] is another graphical interface of the EnergyPlus simulation engine. It is available as a plug-in for the 3D modeling program SketchUp, with the particularity of being free. Being a SketchUp plug-in, its modelling is more intuitive than other programs with the same function (the IFC certificate is missing), while the analysis tools are not as intuitive as those of other software.

4.7. Simergy

Simergy [39] was developed as an independent program. It uses the EnergyPlus simulation engine, thus operating at a dynamic speed. Its user-friendly graphic interface is particularly effective for its use as a calculation tool, while the 3D modelling integrated in it is not easy and immediate to use. A peculiarity of the software is the possibility of comparing different project hypotheses with relative analyses. Additionally, in this case, since it is not a tool for parametric modelling, the IFC certification is absent.

4.8. TermoLOG

TermoLOG [40], by Logical Soft, is independent software that integrates parametric modelling tools and energy diagnosis. There is the possibility to import models in IFC format built with other programs. As a parametric modelling tool, it does not have buildingSMART certification. According to the standards dictated by UNI EN ISO 52016 [41], and validated by the Politecnico di Milano according to ASHRAE 140-2017 [32], it operates with a dynamic hourly engine (CENED + 2.0).

4.9. EcoDesigner Star

EcoDesigner Star [42] is a plug-in integrated into the ArchiCAD software. It is a graphical interface of the VIP-Core calculation engine optimized to work in harmony with the design tool in a BIM environment. This ArchiCAD extension was created with the aim of facilitating the design of buildings, directing them immediately towards more sustainable solutions. It is therefore a design tool and it does not allow the certification of buildings according to Italian standards. So, in this case, certification must be carried out using analysis software mainly dedicated to it. However, it is specified that the software calculations are not to be considered incorrect or non-compliant with current standards. In fact, they are based on data and specific parameters relating to the ASHRAE standard, and in the input phase, these parameters can be modified in order to obtain results in line with current legislation. It is not possible to carry out immediately and automatically to check required by the regulations.

The novelty proposed with EcoDesigner STAR is to have, within a software in a BIM environment, a powerful energy analysis tool. It is possible to work on a single Building Information Model and transform it almost instantly into a Building Energy Model ready to be analyzed. Moreover, it is also possible to orient the design of a building towards more sustainable ways both from the point of view of energy consumption and that of energy production through renewable sources. From the early stages of the project, an energy analysis can be obtained by defining the parameters necessary for the calculation. Therefore, an overview of the performance of the building and guide the designer towards the best-integrated design solution can be carried out. Using this tool, the path to be explored is established, and after having decided all the details of the building envelope and systems, it is possible to proceed with a further analysis, this time more detailed, to know the behavior of the building through a calculation dynamic, on an hourly basis. The integrated plug-in has many advantages, e.g., the possibility to manage the workflow in an optimal manner, and guarantees, both in terms of parameter input and in the output phase, high versatility and a high degree of data customization. EcoDesigner STAR, through the tools already present in ArchiCAD and connected to it, in the input phase allows you to:

- To build a BEM model from the BIM model, with the definition of thermal zones. They are automatically detected, as well as all the constructive elements that delimit them (walls, floors, doors, windows, beams, pillars, etc.); they can be viewed in 3D, both as an overall volume and as specific elements with their properties (e.g., by selecting a wall, it is possible to access all the parameters that define it, including the thermo-physical properties and orientation);
- to select built materials, with related thermophysical properties, from a large catalogue, or insert new ones with customized parameters, while maintaining the possibility of changing the assigned parameters; these changes are automatically sent to the BEM;
- to geo-localize the building by entering its geographical coordinates, to set the north and the elevation with respect to the sea level with the identification (automatic or manual) of the quote 0; the information entered can be verified through a link with Google Maps that indicates the position just defined;
- to define precisely both the surroundings of the building, by entering the type of land with its thermo-physical properties, the wind and sun shields present, the climatic data, such as air temperature, relative humidity, solar radiation, and wind speed; they can be displayed graphically through monthly, weekly, daily, or hourly charts; (downloading them directly from the dedicated server);
- to add thermal blocks to defined thermal zones that can be inserted; they are characterized not only by the zones, but also by heating, cooling, and ventilation systems, and by operation profiles. The systems that can be inserted are different, all already present in a plug-in catalogue, but new ones can be added based on the type of those present; the operation profiles are also already present in good quantity in the plug-in library, but new ones can be added, customizing them in each of their parameters;
- to define the energy vectors and their costs;
- to calculate according to a finite element approach (FEM) the thermal bridges present in the construction. It is possible using the "Detail" tool of ArchiCAD, which extracts a 2D drawing from a plan or section of the project. In addition, it is possible to make changes both in terms of geometry and materials and finally calculate the thermal bridges through a special window, saving both the numerical data and the temperature or heat flow graphs. Each thermal bridge can then be connected to the thermal block to which it refers;
- to perform a solar analysis on each frame of the building with the creation of a graph; it allows us to understand when the frame in question is exposed to the sun and in what percentage;

In the output phase, it is possible to obtain:

- An illustrative report of the performed energy analysis. It includes a first part that includes all the data common to the entire project (e.g., general graphs on the building's consumption and energy inputs);
- an excel spread sheet containing, in an extensive and detailed way, all the information that makes up the report;
- a file format different from the ArchiCAD ones, such as .gbXML, PHPP, and a format compatible with VIP-Energy, for exporting the Building Energy Model within other programs for energy analysis or certification. It is also possible to export the project as a "reference building" in order to make a comparison, during the energy simulation phase, between two alternatives of the same project.

4.10. Selection Criteria and Choice of Software

The criterion that led to the choice of the specific software is based on the following characteristics:

- Versatility, i.e., the presence of integrated functions that allow BIM modelling and BEM modelling; the software for BIM modelling can be combined with a well-integrated plug-in for energy analysis. In this way, it is possible to import and export building 2D drawings, BIM modelling and 3D visualization, quasi-static energy diagnosis, dynamic energy diagnosis, calculation of thermal bridges, and calculation of renewable energy sources. The importance of this parameter is given by the significant limitation of errors and simplifications that may arise from the management of the model with different software;
- certification: The BIM software must meet the validity requirements required by buildingSMART for IFC certification;
- in compliance with standards: The software or plug-in for energy analysis meets the requirements of the most advanced energy diagnosis standards, such as UNI EN ISO 52016-1 [41] for the calculation in dynamic hourly regime, and ASHRAE 140-2017 [32] and UNI/TS 11300 [35] on the monthly average stationary calculation.

The following table shows the BIM and BEM software and plug-ins already mentioned in the previous paragraph. For each of them, the greater or lesser compliance with the criteria described above is reported.

Table 1 shows the comparison of the eight examined software.

The first important difference concerns the IFC certifications. Indeed, it can be seen that only the BIM software combined with the plug-in has the third requirement (e.g., the buildingSMART certification).

The stand-alone BEM software, even if equipped with tools for the construction of a BIM model, cannot match the software dedicated to BIM in terms of functions, interoperability, and complexity. The adoption of the BIM Software + plug-in combination can guarantee a faster workflow. It is free from possible simplifications or misinterpretations of the data, resulting from exporting to external software from a different software house. Moreover, regarding the compliance of the software with the parameter relating to compliance with current legislation, many of them do not operate according to the Italian guidelines for stationary and dynamic calculation. Nevertheless, the 8 software examined comply (except for Allplan) with the ASHRAE 140-2007 standards relating to the validity of the calculation adopted. This does not mean that the calculation tools are wrong, but that these programs can only be used for energy diagnosis. For the compilation of energy performance certificates (APE) and other certification documents, different software has to be used. From the examination of these two parameters, the choice can be restricted to a more limited number of software. The BIM + plug-in software solutions that meet the requirements include:

- Termus (even if the latter is not really a plug-in but a program to complement the first and completely compatible), which allows you to produce documents valid in Italy;
- ArchiCAD + Ecodesigner STAR and Revit + Energy Analysis, which have a similar range of features, but do not produce documents conforming to Italian standards.

Table 1. Comparison of eight examined tools.

	Tool	Versatility					In Compliance with Standard			IFC Certification (buildingSMART)
		2D Import	3D/IFC Import	Energy Diagnosis	Thermal Bridges	Renewable Energy	UNI/TS 11300	EN ISO 52016-1	ASHRAE 140-2007	
SoftwareBIM + plug-in	Allplan + AX-Energia	✓	✓		✓	✓	✓			✓
	ArchiCAD + EcoDesigner STAR	✓	✓	✓	✓	✓			✓	✓
	Edificius + Termus	✓	✓	✓	✓	✓	✓	✓	✓	✓
	Revit + Energy Analysis	✓	✓		✓	✓			✓	✓
Stand-alone software	Design Builder	✓	✓	✓	✓	✓	✓	✓	✓	
	Open studio	✓	✓	✓			✓		✓	
	Simergy	✓	✓	✓			✓		✓	
	Termolog	✓	✓	✓	✓	✓	✓	✓	✓	

Among the stand-alone plug-in, TermoLOG appears to be the more complete than the competitors Design Builder, Open Studio, and Simergy, based on the EnergyPlus calculation engine, and therefore quite similar. Therefore, the main parameter to choose it is the versatility, or the presence of all functions in a single work environment, in order to limit the use of other software.

The versatility suggested adopting an integrated BIM + BEM plug-in solution. It offers a much larger package of features and greater interoperability than the stand-alone solution offered by TermoLOG. Furthermore, Revit + Energy Analysis package is not the best choice according to its versatility. Indeed, it needs to be accompanied by other plug-ins (such as Insight for solar analysis) to work. Finally, for these reasons, the ArchiCAD + Ecodesigner STAR combination has been selected. The parametric modelling of ArchiCAD has been implemented and updated. The model can guarantee full compatibility with the EcoDesigner Star plug-in, making the transition from the BIM model to the BEM model almost instantaneous. The adopted solution solves one of the most common problems in the integration between energy and BIM. Indeed, if the BIM model is imported and interpreted without errors or excessive simplifications, the BEM can be built and obtained by simply enriching the information present in BIM from the Energy Analysis Program. If the import/verification step does not take place correctly, it will be necessary to perform a specific BEM modelling. It causes longer time of work and more effort by the designer. Thus, it nullifies the advantages of the BIM workflow. If the model was correctly set, the ArchiCAD + Ecodesigner STAR solution automatically performs the transition from the BIM model to the BEM model. It has the great advantage of not losing any information present in the BIM and recording in real time in the BEM model all the changes made to the BIM model. It enhances the aim of BIM design.

These considerations can be summarized in graphs in Figure 2.

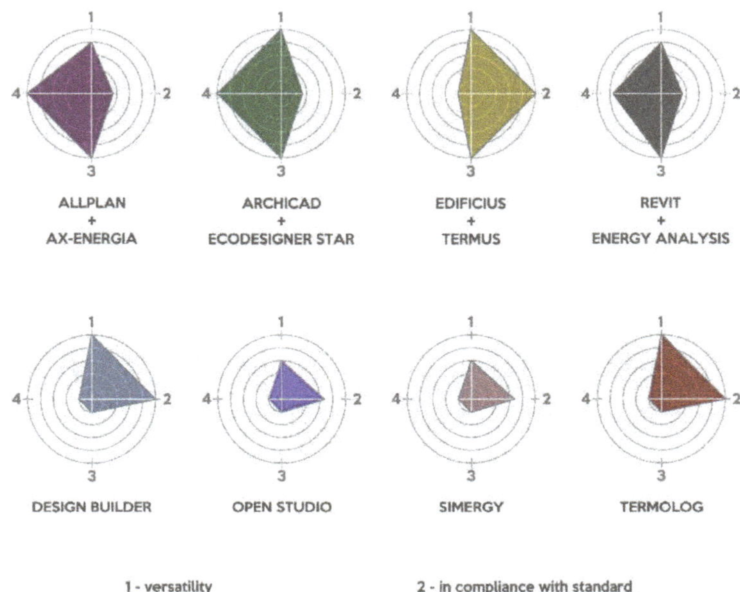

Figure 2. Comparison of the analyzed tools.

Each figure represents a tool. The vertexes are four main characteristics: Versatility, compliance with the standards, certification, and the "workflow-continuity". This latter identified the advantage of not having to open another software to conduct energy analysis. It leads to less interoperability since the changes made to the BIM will not be directly reported in BEM. The value related to the "versatility" was associated according to the number of features available (e.g., Open Studio has 3/5 features, so the value of its versatility is 0.6). The value related to the "compliance with the standard" was calculated according to the number of standards complied with, e.g., Allplan + AX-Energia complies with 1/3 standards, so the value is 0.33. The value of the certification is equal to 1, if the tool is certified, and 0 if it is not certified. The value of the workflow-continuity is equal to 1, if the tool has this characteristic, and 0 if it does not.

5. Modelling and Pre-Analysis of a Simplified Building

As a first step, the energy analysis of an apartment in a three-story building was carried out to study the characteristics of the software using EcoDesigner STAR, a plug-in for ArchiCAD. It allowed highlighting the main calculation characteristics and detecting the first advantages and disadvantages of using this tool.

The choice was determined by the value and also by the simplicity of the building, which allows us to control the result of the simulation process more precisely.

The structure is made up of a reinforced concrete frame made of rectangular section beams and pillars (30 $\times$ 60 cm). The indoor walls are composed of non-insulated brick blocks. It was geo-located in the city of Palermo. As regards the immediate surroundings, it was decided to consider it not bordering other buildings.

The construction of the BIM model was carried out using the construction components and materials already present in the program library. In this way, possible conflict situations were avoided to better control the process. Figure 3 shows the three-dimensional view of the building with the apartment examined in evidence.

Figure 3. Three-dimensional view of the building with the apartment examined in evidence.

5.1. Switching to the Building Energy Model

Once the BIM model was obtained, the missing information was implemented for the construction of the BEM model. In particular, data related to the project site and its location with related climatic data (air temperature, relative humidity, solar radiation, and analysis of wind speed and direction) were included. The software can download automatically the information from the Strusoft Climate Server. The Strusoft server bases its climate data on those provided by "Reanalysis NCEP" available on the website of the "NOAA-Cires Climate Diagnosis Center". The information obtained was compared with the climatic data used by the EnergyPlus calculation engine. It is based on the data collection commonly known as "IGDG—Climatic data G. De Giorgio" [43] (Figure 4).

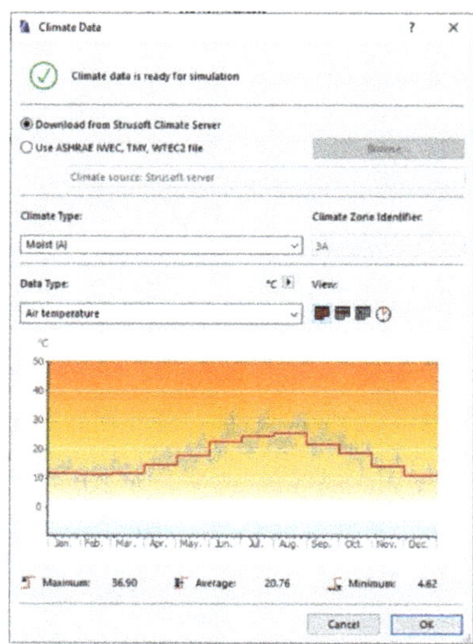

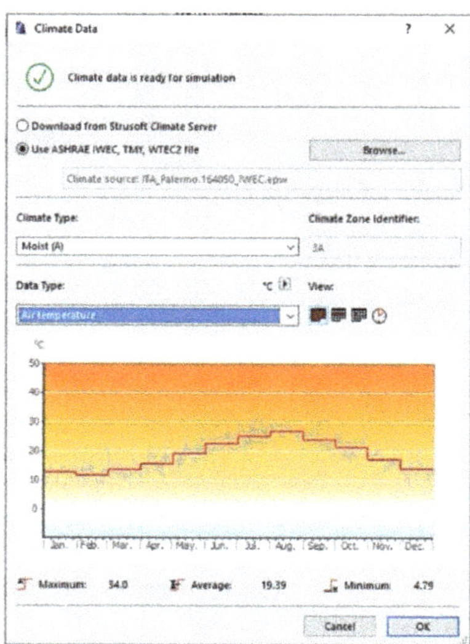

Figure 4. On the left the climatic data obtained from the Strusoft server, on the right the climatic data G. De Giorgio.

The comparison shows that the climatic data used by the EcoDesigner STAR calculation engine are in compliance with those used by the EnergyPlus calculation engine, with maximum and minimum temperatures very close to each other:

- Maximum temperature: 36.9 °C—minimum temperature 4.62 °C (Strusoft);
- maximum temperature: 34.0 °C—minimum temperature 4.79 °C (EnergyPlus).

It was supposed that the slight deviations between the monthly temperatures were due to the different time intervals relied on for data collection. In particular, the EnergyPlus data refer to a period ranging from 1951 to 1970; while those of the Strusoft servers are updated from 1948 to today. Then, the areas of the building characterized by the same orientation, by the same usage profile and above all by the same system (thermal blocks) were defined (Figure 5). It was possible to identify only two thermal blocks (Figure 4): that of the heated rooms and that of the unheated rooms.

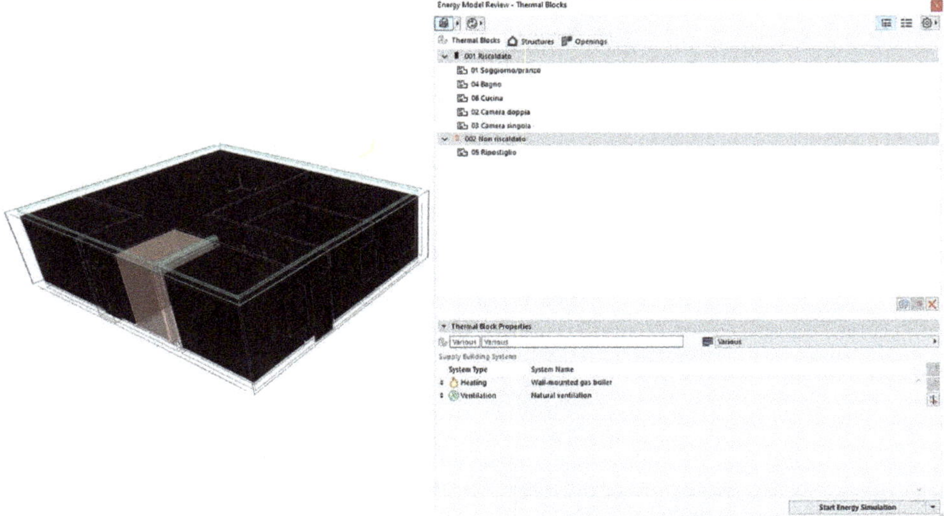

Figure 5. Screen of the thermal blocks window with 3D visualization of the selected zones.

For each block, an operating profile and a plant system (heating, air conditioning, and ventilation) were set. The software includes different operating profiles (residential, commercial activity, hotel, cinema, museum, and others), but it was chosen to have a more complete overview of the program and its potential, to create a new one. As for the most common software, it is possible to select different schedules for different seasons: One for the summer season (for the cooling plant) and one for the winter season (for the heating plant).

It is possible to assign several thermal blocks to a single system or to assign a specific one to a single thermal block. In particular, it is possible to set the use of a boiler for heating and the production of domestic hot water to several thermal blocks; while the cooling system is, if made with single units serving only one room at a time, it must be assigned for each air-conditioned block. In this case, there is only a natural gas boiler for heating and the production of domestic hot water. Furthermore, there is no type of summer air conditioning and the ventilation is natural as it is normally the case in common homes.

The data relating to the autonomous heating system are very simplified. It is possible to select the nominal power of the element, the type of control (with internal/external sensor or by manual ignition), the type of energy source used, the cost of the energy used, in order to obtain an estimate of the costs and consumption of that particular system. On the other hand, the items relating to terminals are completely missing.

It is not possible to set the number of terminals and their performance. They can be inserted as elements within the model, but they do not interface with the plug-in during the calculation. For a more accurate analysis, it should be necessary to use an external program that includes these attributes. In general, the number of parameters that can be selected is less than the parameters available in the most common software for energy analysis.

For natural ventilation, the parameters to be set are accompanied by an hourly schedule. In this case, it was set to keep the "system" active all year round. In the case of mechanical ventilation, it is certainly useful for calibrating the best usage profile. In addition, it is also necessary to specify the air changes by choosing from four different units of measurement. Furthermore, it is possible to select a function that uses automatically the standard ASHRAE values.

Once all the elements of the building and their materials are set, it is possible to conduct the calculation of the thermal bridges. Before proceeding with the calculations, the plug-in updates the model with the latest changes and automatically detects errors or warnings, which, if not resolved, do not allow it to continue with the simulation. Solving any errors and starting the simulation, EcoDesigner STAR compiles a final report of the simulation.

5.2. First Results of the Analysis

According to the aim of this paper, some advantages and disadvantages were detected already in this step. The main advantages are:

- Automatic construction of the BEM starting from BIM: The model obtained contains in itself a lot of fundamental information for energy analysis, so a few other parameters must be integrated;
- constant updating: Any type of modification made on the BIM model is automatically reflected in the BEM and in the EcoDesigner STAR cards, without any loss of information;
- reliability: The climatic data of the Strusoft server comply with other types of data used by certified analysis programs (see EnergyPlus); furthermore, the calculations are consistent with the parameters set and the result, for example, in terms of consumption, and is plausible if compared with similar buildings;
- good definition of the parameters: The thermo-physical attributes on which one can act are many, allowing us to represent even the most complex elements;
- optimized workflow: Thanks to the use of a single software and a single type of file, it is possible to work quickly and accurately without the loss of data that may occur in the passage of the file from one program to another. Furthermore, all EcoDesigner STAR tools are best calibrated to operate without conflicts with those present in ArchiCAD;
- design support: This tool is not to be used a-posteriori, or after the project is finished, to know only its behavior from an energy point of view. It must be used during the design phase to ensure the maximum result in terms of performance; so, the final simulation refers to a building designed in a truly sustainable way.

Even if they are not many and do not affect the use of the program, some constraints were found and listed following:

- Poor definition of the parameters relating to the systems (especially heating);
- and therefore, it is not possible to insert the calculation of terminals of any kind or the performance of the heating systems. It makes the calculation of these aspects more limited. For this reason, further investigations with other programs should be carried if necessary.
- A separate reasoning must be made for thermal bridges. They are connected to the ArchiCAD Detail tool and they are configured as separate 2D drawings. If a material or a type of construction element changes, the individual modification must be made on the detail and the thermal bridge recalculated. Generally, this operation is not long and it is easy, but for large architectural complexes with numerous thermal bridges where changes need to be made, it is long and hard.

6. Application of the Methodology on an Existing Case Study

In order to verify their correctness, the procedures for the energy analysis in the BIM environment, developed on a sample building, have been tested on a sample building. Indeed, only a building complete in all its parts can provide the necessary information, especially from the plant engineering point of view. Furthermore, in an existing building, the systems have already been measured and have precise characteristics. They can be traced back by recovering the technical sheets drawn up by the manufacturers. The research of the case study was conducted by preliminarily defining some characteristics that the case study building must possess. The aim was to validate the methodology and the final results of the energy simulation. These features are:

- End use: The residential end-use is the one of greatest application interest, as well because many policies aimed to reduce consumption of residential architecture;
- simplicity of the case study: Both in terms of geometric or constructive characteristics, in terms of thermal zones and for the absence of peculiar characteristics (e.g., underground habitable rooms, large glazed areas, or systems designed ad hoc and different from the more common types);
- size: It is a medium size building. It implies the exclusion of individual housing units exposed to the external environment on all sides (this condition would prevent testing all the potential of the software), i.e., large apartment complexes consisting of a few thermal zones, but repeated for several floors (in this case there would be a large amount of data, not very significant from the point of view of the calculation), i.e., considering a 7-storey condominium, in the calculation phase, the significant floors are the ground floor that exchanges heat with the ground, the floor that borders two heated rooms both above and below, and the top floor that exchanges heat with the outside through the cover. For the purposes of this study, it would be enough to consider only one of the 5 floors bordering heated rooms. A good compromise is therefore offered by buildings with two (or rather three) elevations above ground;
- building envelope and thermal plants: All the information about stratigraphy, fixtures, and materials were known or can be inferred with a good approximation;
- it is an existing building.

6.1. Case Study Modelling

As said, this case study was selected as an important sample of existing architecture characterized by the main data available, and, as well in this case, for simplicity. This latter concerns both the spatial and architectural quality of the building or housing, and the relationships with the ground and the characteristics of use that allow more effective control of the BEM energy analysis. The chosen building, in fact, has simple thermal zones, with no particularities (of use or construction) that can alter the calculation of the energy simulation. It is characterized by a simple structure, and a regular plan, with essential thermal zones (with a common residential type of user profile). It is a residential building and has three apartments distributed over three floors above ground connected by an external staircase. Only the apartment on the first floor was chosen.

It must be specified that some simplifications regarding the articulation of the architectural and climate artefact can be made, where necessary, to ensure greater control over data processing by the software. Moreover, some simplifications, e.g., regarding the articulation of the architectural and climate artefact, were made. It endured greater control over data processing by the software.

The selected building met all the necessary characteristics: The Langham House Close residential complex in Richmond (England), designed by James Stirling.

The apartment building consists of 18 residential units. They are spread over three elevations above ground (6 per floor). There are 3 different types of apartments, all based on the same floor plan, which differ in the number of rooms:

- 3 apartments with one bedroom (approx. 65 m^2);

- 9 apartments with two bedrooms (approx. 75 m^2);
- 6 apartments with three bedrooms (approx. 85 m^2).

The three smaller apartments, all on the ground floor, are identical to the apartments with two bedrooms. One of the rooms is intended for the service of condominiums, as an accessory storage. There are also three two-bedroom apartments on the ground floor. On the next floor, there are three two-bedroom apartments and three three-bedroom apartments; the same distribution is repeated on the second and last floor.

At each level, a pair of housing units are served by a common stairwell; one block includes 6 apartments on 3 levels, served by a stairwell; the residential complex consists of 3 blocks (Figures 6 and 7).

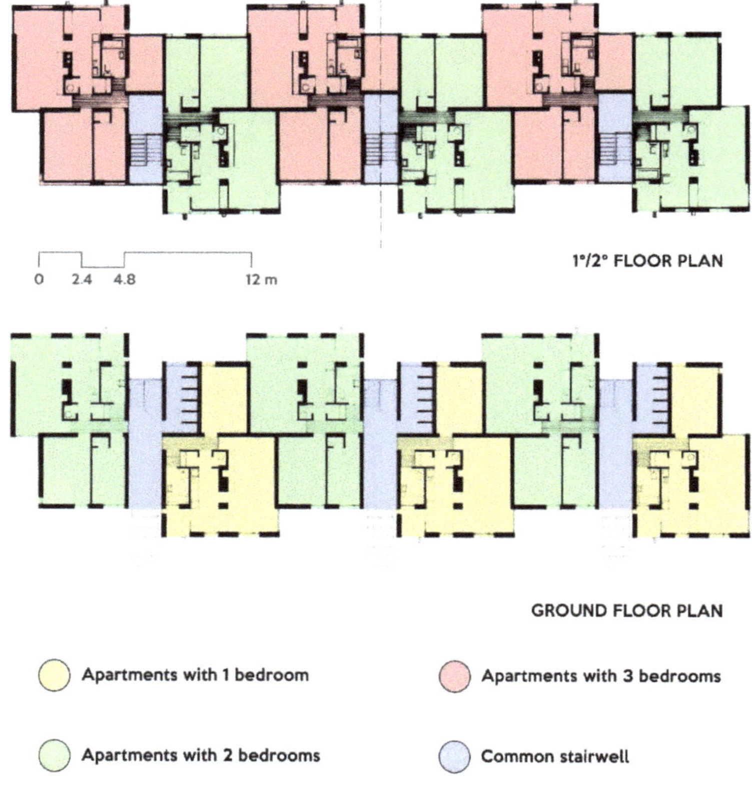

Figure 6. Plans of the building.

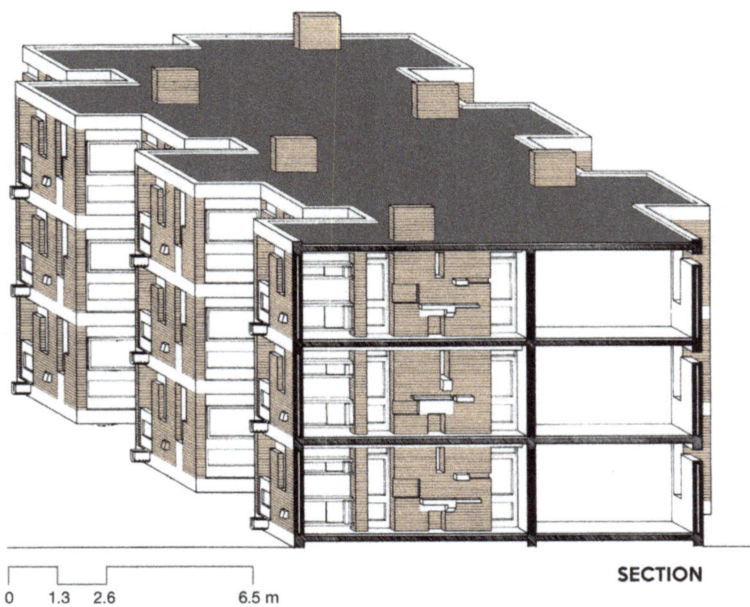

Figure 7. Section of the building.

The complex is located in a suburban area of London, characterized by a low population density and a strong natural presence; the residential complex is open on all fronts and is surrounded by trees [44].

The entire structure is in load-bearing masonry, consisting of solid bricks (215 × 102.5 × 65 mm) and 10 mm lime mortar joints. The wall structures are differentiated into three types:

- The external perimeter walls are cavity walls, very recurrent in English houses of the first half of the twentieth century; they have an overall thickness of about 270 mm (solid brick of 102.5 mm—air gap of about 65 mm—solid brick of 102.5 mm);
- the internal perimeter walls separate the apartments from each other and from the staircase body; they are entirely in solid bricks without cavities and have a thickness of 215 mm (length of a single brick);
- the internal dividing walls are the partitions of the houses, formed by individual courses of bricks (102.5 mm thick).

Each apartment is characterized by the presence of a central block, where the fireplace is located, and the plant room; this block is in load-bearing masonry and plays a decisive role in the load-bearing structure of the building.

Regarding the structure, the load-bearing masonry is combined with reinforced concrete beams characterized by a rectangular section (27 × 35 cm). They work as curbs for the distribution of loads of the upper floors. Outside, the elevations are characterized by the alternation between brick and concrete. In addition to the beams, visible directly from the outside, there are other reinforced concrete elements, such as the U-shaped gargoyles, the ventilation openings, and the panels under the windows. These latter serve to further stiffen the floors, linking the masonry with the reinforced concrete of the beams and floors [45].

The floors are reinforced slabs, made together with the beams and panels under the windows. Their stratigraphy, described from bottom to top, varies according to the reference plane:

- Ground floor slab: This slab rests directly on the ground and is composed of a bed of compacted materials (generally stones, bricks, concrete) of about 20 cm thick, by a 15 cm cast-in-situ reinforced concrete slab, by a layer of bitumen, from a 3 cm screed, from a 1 cm floor, mainly made of wood and stoneware;
- floor slab: 1–1.5 cm lime-based plaster, 15 cm cast-in-situ reinforced concrete slab, bitumen layer, 3 cm screed, 1 cm flooring (wood or stoneware);
- roof slab: 1–1.5 cm lime-based plaster, 12.5 cm cast-in-situ reinforced concrete slab, 3.5 cm screed, bituminous sheath;

The fixtures are very similar to those envisaged in the project and consist of wooden frames, painted white, with single glass.

Regarding the plants, it must be noted that this building, like many others built in the 1950s, was not originally equipped with heating or cooling systems. The control of the internal temperature was therefore obtained through natural ventilation. In the winter season, the heating of the rooms was performed by a wood-burning fireplace in each single house in the living area. Over the years, with the change in technology, product costs, and lifestyles, each home has been equipped with a heating system. The individual owners carried out the construction of the systems independently. For this reason, the components of the systems (boiler, radiators) vary from apartment to apartment. The solution adopted provides, in general, the installation of an autonomous internal boiler of about 24 kW in the central masonry body. The boiler allows the production of domestic hot water and the power supply of the terminals located in each room of the apartment. The terminals are standard radiators. Given the differences between the apartments, a schematic was adopted in the calculation phase, considering the system of a typical accommodation and then applying it to the other apartments examined. Therefore, small differences (model or commercial brand of the radiator, for example) were eliminated, given that they did not affect the calculation results.

The fireplace is present in all apartments, but its function changes, e.g., in some cases, it is not used. Its function is performed by the heating system (Figure 8). In other cases, residents decided to continue to use it in combination with the heating system, maintaining the wood supply, and others replaced the wood-burning fireplace with a gas fireplace that replaces the radiators in the living area.

Figure 8. Pictures of the flat with a gas-fireplace and a wood-fireplace (**above**) and a picture of the external part of the building with the analyzed part (**below**).

Figure 8 shows pictures of the flat with a gas-fireplace and a wood-fireplace and a picture of the external part of the building with the analyzed part. The cooling systems have never been installed because they are not necessary because the climate in London in the summer is not very hot. Furthermore, it should also be considered that the building is located in a well-ventilated area, far from the densely built urban center.

The present study examined a single block consisting of 6 apartments, distributed in pairs, on three elevations, and served by a common staircase. The limitation to a single block of apartments does not affect the search results. It respects the modularity desired by Stirling and excludes the repetition of identical elements, superfluous for the purposes of the calculation. The delimitation required the modification of the perimeter walls of the housing: The walls that previously bordered other apartments, in solid bricks, were transformed into cavity walls bordering the external environment, as is already the case for the rest of the construction.

The modelling process was conducted on the basis of two-dimensional graphic references (plans, sections, elevations) produced through the redesign of the project drawings and the verification of the relative congruence (between plan and section, for example); the modelling phase was conducted, as required by BIM, specifying the material and construction characteristics of the individual elements and also the parameters useful for the energy simulation. An accurate BIM model of the building, defined in its architectural-construction aspects, was developed. From the model, it is possible to extract plans and sections, or inspect the building in three-dimensional views.

To export the model from the Building Information Model (Figure 9) and to import it into the Building Energy Model, it is necessary to create the thermal zones to which each room is assigned. This operation, easy and immediate from an operational point of view, requires particular attention from a conceptual point of view.

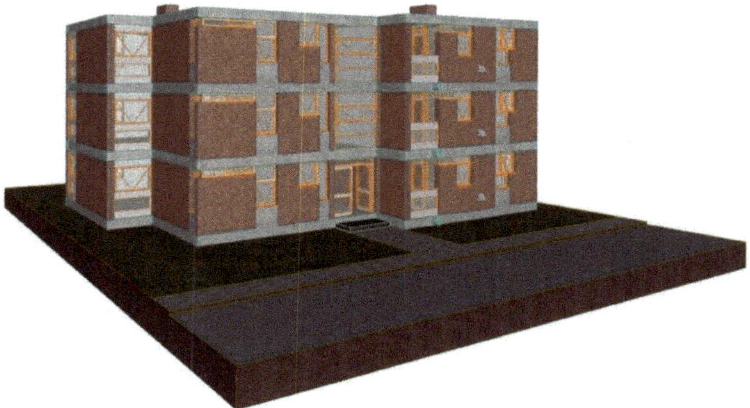

Figure 9. Navigable virtual model, view of the main front.

For the Stirling building, 7 thermal blocks were identified (Figure 10), one for each of the 6 apartments, and one for the common areas. The thermal block referred to an apartment, contains within it as many thermal zones as there are rooms that compose it. Indeed, all the rooms are heated by the same system and therefore share the same internal temperature. Furthermore, since this is a residence, the occupancy will also be homogeneous throughout the apartment.

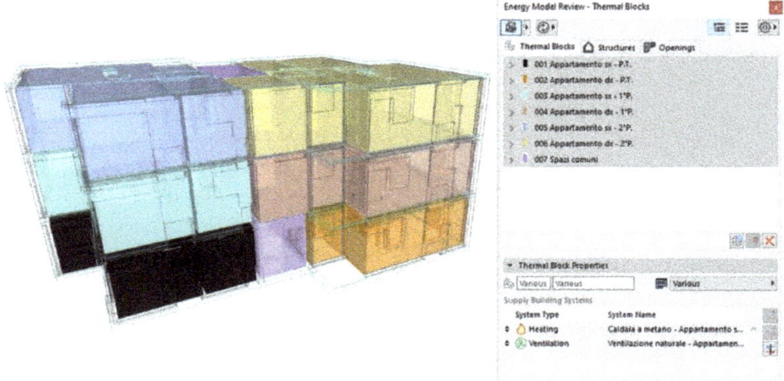

Figure 10. Graphic visualization of the Building Energy Model (BEM) model with chromatic distinction of the thermal blocks.

Common areas do not have any type of heating and have a different occupancy profile from that of the apartments. After verifying the correct definition of the Building Energy Model, it is possible to proceed to enrich its information content. The operations carried out within the three EcoDesigner STAR tabs are: Thermal Blocks, Structures, and Openings. Within the thermal block section, it is possible to modify the operation profiles of each thermal block with related heating and ventilation systems, to determine, during the calculation phase, the thermal inputs deriving from external and internal factors. The analysis on the operation profiles was calibrated on the basis of a typical English family. The schedules were set including the differentiation of the types of use over the different seasons and working and non-working days. To identify the period of operation of the heating system, the graphs on the climatic data generated by the software were examined. It is thus determined that the apartment is inhabited for a few hours a day during work and school days and that, consequently, systems, lights, and appliances will be active for a few hours. On the other hand, the occupation during non-working days is different, when the apartment is occupied for most of the day, generating a more intensive use of systems, lights, and appliances.

A deep study made it possible to trace in detail the characteristics and technical data sheets of the elements that compose it (Figure 11). Similarly to the operation profiles setting, heating and ventilation systems must be assigned to each thermal block.

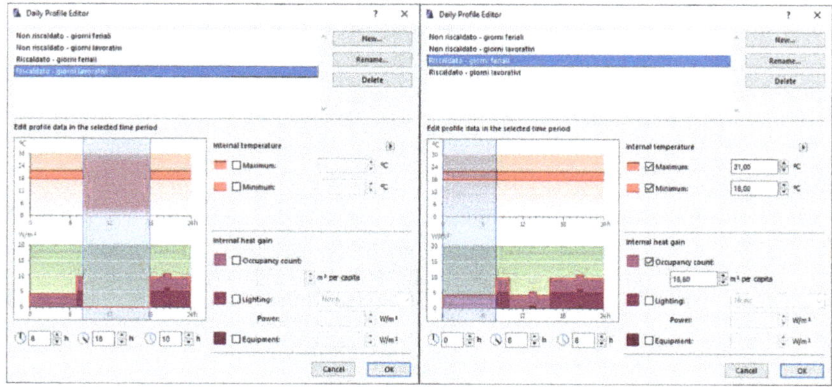

Figure 11. Schedules of the operation profiles. Note how the occupation of the houses changes between working and school days (**left**) and non-working days (**right**) within the period of the year in which the heating is activated.

EcoDesigner STAR, as already specified above, does not provide for the insertion of the heating system terminals (radiators); therefore, only the parameters relating to the boiler (nominal power and flow and return temperatures) and the production of domestic hot water have been entered. Figure 12 shows the window in which the flow and return temperature of the heating system water is set and the energy source used to power the boiler.

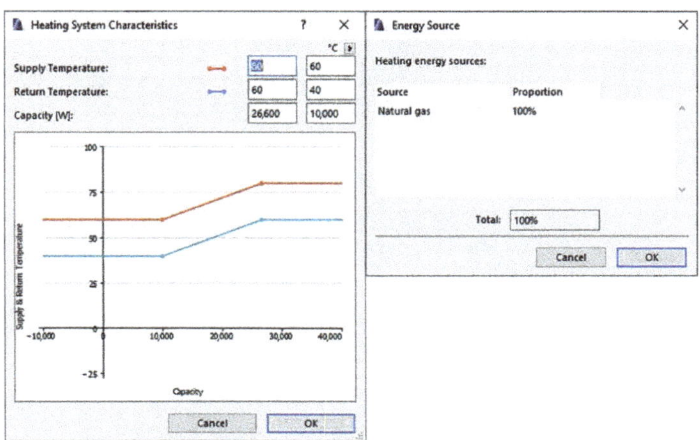

Figure 12. On the left, the window in which the flow and return temperature of the heating system water is set; on the left, the energy source used to power the boiler.

The ventilation of the apartments is natural. Therefore, it was possible to set parameters such as the number of air changes per hour (defined as ACH—Air Changes per Hour) and the program for using the fixtures. The air exchange has been set with a value of 0.5 ACH. For the attribution of this value, the harsh climate was considered.

The tab called "Structures" can be used to insert and calculate thermal bridges.

The software manages the calculation of the thermal bridge in an extremely intuitive way:

- With the Detail tool, it is possible to isolate the desired portion of the drawing in an independent tab;
- the 2D drawing is graphically improved and the materials are attributed to the various screens representing the construction elements;
- through a dedicated command, the calculation of the thermal bridge is started and takes place in several phases: (i) In the first phase, the area relating to the external air is selected with its temperature (already calculated based on climatic data, but possibly editable); (ii) in the second phase, the same operation is carried out, this time for the indoor air; (iii) we then move on to the identification of the foundation soil using a net (if it is a matter of structures in direct contact with the ground, otherwise we move on to the next phase); (iv) in the fourth phase, the thermo-physical characteristics of the building materials are checked.

This calculation provides an interactive graph of the temperatures (or a graph of the heat flow). The main thermal bridges identified in the analyzed case study concern the combination of bricks and concrete beams/panels of reinforced concrete, and the material discontinuities at the windows and the corners of the structure. The window fixtures required a detailed study for each frame. Each window was decomposed optimally to not distort the performance of the building envelope. In order to conduct this analysis, the factors considered are: The juxtaposition between the concrete panel and the masonry wall, the particular shaped frame and its contact with the bricks and concrete, and the angle of the structure, which is identified as a thermal bridge in shape. Normally, these aspects should be taken into account separately, but their positioning within a very small area does not allow this procedure. In doing so, thermal bridges would be calculated two or more times, in the elements to which they belong and in the adjacent ones. It should distort the performance of the building envelope, which would be worse than they really are. It was, therefore, decided to break down and schematize these factors. In this way, they were considered independent, and a correct calculation was obtained.

The first thermal bridge calculated was that between the reinforced concrete panel under the window and the adjacent masonry. In this case, the thermal bridge of the shape deriving from the angle formed by the structure (the wall to the right of the panel) is calculated simultaneously. The thermal bridge was divided into two parts: One between window and brick and one between window and concrete. The resulting thermal bridge value has attributed a length equal to the perimeter of the window in contact with the bricks, excluding that part in contact with the wall in the right corner. It was because the effect of this thermal bridge was calculated in the case of the previous step.

The thermal bridge of the window was calculated in contact with the concrete, considering a vertical section of the frame (Figures 13 and 14). The length to be attributed to the thermal bridge is therefore the perimeter of the frame in contact with the concrete.

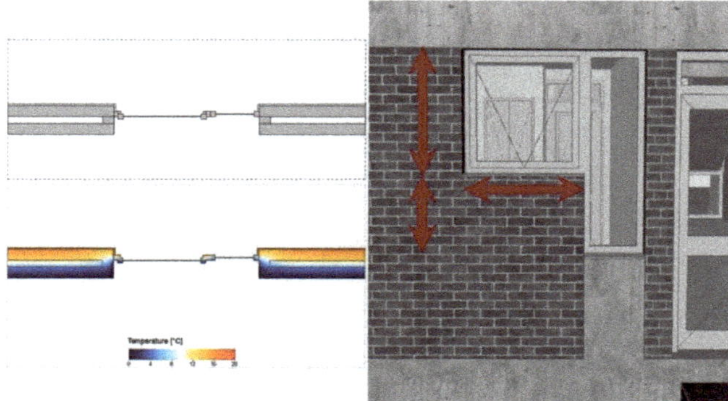

Figure 13. Output of the calculation of the thermal bridge of the window inside a fictitious wall (**left**) and dimensions considered for the length to be attributed to the thermal bridge (**right**).

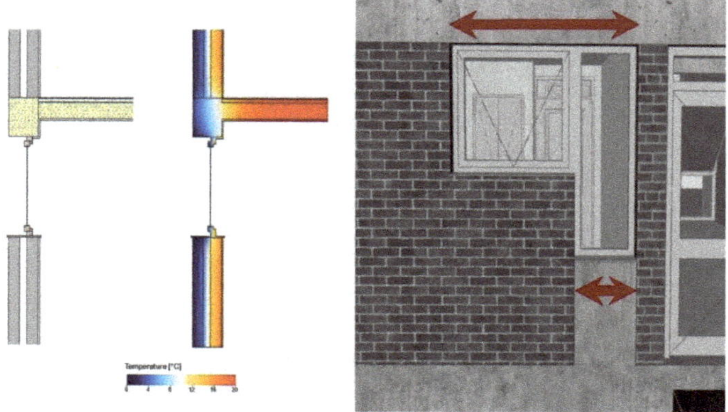

Figure 14. Output of the calculation of the thermal bridge of the window in contact with the concrete (**left**) and dimensions considered for the length to be attributed to the thermal bridge (**right**).

In this calculation, the floor in contact with the beam was also included because it gave rise to another thermal bridge. By calculating the thermal bridge of the concrete beam in contact with the floor, the lengths already taken into consideration for the frames were excluded. Once all thermal bridges were calculated, they were attributed to each thermal block via the structures table. Alternatively, it is also possible to enter a table value of the thermal bridge, but the specific calculation for each element is always to be preferred. In this last sheet, the characteristics of the frame are specified. They can be selected from a vast library inside the plug-in, divided into glass type and frame material. The alternatives made available by the library are numerous and can satisfy even the cases of fixtures with particular performances. However, if the characteristics to be entered do not correspond to those present in the library, it is always possible to manually overwrite them for each frame or groups of frames. It is also necessary to start the calculation of the solar analysis (Figure 15) for all external frames (all internal doors will be automatically excluded). This operation has a double advantage. The first is that the data obtained can be used by the program in the calculation phase, while the second is that the professional receives support, during the design, from the interactive graph produced as a result of the calculation. This graph offers the possibility to investigate, day-by-day, hour-by-hour, the irradiation conditions of a given frame. So, it is possible to instantly evaluate the effectiveness of the positioning, dimensions, or shielding system adopted (Figure 16).

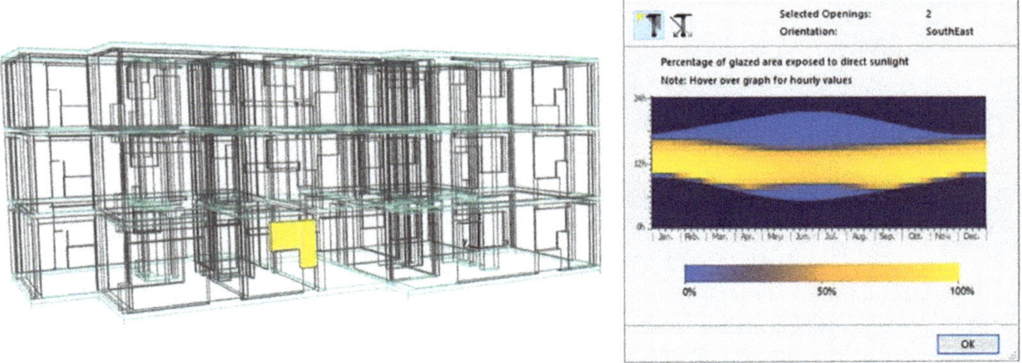

Figure 15. Three-dimensional display of the selected window element and relative solar analysis.

Before proceeding with the energy simulation of the building, it is possible to select a "Reference building" in the calculation phase. This building is another virtual model that serves as a benchmark. Thanks to it, the advantages and disadvantages of the two alternatives of the same project can be immediately highlighted. Moreover, it is possible to compare the building under consideration with a similar one whose performance we already know. This is a completely optional operation. The simulation can very well proceed with the building data without any reference building.

However, in this case, it was decided to build a reference building model both for completeness in the study of the possibilities of this software, and to highlight considerations on the design applications extensively presented in the next paragraph. The construction of the reference building was based on the same geometric-architectural model of the case study building. The elements and climatic data used are completely identical; while thermo-physical properties of the elements are different.

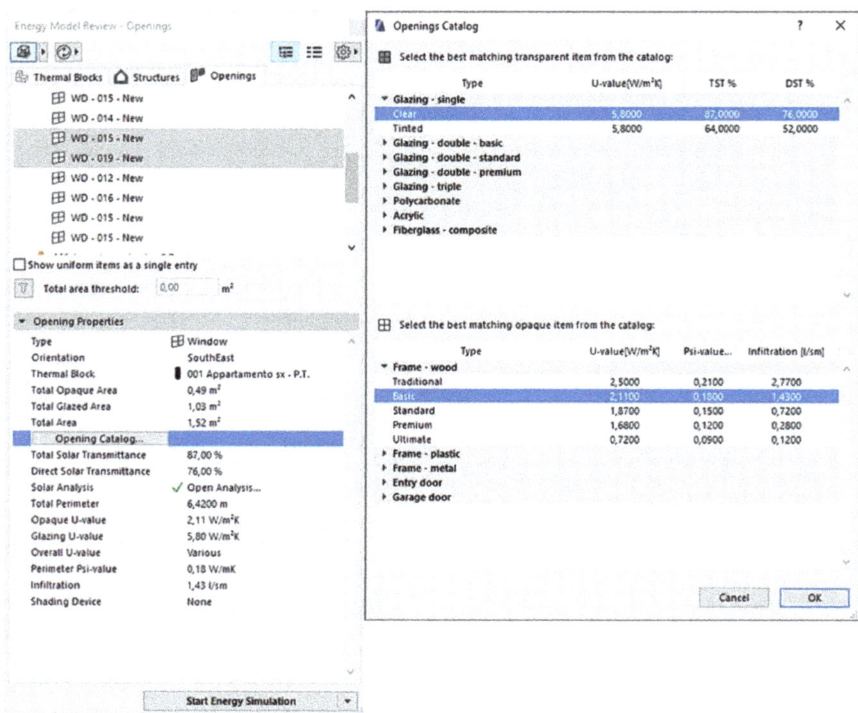

Figure 16. Window for selecting openings/frames on the left, library of the glass type and frame material on the right.

In particular, the calibration of the building envelope was based on the guidelines dictated by the "Interministerial Decree of 26 June 2015—A" and by the "Interministerial Decree of 26 June 2015—B" [46]. The procedure adopted is purely for study purposes, as the building was not designed in Italy nor is it subject to Italian regulations. The parameters are classified in the standard according to the climatic zones of the locality. Italy is divided into 6 climatic zones ranging from zone A to zone F [47]. They differ in the value of degree days (GG). The "degree-day" is defined as the sum, over one year, of the (positive) difference between the internal ambient temperature and the average daily external temperature [48]. The indoor temperature in Italy has been set at 20 °C, so the degree days are calculated based on this temperature. It is evident that the Stirling building cannot be placed in the Italian climatic zones, therefore the problem arises of which climatic zone to choose to obtain the parameters of the reference building. It was decided to calculate the degree days near London (Ham Common, Richmond, BC, USA). In doing so, also the fictitious climatic zone according to Italian parameters was found. Although in England the degree days (HDD and CDD, respectively, Heating Degree Day and Cooling Degree Day) are used, they are calculated differently than the Italian ones, and in particular, they refer to an internal temperature of 15.5 °C instead of 20 °C. Comparing the GG calculated with two different temperatures is an operation that distorts the results at the start, it was necessary to calculate them through an online application [49]. It was used to choose the internal temperature to be included in the calculation. The output of this process is a spreadsheet in which the following are entered: The period of time considered and temperature of the reference indoor environments, source, accuracy of the climatic data for that area, weather station used, a table with monthly degree day values, and finally the total. A value of 3063 GG was considered. It was compared with the parameter suggested by the Italian standard. So, the Stirling building was ideally placed in the "climate zone F". It indicates the period of heating system operation and it allows identifying the parameters that the building envelope must have to be considered as a reference building. EcoDesigner STAR can overwrite the new parameters on the old, as a finished element, to quickly update the Building Energy Model in a few steps. This avoids replacing the elements built previously (walls, floors, windows, etc.) and having to model the building again. The resulting building is better performing than the real building. It is because the building envelope is made up of elements with high thermal efficiency. After starting the energy simulation calculation, it is possible to save the building as a reference building. Finally, it can be used as a term of comparison when analyzing the building with real parameters.

6.2. Simulation Results

All the aspects studied and exposed were useful for the correct construction of the BEM model and a valid setting of the data for energy simulation. The previously built reference building was included in the appropriate tool section. At the end of the calculation, a final report was obtained.

The first section reports all the general data of the entire building, including geometric ones, and some average values for all the thermal zones. The data concerning the energy supplied to the building (heating system, lighting systems, internal heat inputs due to the presence of people, etc.) are shown in the form of a weekly chart. It provides an immediate understanding of which system requires the most energy. Therefore, it predicts which of these fields can be the most expensive in economic terms. The first section contributes to giving an overall view of the analyzed construction. The second section contains the same information as the first, but this time concerning the individual thermal blocks. Each block is associated with a weekly graph of the energy supplied and a graph on the energy emitted. In the specific case of the Stirling building, the 6 apartments are shown (each definitive with a thermal block) and a single common area (represented with a single thermal block). It was noted that the apartments that require less energy in a year are those on the first floor. It is because they exchange heat with the external environment only on three sides. On the fourth side, the apartments border an unheated internal environment. Both at floor and ceiling level they border with other heated rooms. Therefore, the heat losses are less. The accommodations that require more energy to maintain an optimal internal temperature are those on the top floor. Indeed, they exchange heat with the external environment even from the ceiling. Moreover, the roof slab, compared to the floor slabs, has a lower thickness and there is not any type of insulating layer. The apartments on the ground floor, on the other hand, have energy consumption closer to those on the first floor. Another aspect that can be immediately noticed is that the apartments on the right of the entrance (and therefore facing north) have slightly higher energy consumption than those facing south. The third section relates to the daily temperature profiles. It is possible to insert graphs for each day of the year of any thermal block, to show the curves of internal and external temperatures. During the design phase, these graphs are very useful to better calibrate the systems and know when it is necessary to heat or cool the rooms. The fourth section is dedicated to the energy consumption, environmental impact, and energy production (if renewable energy sources are present). Energy consumption is shown both as a table and as a graph, and is divided into categories (heating, cooling, domestic hot water production, consumption due to mechanical ventilation if present, lighting, and equipment). If the prices of the various energy sources are also set, the cost of the various systems can be known. The environmental impact is instead calculated through the kg/y of CO_2 emitted, or the carbon dioxide expressed in kg emitted in a year.

The last section focuses on comparing the consumption of the building to be analyzed and a reference building. It is possible, through this section, to compare two variants of the same project, in order to know both the consumption of the two buildings. They are compared in economic terms.

In Table 2, the results of the simulation were reported.

Table 2. Geometric characteristic of the analyzed apartment and simulation result ante and post ideal retrofit action.

			Ground Floor	1st Floor	2nd Floor
Geometry Data	Gross Floor Area [m^2]		67.68	67.67	67.7
	Treated Floor area [m^2]		56.3	56.3	56.3
	Building shell area [m^2]		78.09	75.74	132.41
	Ventilated Volume [m^3]		135.41	135.41	135.41
	Glazing ration [%]		16	16	9
Heat Transfer Coefficients	Floors [W/m^2K]	ante	2.18–2.18	-	-
		post	-	-	-
	External [W/m^2K]	ante	1.04–3.74	1.04–3.74	1.04–3.74
		post	0.43–3.74	0.43–3.74	0.43–3.74
	Openings [W/m^2K]	ante	2.35–5.70	2.35–5.70	2.35–5.70
		post	1.81–2.48	1.81–2.48	1.81–2.48
Internal Temperature	Min (06:00, 29) [°C]	ante	6.59	6.3	4.52
		post	6.07	6.61	5.96
	Annual Mean [°C]	ante	14.3	15.32	15.57
		post	14.33	14.62	14.37
	Max (19:00 05 August) [°C]	ante	20.36	21.83	27.99
		post	18.6	18.65	18.91
Unmet Load Hours	Heating [hrs/a]	ante	127	126	124
		post	128	126	128
Annual Supplies	Heating [kWh]	ante	16,249.43	14,766.44	22,253.95
		post	12,243.32	11,741.08	13,873.37
Peak Loads	Heating (19:00 14 December) [kW]	ante	10.22	10.05	15.03
		post	9.11	10.27	10.06

The proposed interventions are to be considered the hypothesis that complete and conclude the entire process outlined above. It has to be remembered that the aim of this paper is not the design of improvement solutions for the James Stirling building, but the definition of a methodology that integrates the simulation energy in the BIM environment, thus identifying a valid design support tool. Anyway, some possible interventions aimed at reducing the consumption of individual apartments and improving the performance of the entire building are listed below:

- Addition of an insulating layer in the roof slab; the second and last level are those affected by greater heat losses; this occurs at the roof level, as can be seen from the thermography; thus acting on the attic, inserting an insulating layer that does not alter the aesthetic and structural components of the building, the consumption and costs for heating these apartments would be reduced;
- of the ground floor slab in contact with the ground; the consumption of the ground floor apartment, compared with the consumption of the apartment on the first level, is higher; it is hypothesized that improving the contact between the ground and the attic through the insertion of a crawl space, consumption can be reduced and made similar to that of the apartment on the first level;
- strengthening of the insulation package of the external cavity walls; currently the walls have an air gap and by blowing in insulating material their thermal transmittance values could be improved, thus reducing heat loss;
- replacement of fixtures; for this intervention two solutions could be opted for: (i) The first involves the complete replacement of the current fixtures with fixtures, similar in materials, but with double glass and thermal break; (ii) the second, if the first cannot be carried out due to the protected nature of the building, provides for the maintenance of the current wooden frames and for the replacement of the single glass with a double glass, capable of guaranteeing an improvement in the performance of the fixture.

By updating the BIM model with the new parameters, the BEM model will also be updated automatically. So, the energy simulation calculation can be quickly started.

The consumption of the apartment at the ground floor after the application of the retrofit actions is 12,243.32 kWh; while before the ideal retrofit actions was 16,349.43 kWh.

The apartment on the first floor adjoins two heated rooms, both at floor and ceiling level, with an annual consumption of 14,766.44 kWh, before the ideal retrofit actions, and of 11,741.08 kWh, after the ideal retrofit actions.

The consumption of this apartment is slightly lower than the ground floor apartment and significantly lowers. The apartment on the second level borders on a heated room at floor level and with the external environment at ceiling level. The annual consumption calculated was 22,539.95 kWh before the ideal retrofit actions, and 13,873.37 kWh after the ideal retrofit actions. Furthermore, it is possible to obtain the perspective sections of the apartments in which it is also possible to read the distribution of temperatures within the building envelope; this type of paper, following a correct interpretation, is particularly useful for identifying the areas of intervention.

Looking at the thermography of the new configuration of the building, it can be seen how the roof slab disperses less heat than the real configuration and how the windows and cavity walls break down the heat flow.

7. Results

In the previous section, the results of the process were reported. By conducting this study, it was possible to highlight the advantages and limitations of the tool EcoDesigner STAR application and to outline a clear picture of the potential and criticality of the chosen. The greatest advantage that is obtained from the use of ArchiCAD associated with EcoDesigner STAR is the overcoming of one of the major problems in this field, namely the transition from BIM to BEM. The modelling software allows the construction of a BIM model according to certified tools. Plug-in allows its interpretation in BEM with an almost instantaneous operation.

It is clear that this transition from BIM to BEM is error-free. Moreover, it is configured as an immediate operation when the designer builds the BIM to then conduct an energy analysis. Indeed, during the design and modelling process, it is necessary to better calibrate the data entered in the BIM, the used elements and the relationships established with other elements. It is also useful to facilitate the subsequent transformation of the Building Information Model Building Energy Model. It has to be reminded that, according to the BIM goals, the interoperability must take place from the earliest stages of design. To do this, all the professional figures required should be involved to operate with the same objectives and making the work faster and more effective.

If the BIM model is built to transform it into BEM, the operation is easy. In this way, it is possible to apply an energy simulation in all the design phases of a building. Clearly, this action occurs with degrees of detail and accuracy based on the progress of the project. So, better integration between design and energy analysis can be achieved. This latter is not relegating to the final phases of the project. As for typical energy analysis, at the end of the simulation process, many data can be obtained (both directly and indirectly). From them, it is possible to consider possible interventions aimed at reducing the consumption of apartments.

Another advantage of this tool is the feasibility to compare two or more variants of one of the same building. They can be compared both in the early stages of the project and in the final stages, managing to choose the best solution according to needs.

The output data looks user-friendly. It is an important aspect for the designer, who can know the advantages and disadvantages of a design choice almost immediately. For example, it is possible to choose the orientation and materials of the building envelope and start an initial energy simulation by mentioning the other fundamental data. Furthermore, it is possible to propose a different orientation and different types of materials for the building envelope, keeping the other data completely identical to the previous variant. Finally, it proceeds with the simulation of a second alternative and automatically compares it with the first, in order to obtain graphs on the savings and consumption of both and choose the most suitable solution. In order to support this consideration and to check the possible improvements of the tool, further comparison with a stand-alone software was performed. The main results are reported in Table 3. Regarding the building systems, "basic" means that there are just a few options to set the plants; "detailed" means that it is possible to set all the parameters.

Table 3. Comparison with Termolog software.

				ArchiCAD + EcoDesigner STAR	Termolog
Material physical properties			Density [ρ]	✓	✓
			Thermal conductivity [λ]	✓	✓
			Vapour diffusion resistance factor [μ]	✕	✓
			Heat Capacity	✓	✓
			Embodied energy	✓	✕
			Embodied carbon	✓	✕
Building Systems	heating and cooling	Boiler	Nominal capacity	✓—basic—	✓—detailed
			Efficiency [η]	✕	✓—detailed
			Emission system	✕	✓
			Thermoregulation	✓—basic	✓—detailed
		AC unit	Nominal capacity	✓—basic	✓—detailed
			Efficiency [COP/EER]	✓—basic	✓—detailed
			Emission system	✕	✓
			Thermoregulation	✓—basic	✓—detailed
	lighting		Power	✓	✓
			Sources typologies	✓	✓
			Control system	✕	✓

8. Discussion

The main goals of this study are the analysis of the interaction between BIM and energy simulation, through a review of the main existing commercial tools, and the identification and application of a methodology in a BIM environment by using Graphisoft's BIM software Archicad and the plug-in for dynamic energy simulation EcoDesigner STAR. The application on a case study gave the possibility to explore advantages and limits of this commercial tools and, consequently, to provide some possible improvements.

As said in Section 4, the selection of the analyzed tool was based on some main criteria. The first characteristics are the versatility and the possibility to combine BIM modelling with well-integrated plug-in for energy analysis, to import and export building 2D drawings, BIM modelling and 3D visualization, quasi-static energy diagnosis, dynamic energy diagnosis, calculation of thermal bridges, and calculation of renewable energy sources. The second characteristic is that it meets the requirements set by the most advanced energy diagnosis standards, such as UNI EN ISO 52016-1 for the calculation in dynamic hourly regime, and ASHRAE 140-2017 and UNI/TS 11300 on the monthly average stationary calculation. Finally, it meets the validity requirements required by buildingSMART for IFC certification. All these requirements, according to the conducted study and the results reported above, were confirmed.

Although the results of this study are satisfactory, some critical issues and disadvantages were found within the application and possible fields. They can be addressed in future research, in order to improve digital tools and achieve perfect integration between BIM and BEM, without the passage of digital models in third-party analysis software. As shown in Table 3, the major limitation that has been detected is the poor personalization of data relating to heating systems. The type of data that can be set is limited and inherent to the fundamental characteristics of the system, such as the nominal power, the thermoregulation, the nominal capacity, and the COP/EER. It is not possible to specify the efficiency and to insert the heating terminals with their technical specifications. Furthermore, it is not possible to set the vapor diffusion resistance factor of the material, the efficiency and the emission of the system.

Thus, as improvements of the tools there is the implementation of the possibility to set these latter parameters. In particular, giving the possibility to set the vapor diffusion resistance factor of the materials, it will improve the calculation of possible surface and interstitial condensation. Regarding the HVAC system, in order to provide a more precise energy analysis, the tool could be improved by giving the possibility to set the efficiency of the boiler and the emission system characteristics and typology. Furthermore, it could be necessary to detail the parameters to be inserted regarding the nominal capacity, the thermoregulation and the COP/EER. Finally, some details about the lighting system and mainly the control system should be added.

9. Conclusions

The purpose of the paper is to study a new integrated energy simulation methodology in a BIM environment by using Graphisoft's BIM software Archicad and the plug-in for dynamic energy simulation EcoDesigner STAR. They have been selected after a careful study of the rules on energy analysis, an examination of the operational potential of different software on the market, and research conducted by a wide scientific community interested in various capacities in issues related to the interaction between architecture and energy analysis. Thanks to the application on the case study, the advantages and disadvantages of the existing tools were highlighted and compared. The greatest advantage that is obtained from the use of ArchiCAD associated with EcoDesigner STAR is the overcoming of one of the major problems in this field, namely the transition from BIM to BEM. The modelling software allows the construction of a BIM model according to certified tools. Plug-in allows its interpretation in BEM with an almost instantaneous operation when the designer builds the BIM with the intention of conducting an energy analysis. Another advantage of this tool is the feasibility to compare two or more variants of one of the same building, both in the early stages of the project and in the final stages, managing to choose the best solution according to needs. Furthermore, the output data looks user-friendly. It is an important aspect for the designer, who can know the advantages and disadvantages of a design choice almost immediately. Moreover, some critical issues and disadvantages were found within the application and possible fields. The major limitation that has been detected is the poor personalization of data relating to heating systems. The type of data that can be set is limited and inherent to the fundamental characteristics of the system, such as the nominal power, the thermoregulation, the nominal capacity, and the COP/EER. It is not possible to specify the efficiency and to insert the heating terminals with their technical specifications. Furthermore, it is not possible to set the vapor diffusion resistance factor of the material, the efficiency, and the emission of the system.

These lacks can be used as a starting point for further improvements to the tool. The possibility to set the vapor diffusion resistance factor of the materials would improve the calculation of possible surface and interstitial condensation. Regarding a more precise analysis, the tool could be improved by giving the possibility to set the efficiency of the boiler and the emission system characteristics and typology and to detail the parameters to be inserted regarding the nominal capacity, the thermoregulation, and the COP/EER. Finally, some details about the lighting system and mainly the control system should be added.

Author Contributions: Conceptualization, M.B., S.D.L. and G.L.; methodology, M.B., S.D.L. and G.L.; software, S.D.L.; validation, M.B., S.D.L. and G.L.; formal analysis, M.B., S.D.L. and G.L.; investigation, M.B., S.D.L. and G.L.; data curation, S.D.L.; writing—original draft preparation, M.B., S.D.L. and G.L.; writing—review and editing, M.B., S.D.L. and G.L.; visualization, M.B., S.D.L. and G.L.; supervision, M.B. All authors have read and agreed to the published version of the manuscript.

Funding: This research received no external funding.

Institutional Review Board Statement: Not applicable.

Informed Consent Statement: Not applicable.

Data Availability Statement: The data presented in this study are available on request from the corresponding author.

Acknowledgments: This study starts from the degree thesis [50], carried out by Simone Di Lisi with the fundamental support of Arch. Fabrizio Agnello and of the Ing. Marco Beccali.

Conflicts of Interest: The authors declare no conflict of interest.

References

1. Wong, J.H.; Rashidi, A.; Arashpour, M. Evaluating the Impact of Building Information Modeling on the Labor Productivity of Construction Projects in Malaysia. *Buildings* **2020**, *10*, 66. [CrossRef]
2. Tu, K.J.; Vernatha, D. Application of Building Information Modeling in energy management of individual departments occupying university facilities. *Int. J. Archit. Environ. Eng.* **2016**, *10*, 225–231.
3. Alahmad, M.; Nader, W.; Neal, J.; Shi, J.; Berryman, C.; Cho, Y.; Lau, S.-K.; Li, H.; Schwer, A.; Shen, Z.; et al. Real Time Power Monitoring & integration with BIM. In Proceedings of the IECON 2010—36th Annual Conference on IEEE Industrial Electronics Society, Glendale, AZ, USA, 7–10 November 2010; pp. 2454–2458.
4. Woo, J.H.; Diggelman, C.; Abushakra, B. BIM-based energy monitoring with XML parsing engine. In Proceedings of the 28th ISARC, Seoul, Korea, 29 June–2 July 2011; pp. 544–545.
5. Maltese, S.; Tagliabue, L.C.; Cecconi, F.R.; Pasini, D.; Manfren, M.; Ciribini, A.L.C. Sustainability Assessment through Green BIM for Environmental, Social and Economic Efficiency. *Procedia Eng.* **2017**, *180*, 520–530. [CrossRef]
6. Shin, M.; Haberl, J.S. Thermal zoning for building HVAC design and energy simulation: A literature review. *Energy Build.* **2019**, *203*, 109429. [CrossRef]
7. Waibel, C.; Evins, R.; Carmeliet, J. Co-simulation and optimization of building geometry and multi-energy systems: Interdependencies in energy supply, energy demand and solar potentials. *Appl. Energy* **2019**, *242*, 1661–1682. [CrossRef]
8. Liu, S.; Ning, X. A Two-Stage Building Information Modeling Based Building Design Method to Improve Lighting Environment and Increase Energy Efficiency. *Appl. Sci.* **2019**, *9*, 4076. [CrossRef]
9. Jeong, W.; Yan, W.; Lee, C.J. Thermal Performance Visualization Using Object−Oriented Physical and Building Information Modeling. *Appl. Sci.* **2020**, *10*, 5888. [CrossRef]
10. Clarke, J.; Hensen, J. Integrated building performance simulation: Progress, prospects and requirements. *Build. Environ.* **2015**, *91*, 294–306. [CrossRef]
11. Ramaji, E.; Memari, A.M. Review of BIM's application in energy simulation: Tools, issues, and solutions. *Autom. Constr.* **2019**, *97*, 164–180.
12. Andriamamonjy, A.; Saelens, D.; Klein, R. A combined scientometric and conventional literature review to grasp the entire BIM knowledge and its integration with energy simulation. *J. Build. Eng.* **2019**, *22*, 513–527. [CrossRef]
13. Bracht, M.; Melo, A.; Lamberts, R. A metamodel for building information modeling-building energy modeling integration in early design stage. *Autom. Constr.* **2021**, *121*, 103422. [CrossRef]
14. Kamel, E.; Memari, A.M. Automated Building Energy Modeling and Assessment Tool (ABEMAT). *Energy* **2018**, *147*, 15–24. [CrossRef]
15. Kim, J.B.; Jeong, W.; Clayton, M.J.; Haberl, J.S.; Yan, W. Developing a physical BIM library for building thermal energy simulation. *Autom. Constr.* **2015**, *50*, 16–28. [CrossRef]
16. El Sayary, S.; Omar, O. Designing a BIM energy-consumption template to calculate and achieve a net-zero-energy house. *Sol. Energy* **2021**, *216*, 315–320. [CrossRef]
17. Xu, X.; Mumford, T.; Zou, P.X. Life-cycle building information modelling (BIM) engaged framework for improving building energy performance. *Energy Build.* **2021**, *231*, 110496. [CrossRef]
18. Utkucu, D.; Sözer, H. Interoperability and data exchange within BIM platform to evaluate building energy performance and indoor comfort. *Autom. Constr.* **2020**, *116*, 103225. [CrossRef]
19. Rodrigues, F.; Isayeva, A.; Rodrigues, H.; Pinto, A. Energy efficiency assessment of a public building resourcing a BIM model. *Innov. Infrastruct. Solut.* **2020**, *5*. [CrossRef]
20. Tushar, Q.; Bhuiyan, M.A.; Zhang, G.; Maqsood, T. An integrated approach of BIM-enabled LCA and energy simulation: The optimized solution towards sustainable development. *J. Clean. Prod.* **2021**, *289*, 125622. [CrossRef]
21. Alam, J.; Ham, J.J. Towards a BIM-based energy rating system. In Proceedings of the 19th International Conference on Computer-Aided Architectural Design Research in Asia: Rethinking Comprehensive Design: Speculative Counterculture, Kyoto, Japan, 14–16 May 2014; pp. 285–294.
22. Jalaei, F.; Jrade, A. Integrating BIM with Green Building Certification System, Energy Analysis, and Cost Estimating Tools to Conceptually Design Sustainable Buildings. In Proceedings of the Construction Research Congress 2014, Atlanta, GA, USA, 19–21 May 2014; American Society of Civil Engineers (ASCE): Reston, VA, USA, 2014; pp. 140–149.
23. IESVE. Available online: https://www.iesve.com/software/BIM (accessed on 10 January 2021).
24. ENERGY PLUS. Available online: https://energyplus.net/ (accessed on 10 January 2021).
25. Hirsch, J.J.; Lawrence Berkeley National Laboratory. eQUEST. Available online: http://www.doe2.com/eQuest/ (accessed on 12 January 2013).
26. Gao, H.; Koch, C.; Wu, Y. Building information modelling based building energy modelling: A review. *Appl. Energy* **2019**, *238*, 320–343. [CrossRef]
27. Autodesk. Available online: https://www.autodesk.com/products/revit/architecture (accessed on 10 January 2021).

28. Graphisoft. Available online: https://graphisoft.com/ (accessed on 10 January 2021).
29. Allplan. Available online: https://www.allplan.com/us_en/ (accessed on 10 January 2021).
30. Muñoz, J.N.; Manzanares, F.V.; Gonçalves, M.M. BIM Approach to Modeling a Sports Pavilion for University Use. *Appl. Sci.* **2020**, *10*, 8895. [CrossRef]
31. Available online: https://www.acca.it/bim-modeling-software?gclid=CjwKCAiAxeX_BRASEiwAc1Qdkb_-Th144lYXAi8UFc8tsWEIJt745zXabINlVxZUYBia5CbHvS0zIBoCCYkQAvD_BwE (accessed on 3 March 2021).
32. ASHRAE 140-2017. Available online: https://xp20.ashrae.org/140-2017/ (accessed on 3 March 2021).
33. Decreto Legislativo 29 Dicembre 2006, n. 311 "Disposizioni Correttive ed Integrative al Decreto Legislativo 19 Agosto 2005, n. 192, Recante Attuazione della Direttiva 2002/91/CE, Relativa al Rendimento Energetico Nell'edilizia". Available online: https://www.camera.it/parlam/leggi/deleghe/testi/06311dl.htm (accessed on 3 March 2021).
34. Decreto Legislativo 30 Maggio 2008, n. 115 "Attuazione della direttiva 2006/32/CE Relativa All'efficienza Degli Usi Finali Dell'energia e i Servizi Energetici e Abrogazione Della Direttiva 93/76/CEE". Available online: https://www.camera.it/parlam/leggi/deleghe/08115dl.htm (accessed on 3 March 2021).
35. UNI/TS 11300-1:2014, Prestazioni Energetiche Degli Edifici—Parte 1: Determinazione del Fabbisogno di Energia Termica Dell'edificio per la Climatizzazione Estiva ed Invernale. Available online: http://store.uni.com/catalogo/uni-ts-11300-1-2014 (accessed on 3 March 2021).
36. UNI/TS 11300-2:2019, Prestazioni Energetiche Degli Edifici—Parte 2: Determinazione del Fabbisogno di Energia Primaria e dei Rendimenti per la Climatizzazione Invernale, per la Produzione di Acqua Calda Sanitaria, per la Ventilazione e per L'illuminazione in Edifici non Residenziali. Available online: http://store.uni.com/catalogo/uni-ts-11300-2-2014 (accessed on 3 March 2021).
37. Design Builder. Available online: https://designbuilder.co.uk// (accessed on 10 January 2021).
38. Open Studio. Available online: https://www.openstudio.net/ (accessed on 10 January 2021).
39. Simergy. Available online: https://d-alchemy.com/products/simergy (accessed on 10 January 2021).
40. Logicalsoft. Available online: https://www.logical.it/software-per-la-termotecnica (accessed on 10 January 2021).
41. UNI EN ISO 52016-1:2018, Prestazione Energetica Degli Edifici—Fabbisogni Energetici per Riscaldamento e Raffrescamento, Temperature Interne e Carichi Termici Sensibili e Latenti—Parte 1: Procedure di Calcolo. Available online: http://store.uni.com/catalogo/uni-en-iso-52016-1-2018 (accessed on 3 March 2021).
42. EcodesignStar. Available online: https://www.bimtechla.com/pdf/Brochure-EcoDesigner_STAR_Brochure.pdf (accessed on 3 March 2021).
43. Mazzarella, L. Dati climatici G. De Giorgio. In *Giornata Di Studio Giovanni De Giorgio*; Politecnico di Milano: Milano, Italy, 18 November 1997.
44. Stirling, J.; Gowan, J. Flats at Langham House, Ham Common, Richmond. In *Architectural Design*; 1958; Available online: https://historicengland.org.uk/listing/the-list/list-entry/1051027 (accessed on 3 March 2021).
45. Izzo, A.; Gubitosi, C. James Stirling, Officina Edizioni, Roma. 1976.
46. Decreto Interministeriale 26 Giugno 2015—Applicazione Delle Metodologie di Calcolo Delle Prestazioni Energetiche e Definizione Delle Prescrizioni e dei Requisiti Minimi Degli Edifici. Available online: https://www.gazzettaufficiale.it/eli/id/2015/07/15/15A05198/sg (accessed on 3 March 2021).
47. D.P.R. 26 August 1993, n. 412 (Annex A), on the subject of the "Regulation containing rules for the design, installation, operation and maintenance of the thermal systems of buildings for the purpose of containing energy consumption".
48. UNI EN ISO 15927-6: 2008, in Terms of Performance Hygrometric Temperature of Buildings—Calculation and Presentation of Climatic Data—Part 6: Cumulated Temperature Differences (Degree Days). Available online: https://standards.iteh.ai/catalog/standards/sist/3da76bd7-e17c-4e3c-a10e-4f0be76d017d/sist-en-iso-15927-6-2008 (accessed on 3 March 2021).
49. Degreedays. Available online: https://www.degreedays.net/#generate (accessed on 10 January 2021).
50. Di Lisi, S. Simulazione Energetica in Ambiente BIM Per Il Progetto di Architettura. Master's Thesis, University of Palermo, Palermo, Italy, 2020.

Article

An openBIM Approach to IoT Integration with Incomplete As-Built Data

Nicola Moretti [1,*], Xiang Xie [1,*], Jorge Merino [1], Justas Brazauskas [2] and Ajith Kumar Parlikad [1]

[1] Institute for Manufacturing, Department of Engineering, University of Cambridge, 17 Charles Babbage Road, Cambridge CB3 0FS, UK; jm2210@cam.ac.uk (J.M.); aknp2@cam.ac.uk (A.K.P.)

[2] The Computer Laboratory, Department of Computer Science and Technology, University of Cambridge, 15 JJ Thomson Ave, Cambridge CB3 0FD, UK; jb2328@cam.ac.uk

* Correspondence: nm737@cam.ac.uk (N.M.); xx809@cam.ac.uk (X.X.)

Received: 2 November 2020; Accepted: 19 November 2020; Published: 23 November 2020

Featured Application: The proposed methodology has been developed to support the Asset Management (AM) decision making according to an open Building Information Modelling (openBIM) approach. Within the context of the West Cambridge Digital Twin Research Facility, a real case scenario has been considered, where the as-built data is imprecise or absent. The methodology is well suited to dealing with incomplete data on existing buildings, when the objective is integration among AM, the Internet of Things (IoT) and BIM information.

Abstract: Digital Twins (DT) are powerful tools to support asset managers in the operation and maintenance of cognitive buildings. Building Information Models (BIM) are critical for Asset Management (AM), especially when used in conjunction with Internet of Things (IoT) and other asset data collected throughout a building's lifecycle. However, information contained within BIM models is usually outdated, inaccurate, and incomplete as a result of unclear geometric and semantic data modelling procedures during the building life cycle. The aim of this paper is to develop an openBIM methodology to support dynamic AM applications with limited as-built information availability. The workflow is based on the use of the IfcSharedFacilitiesElements schema for processing the geometric and semantic information of both existing and newly created Industry Foundation Classes (IFC) objects, supporting real-time data integration. The methodology is validated using the West Cambridge DT Research Facility data, demonstrating good potential in supporting an asset anomaly detection application. The proposed workflow increases the automation of the digital AM processes, thanks to the adoption of BIM-IoT integration tools and methods within the context of the development of a building DT.

Keywords: BIM; openBIM; IFC; IoT; sensors; cognitive buildings; asset management; digital twin

---

1. Introduction

Asset Management (AM) is a key organisational area in Architecture, Engineering, Constructions and Operations (AECO), being a recognised and effective driver for better sustainability of the built environment, while improving asset condition and performance [1,2]. Moreover, the management of the built environment has entered a new phase characterised by a digital transformation of management processes [3]. This phase concerns the adoption of digital tools that can support the production, storage and update of information during the life cycle of assets [4–6].

1.1. Digital Modelling in Asset Management

Building Information Modelling (BIM) is now widely adopted by the industry as part of their digital toolkit, especially when focusing on building systems and components. BIM for building operation has been standardised internationally by ISO 19650-3 [6], which provides guidance for information management during the use phase of the assets [7]. The benefits of BIM have been studied in multiple domains, for example, maintenance prioritisation [8,9], energy management [10], sustainability assessment [11] and life cycle costing [12]. Advances in BIM are likely to reduce the time needed to update databases in the use phase by 98% [13]. However, as a dynamic system, one of the most relevant contemporary challenges in AM concerns the integration of the static data stored and managed through Asset Management Systems (AMS), with the dynamic data provided by Building Management Systems (BMS) [14] and Internet of Things (IoT) sensor networks deployed for specific building management applications [15]. The concept of Digital Twins (DT) aims to address the integration of static and dynamic data, thereby enabling the creation of a digital replica of the physical building that is always up-to-date through its life cycle [16]. DTs are therefore integrated, multifaceted, and multi-scale digital replicas of physical assets, systems, processes, and buildings, that accelerate the development and benefits of BIM in AECO [17,18].

1.2. Data Integration Management

AM processes are still managed based on outdated procedures in practice, hindering the innovation and adoption of digital technologies that could strongly support information management and contribute to the integrity, validity and interoperability of the process [19]. DTs for built environments are still in their infancy, and there are few applications that integrate static and dynamic data in AECO, which is a laggard economic sector in terms of adopting innovative digital tools [20].

The following issues were identified regarding the lack of integration between static and dynamic information in AM:

- BIM models are often created during the design, manufacturing, and construction phases using unclear procedures, and updated as-built models are hardly accessible or even not available. In addition to the static nature of BIM, outdated and unreliable (i.e., inaccurate and incomplete) building information impedes the full potential of the AM applications during the use phase.
- Even when updated as-built BIM data is available, scarce attention is still paid during the design phase to the information management process across the whole asset life cycle. Consequently, the information requirements (IRs) during the use phase are often not met because of the way information is created and aggregated (e.g., classified), during the design and construction phase. BIM is a flexible modelling approach, which supports the inclusion of geometries, assets and systems as part of the model. However this flexibility may result in chaos if recognisable hierarchies and classification systems are not defined in the design phase and adopted during the assets' life cycle.
- Static and real-time data are managed differently because of their nature. For instance, some asset information is designed to be static (e.g., asset locations and geometries), whereas asset performance is measured in real-time in DTs throughout the use phase. Static data is not updated frequently (or at all) and is stored in passive repositories (e.g., relational data-bases or files to query or in COBie spreadsheets). Real-time data is variable, requiring special storage and management (e.g., actively publishing new data for active subscribers). IRs are clearly different for static and real-time data, leading to AM applications that cannot use both sources of information.

To the three main issues described above, a fourth can be added, concerning the inaccessibility of proprietary data formats: siloed black box systems that vendors use often make data interoperability impossible.

Efforts have been made to enable more flexible data integration for AM. On one hand, several studies are currently being conducted to improve the information exchange during the life cycle of the assets within the openBIM approach [21]. OpenBIM indicates the use of BIM based on open standards and workflows to improve the openness, reliability and sustainability of life-long data and enable flexible collaboration between all stakeholders. An example is ongoing standardisation efforts by the International Organisation for Standardisation (ISO), on the 19650 series of standards [4–6]. On the other hand, quality AM processes are being investigated using incomplete and inaccurate information, particularly on existing assets.

1.3. Aim of the Paper

The aim of this paper is to present an openBIM methodology to overcome the separation of existing static/dynamic information in supporting AM applications with awareness of inaccurate and incomplete as-built data. The benefits of this approach include:

- Improved accessibility of the integrated information;
- Users' profiling and access to the right data at the right moment;
- Dynamic AM application support, with limited as-built information availability and
- Enhanced information quality by better matching with the domain specific requirements from different AM applications.

2. State of the Art

The BIM approach can be broadly defined as a set of digital modelling tools, procedures, and methods that support the effective management of information flows during the life cycle of the asset [22]. The benefits of BIM adoption in AM and Facility Management (FM) are well documented [19,23,24]:

- It improves the quality of building data (e.g., preventing data replication and limiting redundancy and inconsistency);
- It facilitates data integration during the building life cycle;
- It improves communication between stakeholders;
- It enables smoother workflows among involved parties according to standardised procedures;
- It allows a reduction in time and cost in the retrieval of FM related information;
- It enables a faster verification process.

Improved information management (i.e., integration, quality, sharing) is the primary benefit that can be achieved through implementing BIM approaches. Through the incorporation of geometry, spatial locations and semantic properties, BIM provides a high-fidelity representation for buildings. The buildings' interaction with users is captured by IoT sensors which are increasingly deployed in the built environment to collect real-time data on the operational condition of buildings [15]. The integration of BIM and IoT has been identified as the key driver for the realisation of cognitive buildings, smart infrastructure and, eventually, the smart built environment [25,26]. Several applications of BIM and IoT data integration can be found in the literature.

2.1. Uses of the BIM and IoT Technologies

In the manufacturing and construction phase, sensor data and BIM technologies can be used to monitor the construction site schedule and improve the procurement process [27]. The use of Virtual and Augmented Reality (VR/AR), which simulate the reality using either virtual reality headsets or multi-projected environments or simply add digital elements to a live view (e.g., the game Pokemon Go), can support construction operations and prevent issues in the execution process (e.g., interference among systems and structural parts) [28]. Global Positioning System (GPS) technologies and Radio-Frequency IDentification (RFID) sensors are utilised to monitor the positioning of building components against the BIM model [29]. In construction logistics and management, IoT data can be employed to track and improve construction site operations [29] within the context of the lean construction, concerning the digitisation and automation of the construction supply chain [30]. In Health and Safety (H&S) management, VR and BIM data have been employed to improve the training process of workers, and their ability to recognise and assess risks [31]. In the same sector, Ref. [32] propose integrating wireless sensor networks and BIM technologies to monitor the safety status (presence of hazardous gas) of underground constructions sites.

In FM, BIM and IoT data integration have been studied to enrich the condition monitoring of critical assets and real-time assessment of their performance [33]. VR/AR technologies enhance indoor navigation [34], which upgrades maintenance procedures. BIM and energy data integration improves energy management [35]. Ultrasonic sensors can be used with BIM for maintenance service optimisation [36]. Dynamic environmental data can be used to achieve higher user comfort and to adapt system behaviour [37].

The number of applications is growing, and the topic has gained momentum, representing a leading research field, which lays a solid foundation for the identification, collation and curation of operational datasets and demonstrating the great potential of the DT applications in AECO.

2.2. Integration Architectures

Besides the applications that can be developed through fruitful BIM and IoT integration, a critical aspect can be found in the static and dynamic data integration. The development of an effective architecture allows for leveraging of the true potential of static information concerning geometries, location and relations among the building elements and the related semantics stored in the BIM model; AM and FM information is generally collected in an Asset Information Model (AIM) [8]; and the dynamic data streamed through the IoT technologies and managed through the related infrastructures. Different types of architecture can be found in the literature, allowing diverse operations on data. Ref. [15] classifies these architectures according to five methods, as shown in Table 1. These methods fulfil the integration of BIM and IoT by utilising BIM tools' APIs and relational database, transform BIM data into a relational database using new data schema, create new query language, using semantic web technologies and hybrid approach, respectively. Basically, these methods keep contextual information (BIM data) and time-series (sensor collected) data, and integrate them from different angles. Their methodological description, advantages and disadvantages are explained in Table 1.

Table 1. BIM IoT data integration methods [15].

Method	Advantages	Drawbacks
BIM tools' APIs + relational database: Sensor and BIM data are stored in a relational DB. Virtual objects are connected to sensor data through unique identifiers [38,39].	• Extensive software support; • existing of APIs allow the export/import of BIM data in the right format; • easy of using SQL.	• Poor in BIM data export and enrichment capabilities; • insufficient of model change management support.
New data schema creation: Transform BIM data into relational database using new data schema [40,41].	• Flexible in users' customisation; • supporting data federation (no need for conversion); • allow effective data management in large projects.	• Time-consuming in mapping operations; • requires BIM data knowledge and editing skills.
Create a new Query Language (QL): for querying time-series and IFC data [42,43].	• Expressiveness of QL; • optimised for domain-specific applications.	• Scarce dynamic data query capabilities; • need to develop a dedicated platform; • no standardisation.

Table 1. *Cont.*

Method	Advantages	Drawbacks
Semantic web approach: for storing, sharing, using heterogeneous data [44,45].	• Linking data silos; • managing cross-domain information; • effective in projects with broad scope.	• Need to represent data in homogeneous format (RDF); • RDF is not optimal for querying dynamic data; • data redundancy risk; • fix structure and storage consuming.
Hybrid approach: semantic web + relational database: both approaches are used for storing cross-domain data [46,47].	• Data is stored in the most suitable platform; • time saving (no conversion); • storage saving; • better performance; • effective QL.	• RDF conversion still needed.

3. The Proposed Openbim Methodology

In this section the methods employed to develop the proposed openBIM approach for IoT data integration are presented while exposing the limitations. Then, the proposed workflow is depicted addressing these limitations to leverage the full potential of the available static and dynamic data, supporting the development of AM applications, even when as-built data is incomplete. The purpose is to create an approach that is effective when dealing with existing buildings where as-built information is frequently absent or not reliable.

The proposed openBIM approach aims to extend the BIM methods and tools for improved accessibility, usability, management and sustainability of data in AECO [21]. It promotes data sharing and collaboration among parties using open standards, addressing the common BIM issues related to proprietary technologies and software. For this purpose, the Industry Foundation Classes (IFC) data schema has been adopted to support and handle the BIM data [48].

The IFC schema is an object-oriented open standard [49] widely studied as an effective means for interoperability, sharing, collaboration and classification. The IFC schema is extensive and complex, and therefore its usage has been limited to simple software interoperability workflows and visualisation of key information of the BIM model. Nonetheless, IFC offers good support for not only geometry representation, but also semantic data enrichment. In this research, the `IfcSharedFacilitiesElements` schema has been employed to handle geometric and semantic information of both existing and newly created IFC objects, supporting real-time data integration.

Rarely, the level of both geometric and semantic detail and the classification system (the granularity) adopted in developing the BIM model, in the design and construction phase, is adopted in the use phase. According to [50], the `IfcAsset` has been adopted for re-aggregating building components in order to achieve the desired level of granularity used in AM. The `IfcAsset` (the element breakdown is elaborated in Figure 1) is defined as "a grouping of elements acting as a single element that has a financial value", allowing objects that are not spatially connected to be related, through the relationship `IfcRelAssignsToGroup`. Moreover, another artefact that can be leveraged for semantic enrichment of the BIM model is the `IfcAsset`, which allows the objectified relationships `IfcAssignToActor`, `IfcAssignToControl`, `IfcAssignToProcess` and `IfcAssignToResources` to be associated with a wide set of data within the context of the AM domain.

In practice, as-built data is incomplete and not reliable. When integrating with IoT data, this may result in hampering an effective digital representation of both the sensor objects and the spatial elements measured by the sensors. This issue is addressed by modelling non-geometric objects in IFC, which allows a modular updating of the BIM model. This approach leverages both the 3D and semantic potential of the IFC schema, streamlining the integration of BIM, AM and IoT data.

From the perspective of applications, detailed spatial/geometrical information is not always necessary. As typical distributed systems, buildings need to be monitored, managed and controlled. Eventually, buildings' performance can be simulated under known building systems organisation using individual asset components. For instance, for the anomaly diagnosis of Heating, Ventilation and Air-Conditioning (HVAC) systems in a specific building, only the basic mechanical system information is needed: describing the HVAC system configuration and the links between architecture zones and HVAC terminal units [51]. This scenario is used in Section 4 for demonstrating the benefits of the proposed approach. In the development of the `IfcAsset`, in fact, the classification system adopted by asset and facility managers must be considered in order to support inter-operable and flexible AM processes. For querying and modifying the IFC, several Application Programming Interfaces (APIs) can be used. The IfcOpenShell-python (http://ifcopenshell.org/) module and the BIMserver (https://github.com/opensourceBIM/BIMserver) software for IFC visualisation and queries are used in this research.

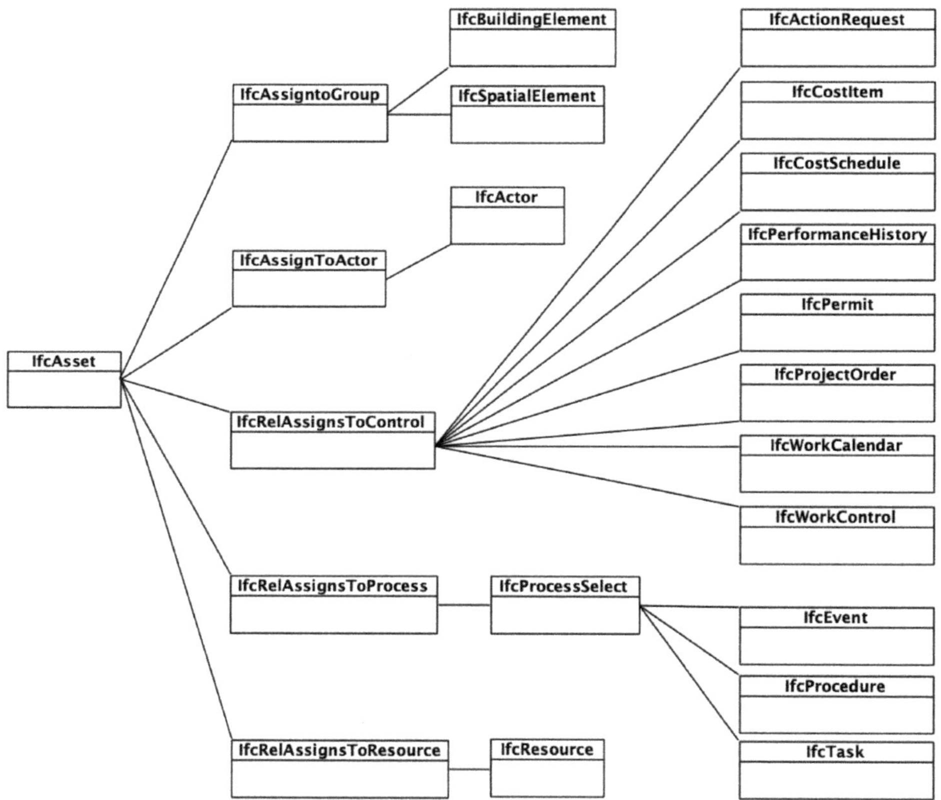

Figure 1. IfcAsset schema.

Figure 2 depicts the proposed methodology. The process starts with the definition of IRs, which is designed to identify the relevant data employed for the operation of the building according to the business and client needs and the outcomes of the application. This step is composed of three sub-tasks regarding the definition of the static and dynamic IRs and the definition of the Service Level Agreements (SLAs).The static IRs correspond to the AM information, characterised by a low frequency of updates and more classical information management. Static IRs include the level of aggregation of the assets (granularity) at which the maintenance interventions are conducted. The dynamic IRs concern the IoT data management, including how to aggregate data into indicators to measure performance through the installed sensors, and how to associate this information with the physical and spatial elements of the building. The SLAs concern the performance agreed for operating the assets at an acceptable service level.

The IRs definition step is crucial to the IFC processing and the creation of the `IfcAsset` assignment relationships. The IFC processing should consider the classification system used for AM (explicitly related to how the static and dynamic data are handled). This would facilitate the adoption of the methodology in practice and increase its usability.

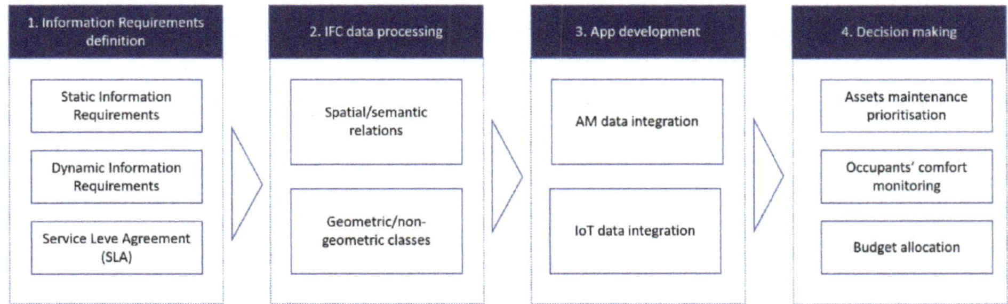

Figure 2. Research schema.

The next step is IFC data processing to meet the IRs. To achieve this, it is necessary to edit the initial IFC file generated through the BIM authoring software. IoT integration requires access to data schemes related to objects that might not be available (modelled) in the IFC file. This is a common issue with existing buildings, where the BIM information is partial or outdated as a result of the modifications of the physical elements and functions of the building during its use. These modifications can be recorded, for example, in the building logbook, but they are rarely collected and managed in BIM models. Furthermore, system components are frequently difficult to access and inspect, and therefore to model correctly in the BIM environment. For this reason, in a streamlined approach to BIM data updates, the geometry of not accessible (or visible) components may be overlooked, focusing mainly on the semantic enrichment. Nonetheless, the modular development of the digital model should always be considered, enabling detailed 3D data integration once available (e.g., through a detailed inspection, after a refurbishment). Additionally, to represent correctly the semantic relationships among the newly created and existing IFC classes, the objectified relationships also need to be modelled. This allows for the connection and effective querying of systematically interdependent IFC classes that are not originally related in the IoT applications. When dealing with existing building data, the aggregation and classification system adopted in design and construction does not match its counterpart used in operations. As a consequence, re-aggregation must be conducted to unlock the real potential of BIM data in AM. The `IfcAsset` class has been employed for this purpose as the grouping entity enabling the collection of homogeneous sets of elements forming the parts of the systems in the building.

The last step concerns IoT data integration. The entire IFC schema is not necessary to handle IFC and IoT integration; rather, it can be achieved by linking the sensor readings to the existing or newly created IFC classes, through the Globally Unique Identifier (GUID) of each IFC object. Therefore, only the relevant IFC subset need to be exported, following a Model View Definition (MVD) approach [52]. The MVD is essentially a filtered view of the IFC, which allows the extraction of specific packages of model information to meet a particular use. The application development and implementation based on this integration must support dynamic decision-making. Some of them are listed in the Step 4 in Figure 2. Step 2, IFC data processing will be discussed in detail in the following section.

4. Case Study

The proposed methodology has been applied to the Institute for Manufacturing (IfM) building located in the West Cambridge campus of the University of Cambridge. It is part of the West Cambridge DT Research Facility [18] and has been equipped with a customised IoT sensor network, comprising a set of Monnit (https://www.monnit.com/Products/Sensor/) wireless sensors measuring indoor environmental and asset parameters, for instance, temperature (°C), relative humidity (%), CO_2 concentration (ppm) of indoor spaces and window open/close status, and HVAC pump vibration frequency (Hz) among others.

A BMS, based on the Trend (http://www.trendcontrols.com/en-GB/) platform, is currently used to monitor the performances of mechanical, electrical and pumping (MEP) systems. This data remains in a different system and is not integrated with the IoT sensor data. Thus, it cannot easily be used together to make informed decisions on assets operations. To demonstrate the capability of the designed scheme, a typical anomaly detection application for the HVAC system monitoring is implemented [35].

Assets responsible for delivering the functionalities of the building determine the quality of the services and the comfort of the spaces that it provides for its inhabitants. Monitoring the working condition of the assets and further revealing the raised anomalies, either environmental or asset-wise, is important for guaranteeing building operational performance. As a result of the limitation in computational resources for buildings, the performances of HVAC system components are monitored individually without considering their interdependence. Even for individual component, the monitoring of assets anomaly detection, for operation and maintenance management requires comprehensive data sources, both static and dynamic, for re-classifying building facilities information. For the definition of the static information requirements, the following information has been considered:

- Geometries and location of the HVAC components, including primary air loop, variable refrigerant flow (VRF), water circulation pumps and radiators;
- Relevant data in the civil components of the building (technical specifications, active contracts, maintenance records, models and producer of the components);
- Sensor location and technical specifications;
- System architecture, that is, the way the HVAC system is organised from multiple components, according to a classification system;
- Interface requirements with the real-time platform.

The real-time information requirements are defined considering the following:

- Set points for the HVAC system (e.g., the temperature of the rooms, relative humidity, CO_2 concentration);
- Data on comfort parameters measurements (BMS, Monnit sensors);
- Data on the BMS and IoT sensors status.

Table 2 collects the information requirements defined for two rooms (labelled G.44 and 1.58) and the related assets in the building. According to the anomaly detection application needs, the level of aggregation of the assets has been defined. This new bespoke classification has been employed to group the relevant building elements to be monitored through the anomaly detection application. Some were already present in the initial version of the IFC file (i.e., Table 2 marked as Existing "yes"), while others had to be created (as non-geometrical classes). In the modelling procedure, the IFC2x3 TC1 [53] version has been used. This version, despite being improved by Version 4, offers wider software support, allowing more accurate visualisation of complex geometries, especially in commercial software. However, the possibility to upgrade the workflow with more recent IFC versions has been considered.

Table 2. Anomaly detection application information requirements definition.

Code	Asset	Sensor Name	Location	Sensor Type	Unit	IFC Entities	Existing
G44	G.44 (lev.0)					IfcSpace	yes
G44_1		Monnit sensor	G.44 room	sensor unit	°C/%	IfcSensorType	no
G44_2		Temperature VRF	Air terminal at VRF 36	sensor unit	°C	IfcEnergyConversionDevice	yes
G44_3		Temperature VRF	Air terminal at VRF 37	sensor unit	°C	IfcEnergyConversionDevice	yes
G44_4		Fan speed VRF	Air terminal at VRF 36	integrated	level (1-n)	IfcFanType	no
G44_5		Fan speed VRF	Air terminal at VRF 37	integrated	level (1-n)	IfcFanType	no
AHU	AHU2					IfcAsset	no
AHU_1		AHU extract air temperature	after the air mixer	sensor unit	°C	IfcSensorType	no
AHU_2		AHU extract fan speed	AHU extract fan	integrated	ls-1/%	IfcFanType	no
AHU_3		AHU extract air filter DPS	AHU extract air filter	integrated	Pa	IfcFilterType	no
AHU_4		AHU supply air filter DPS	AHU supply air filter	integrated	Pa	IfcFilterType	no
AHU_5		AHU supply fan speed	AHU supply fan	integrated	ls-1/%	IfcFanType	no
AHU_6		AHU supply air reheat level	AHU supply air reheat	integrated	%	IfcCoilType	no
AHU_7		AHU supply air temperature	before the air splitter	sensor unit	°C / Pa	IfcSensorType	no
AHU_9		Thermowheel exchange rate	Thermowheel	integrated	% heat	IfcAirToAirHeatRecoveryType	no
WR2	WR2					IfcAsset	no
WR2_1		WR2 supply temperature	before WR2 loop	sensor unit	°C	IfcSensorType	no
WR2_2		WR2 cooling pump DPS	WR2 cooling pump	integrated	Pa	IfcPumpType	no
WR2_3		WR2 return temp	leaving WR2 loop	sensor unit	°C	IfcSensorType	no
DAC_1		Dry air cooler DPS	DAC	integrated	Pa	IfcChillerType	no
DAC_2		DAC on temp	before DAC	integrated	°C	IfcSensorType	no
DAC_3		DAC off temp	after DAC	integrated	°C	IfcSensorType	no
DIAL	1.58 (lev. 1)					IfcSpace	yes
DIAL_1		Space temp	space	sensor unit	°C	IfcSensorType	no
RAD	Radiators					IfcAsset	no
RAD_1		Radiator pump DPS	Radiator pump	integrated	Pa	IfcPumpType	no
RAD_2		VT flow supply temp	radiator inlet	sensor unit	°C	IfcSensorType	no
RAD_3		VT flow return temp	radiator outlet	sensor unit	°C	IfcSensorType	no
RAD_4		VT heat meter		sensor unit	Kwh	IfcFlowMeterType	no

In the definition of the IRs, IoT and BMS sensors are crucial for collecting data on the comfort and function of the spaces and equipment in the building. Different types of sensor had to be handled accordingly: environmental sensors (i.e., temperature, RH and CO_2) have been related directly to the `IfcSpace`, while integrated sensors have been associated with the directly related building components (e.g., AHU_4.AHU supply air filter DPS in Table 2 have been associated to `IfcFilterType`). After creating the missing IFC objects, they have been aggregated to form assets, through the `IfcAsset`. Figure 3 represents the updated IFC including missing building components in Table 2.

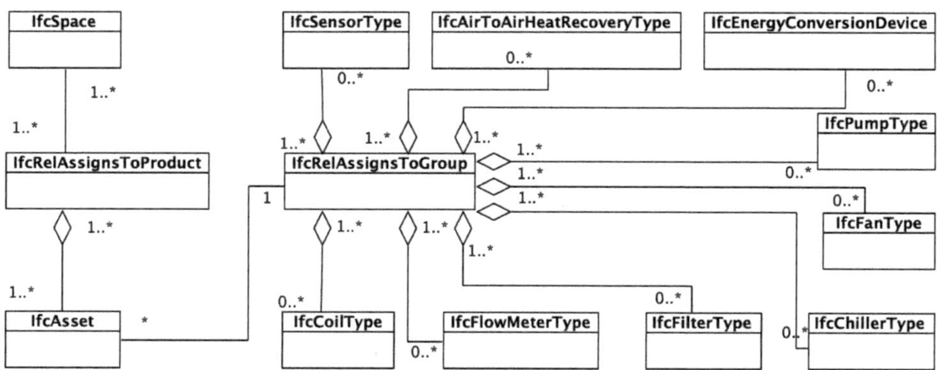

Figure 3. UML schema of the proposed approach implemented in the case study.

The entities necessary to develop the `IfcAsset` classes can be defined in advance through the IfcOpenShell-python software and are used as follows:

```
G44_asset = ifcfile.createIfcAsset(
                        create_guid(),
                        owner_history,
                        'G44 Asset',
                        'Critical components in Room G44',
                        'Asset',
                        Identification,
                        OriginaValue,
                        CurrentValue,
                        TotalReplacementCost,
                        Owner,
                        User,
                        ResponsiblePerson,
                        IncorporationDate,
                        DepreciatedValue
                        )
```

The relationships among the relevant elements and the assets are created and associated with the asset.

```
ifcfile.createIfcRelAssignsToGroup(
                                create_guid(),
                                owner_history,
                                'Asset AHU - group',
                                'Group of objects in AHU asset',
```

```
                                    (AHU_1, AHU_2, AHU_3, AHU_4, AHU_5, AHU_6, AHU_7,
                                    AHU_9),
                                    None,
                                    AHU
                                    )
```

Finally, the assets are connected to the served room.

```
ifcfile.createIfcRelAssignsToProduct(
                                    create_guid(),
                                    owner_history,
                                    None,
                                    None,
                                    (G44_asset, AHU, WR2),
                                    None,
                                    G44)
```

The processed IFC can be visualised and queried in BIMserver as displayed in Figure 4. Furthermore, the model can be queried through JavaScript Object Notation (JSON) queries, and the subset of the original IFC data can be downloaded, allowing the creation of MVDs able to support further data integration. This is relevant in the anomaly detection application since, after running necessary algorithms on dynamic data, it is possible to access the information related to the assets potentially responsible for the detected anomalies.

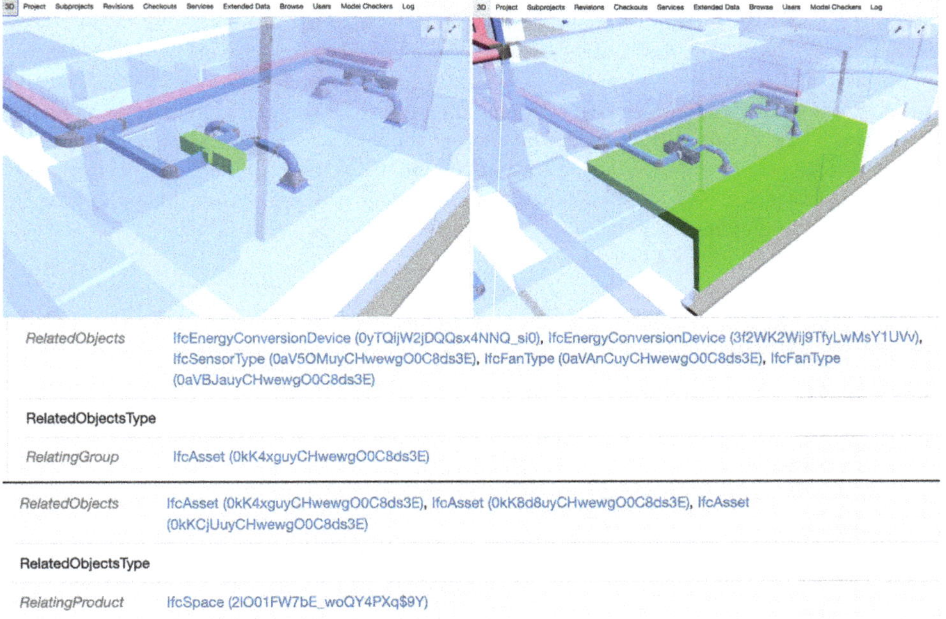

Figure 4. Visualisation of processed IFC through BIMserver.

Taking the asset anomaly detection application on the HVAC system as an example (Figure 5), the processed IFC lays a solid foundation for flexible data integration that supports corresponding AM functions. In this case, the real-time operational data of the HVAC components, such as the WR2 cooling pump and AHU extract fan, and the real-time environmental data of the regulated spaces, are integrated for analysis through the proposed approach (Figure 5).

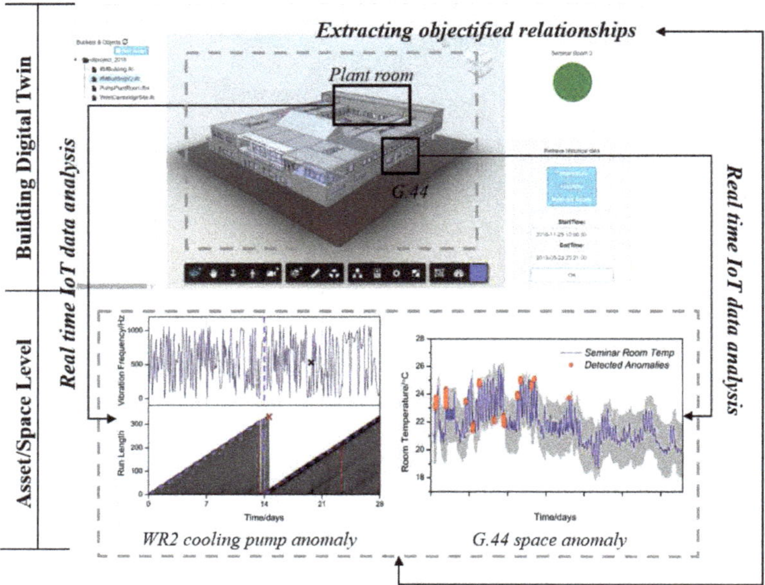

Figure 5. Anomaly detection application.

Contextual anomaly detection algorithms, like cumulative sum (CUSUM) control chart or Bayesian Online Change Point Detection (BOCPD), can be used to dynamically reveal anomalous behaviours that deviate from the anticipation [35]. Subsequently, extracting the objectified relationships among modelled assets and spaces, the causality of found anomalies can be inferred, and the root cause can be identified accordingly. Picking up the correlations of the unexpected anomalies, corresponding local repair, replacement and maintenance operation activities can be triggered to enable preventive maintenance and mitigate the effects of failure risk. In particular, the proposed approach provides a useful tool to back up semantic and geometric data management for AM applications and to facilitate the development of potential application areas [54].

5. Discussion

The proposed approach supports data integration and interoperability in the digital built environment. The methodology enables the effective utilisation of BIM data in the use phase of the assets, supporting the dynamic decision making. Data from the three systems (BIM environment, IoT platform and BMS) were integrated and processed, focusing on the usability of the systems, building a new data structure on top of the existing data sets. Accordingly, information can be accessed and used, integrating and supporting the workflows and operations of the asset management team. The data re-aggregation and processing allows useful insights supporting operations in AM of cognitive buildings'.

The methodology was developed employing open source software for better interoperability and cross-platform usage. In addition, the open standard IFC is used to support information management in the context of the cognitive buildings and smart built environments. IFC also allows the accessibility and integration capabilities to be increased in the development of further applications. The proposed approach, considering the Digital Twin system architecture proposed by [18], sits in the Data/Model integration layer and therefore supports the data integration for the development of multiple AM applications.

IFC artefacts such as `IfcAsset` and `IfcSharedFacilitiesElements` were used in the proposed methodology to enable data integration, including the association with the controls and processes

(Figure 1) capable of supporting the automation of AM processes. Although the enrichment of the IFC schema with relevant AM information (e.g., contractors, resources, economic and financial data, maintenance planning data etc.) may be beneficial in the standardisation of the data collection and process, it can also result in storage and update issues. IFC is a static data format, which poorly supports the dynamic data update. Accordingly, the IRs definition phase in this methodology is vital to the success of this approach, since it includes the potential uses of data during operation.

The case study demonstration was conducted using IFC 2x3, despite Version 4 [48] being available. The methodology can be adapted to the newest version for the data set based on Version 4. The application of the workflow to a digital model with more detailed geometries is possible, since the type entities have been generated in IFC for non-geometric virtual elements. The types can be related to the geometric objects by means of the `IfcRelDefinesByType` relationships, once they become available (e.g., after a refurbishment). After updating the IFC with geometries, further capabilities of the schema can be leveraged for location-enabled AM services (e.g., indoor navigation, H&S and agent-based simulations).

Despite IFC being one of the primary means for interoperability and openBIM standards, it often needs to be converted to be fully usable in the development of software applications. A conversion procedure should be defined. BIMserver offers the possibility to export a sub-set of the imported IFC data, in order to achieve this result. Data can also be exported in JSON format, enabling the development of more generic and cross-domain applications [55]. This process should follow the MVD approach [52] in order to be repeatable and recognisable within the BIM domain, even though this approach is not extensively described in this article. Accordingly, it is possible to enhance the capabilities offered by the IFC model and IFC data, which can be queried through existing technologies and languages after being processed.

We consider that the classification of BIM IoT data integration methods in Table 1 requires some extension with the advent of new technologies. The openBIM methodology proposed in this paper cannot be completely classified into any of those categories, as it shares some characteristics of multiple methods. The characteristics of the openBIM approach that we propose are:

- Flexible schema: Data in the OpenBIM approach is not constraint by a classical relation data schema. A flexible data schema is proposed to facilitate data collection from diverse data sources.
- Standardised metadata: Predefined common metadata attributes to tag data from different sources homogeneously in the data platform. These agreed metadata attributes also enable dynamic data integration and multi-format conversion.
- Real-time perspective: One of the main goals is to enable rapid data transfers by limiting the size of data packages. This reduces the latency of data end-to-end and allows timely decision-making.

We have implemented this data management approach by using JSON Objects throughout the platform. JSON is a Not Only Structured Query Language (NoSQL) approach with a flexible schema. The flexibility is managed through predefined attributes to tag each data message. There is no need for a particular querying language, since JSON Objects can be serialised in plain text. This also supports rapid data transfer given that data coming from dynamic sources send small packages embedding data in individual JSON files. BIM information is extracted from the original IFC files with BIM tools APIs and integrated as required by the AM applications. Thus, it is possible to assert that our approach could be considered as a combination of all the categories in Table 1, leveraging most advantages from each one of them. Further quantitative investigation of the methodology performance is necessary, particularly when comparing it with other integration architectures.

From the application perspective, the designed IFC scheme opens the door for a diversity of AM applications. Through the definition of the IRs at the beginning of the specific application development, it is possible to integrate both real-time and static information from different systems to support conventional and dynamic decision-making in AM. Overcoming the challenge of fragmented data, the use of diverse information collected in the design, construction and, particularly, the use phase, together, can be beneficial for a variety of AM practices, such as commissioning and closeout, quality control and assurance, energy management, maintenance and repair, and space management [54]. Within this context, the proposed methodology will be further tested in applications requiring IFC data to be re-aggregated according to a different criteria, supporting information management and static/dynamic data integration.

6. Conclusions

The proposed approach has shown good applicability to existing buildings, allowing the issues arising from the lack or incompleteness of data to be addressed. Through the proposed methodology, the potential of data usually siloed in their own domain can be accessed more easily, supporting the development of AM applications for cognitive buildings. It offers an effective approach to data integration in the mid-term perspective, providing support for both the integration of static AM information and real-time IoT data. Furthermore, this paper demonstrates the potential of the openBIM approach in built AM, enabling a data-driven approach that can help to reduce the uncertainty arising from the lack of knowledge on the physical and digital assets and automating operations. In future research, its robustness should be tested in the development of additional application case studies.

Author Contributions: Conceptualization, N.M., X.X.; methodology, N.M., X.X.; software, N.M., X.X.; validation, N.M., X.X., J.M., J.B., A.K.P.; formal analysis, N.M., X.X.; investigation, N.M., X.X., J.M., J.B., A.K.P.; resources, A.K.P.; data curation, N.M., X.X., J.M.; writing–original draft preparation, N.M., X.X., J.M.; writing–review and editing, N.M., X.X., J.M., J.B., A.K.P.; visualization, N.M., X.X.; supervision, A.K.P.; project administration, A.K.P.; funding acquisition, A.K.P. All authors have read and agreed to the published version of the manuscript.

Funding: This research forms part of the Centre for Digital Built Britain's (CDBB) work at the University of Cambridge within the Construction Innovation Hub (CIH). The CIH is funded by UK Research and Innovation through the Industrial Strategy Fund.

Conflicts of Interest: The authors declare no conflict of interest.

References

1. International Organization for Standardization (ISO). *ISO 55000:2014 Asset Management Overview, Principles and Terminology*; ISO: Geneva, Switzerland, 2014.
2. Lu, Q.; Xie, X.; Parlikad, A.K.; Schooling, J.M.; Konstantinou, E. Moving from building information models to digital twins for operation and maintenance. *Proc. Inst. Civ. Eng. Smart Infrastruct. Constr.* **2020**, 1–11. [CrossRef]
3. Wong, J.K.W.; Ge, J.; He, S.X. Digitisation in facilities management: A literature review and future research directions. *Autom. Constr.* **2018**, *92*, 312–326. [CrossRef]
4. International Organization for Standardization (ISO). *EN ISO 19650-1:2018. Organization and Digitization of Information about Buildings and Civil Engineering Works, Including Building Information Modelling (BIM)—Information Management Using Building Information Modelling. Part 1: Concepts and Principles*; ISO: Geneva, Switzerland, 2018.
5. International Organization for Standardization (ISO). *EN ISO 19650-2:2018. Organization and Digitization of Information about Buildings and Civil Engineering Works, Including Building Information Modelling (BIM)—Information Management Using Building Information Modelling. Part 2: Delivery Phase of the A*; ISO: Geneva, Switzerland, 2018.
6. International Organization for Standardization (ISO). *EN ISO 19650-3:2020. Organization and Digitization of Information about Buildings and Civil Engineering Works, Including Building Information Modelling (BIM)—Information Management Using Building Information Modelling*; ISO: Geneva, Switzerland, 2018.
7. Royal Institute of British Architects (RIBA). *RIBA Plan of Work 2020 Overview*; Technical Report; RIBA: London, UK, 2020. [CrossRef]
8. Patacas, J.; Dawood, N.; Kassem, M. BIM for facilities management: A framework and a common data environment using open standards. *Autom. Constr.* **2020**, *120*, 103366. [CrossRef]
9. Moretti, N.; Re Cecconi, F. A cross-domain decision support system to optimize building maintenance. *Buildings* **2019**, *9*, 161. [CrossRef]

10. Chen, C.J.; Chen, S.Y.; Li, S.H.; Chiu, H.T. Green BIM-based building energy performance analysis. *Comput. Aided Des. Appl.* **2017**, *14*, 650–660. [CrossRef]
11. Wong, J.K.W.; Zhou, J. Enhancing environmental sustainability over building life cycles through green BIM: A review. *Autom. Constr.* **2015**, *57*, 156–165. [CrossRef]
12. Santos, R.; Costa, A.A.; Silvestre, J.D.; Pyl, L. Integration of LCA and LCC analysis within a BIM-based environment. *Autom. Constr.* **2019**, *103*, 127–149. [CrossRef]
13. Ding, L.; Drogemuller, R.; Akhurst, P.; Hough, R.; Bull, S.; Linning, C. Towards sustainable facilities management. In *Technology, Design and Process Innovation in the Built Environment*; Taylor & Francis: London, UK, 2009; pp. 373–392.
14. Jourdan, M.; Meyer, F.; Bacher, J.P. Towards an integrated approach of building-data management through the convergence of Building Information Modelling and Internet of Things. *J. Phys. Conf. Ser. Inst. Phys. Publ.* **2019**, *1343*, 12135. [CrossRef]
15. Tang, S.; Shelden, D.R.; Eastman, C.M.; Pishdad-Bozorgi, P.; Gao, X. A review of building information modeling (BIM) and the internet of things (IoT) devices integration: Present status and future trends. *Autom. Constr.* **2019**, *101*, 127–139. [CrossRef]
16. Bolton, A.; Butler, L.; Dabson, I.; Enzer, M.; Evans, M.; Fenemore, T.; Harradence, F.; Keaney, E.; Kemp, A.; Luck, A.; et al. *The Gemini Principles*; CDBB: Cambridge, UK, 2018.
17. BuildingSMART International. Enabling an Ecosystem of Digital Twins. A buildingSMART International Positioning Paper. How to Unlock Economic, Social, Environmental and Business Value for the Built Asset Industry. Available online: https://www.buildingsmart.org/wp-content/uploads/2020/05/Enabling-Digital-Twins-Positioning-Paper-Final.pdf (accessed on 20 November 2020).
18. Lu, Q.; Parlikad, A.K.; Woodall, P.; Don Ranasinghe, G.; Xie, X.; Liang, Z.; Konstantinou, E.; Heaton, J.; Schooling, J. Developing a Digital Twin at Building and City Levels: Case Study of West Cambridge Campus. *J. Manag. Eng.* **2020**, *36*, 1–19. [CrossRef]
19. Dixit, M.K.; Venkatraj, V.; Ostadalimakhmalbaf, M.; Pariafsai, F.; Lavy, S. Integration of facility management and building information modeling (BIM): A review of key issues and challenges. *Facilities* **2019**, *37*, 455–483. [CrossRef]
20. Baldini, G.; Barboni, M.; Bono, F.; Delipetrev, B.; Duch Brown, N.; Fernandez Macias, E.; Gkoumas, K.; Joossens, E.; Kalpaka, A.; Nepelski, D.; et al. *Digital Transformation in Transport, Construction, Energy, Government and Public Administration* Technical Report; Publications Office of the European Union: Luxembourg, 2019. [CrossRef]
21. BuildingSMART International. openBIM. Available online: https://www.buildingsmart.org/about/openbim/ (accessed on 20 November 2020).
22. The British Standards Institution (BSI). Building Information Modelling (BIM). Available online: https://www.bsigroup.com/en-GB/Building-Information-Modelling-BIM/ (accessed on 20 November 2020).
23. Bryde, D.; Broquetas, M.; Volm, J.M. The project benefits of building information modelling (BIM). *Int. J. Proj. Manag.* **2013**, *31*, 971–980. [CrossRef]
24. Azhar, S. Building information modeling (BIM): Trends, benefits, risks, and challenges for the AEC industry. *Leadersh. Manag. Eng.* **2011**, *11*, 241–252. [CrossRef]
25. Andrade, R.O.; Yoo, S.G. A comprehensive study of the use of LoRa in the development of smart cities. *Appl. Sci. (Switzerland)* **2019**, *9*, 4753. [CrossRef]
26. Boje, C.; Guerriero, A.; Kubicki, S.; Rezgui, Y. Towards a semantic Construction Digital Twin: Directions for future research. *Autom. Constr.* **2020**, *114*, 103179. [CrossRef]
27. Alizadehsalehi, S.; Yitmen, I. The Impact of Field Data Capturing Technologies on Automated Construction Project Progress Monitoring. *Procedia Eng.* **2016**, *161*, 97–103. [CrossRef]
28. Park, C.S.; Lee, D.Y.; Kwon, O.S.; Wang, X. A framework for proactive construction defect management using BIM, augmented reality and ontology-based data collection template. *Autom. Constr.* **2013**, *33*, 61–71. [CrossRef]
29. Zekavat, P.R.; Moon, S.; Bernold, L.E. Performance of short and long range wireless communication technologies in construction. *Autom. Constr.* **2014**, *47*, 50–61. [CrossRef]
30. Dave, B.; Kubler, S.; Främling, K.; Koskela, L. Opportunities for enhanced lean construction management using Internet of Things standards. *Autom. Constr.* **2016**, *61*, 86–97. [CrossRef]

31. Sacks, R.; Perlman, A.; Barak, R. Construction safety training using immersive virtual reality. *Constr. Manag. Econ.* **2013**, *31*, 1005–1017. [CrossRef]
32. Cheung, W.F.; Lin, T.H.; Lin, Y.C. A Real-Time Construction Safety Monitoring System for Hazardous Gas Integrating Wireless Sensor Network and Building Information Modeling Technologies. *Sensors* **2018**, *18*, 436. [CrossRef]
33. Chang, K.M.; Dzeng, R.J.; Wu, Y.J. An automated IoT visualization BIM platform for decision support in facilities management. *Appl. Sci. (Switzerland)* **2018**, *8*, 1086. [CrossRef]
34. Herbers, P.; König, M. Indoor localization for augmented reality devices using BIM, point clouds, and template matching. *Appl. Sci. (Switzerland)* **2019**, *9*, 4260. [CrossRef]
35. Lu, Q.; Xie, X.; Parlikad, A.K.; Schooling, J.M. Digital twin-enabled anomaly detection for built asset monitoring in operation and maintenance. *Autom. Constr.* **2020**, *118*, 103277. doi:10.1016/j.autcon.2020.103277. [CrossRef]
36. Moretti, N.; Blanco Cadena, J.D.; Mannino, A.; Poli, T.; Re Cecconi, F. Maintenance service optimization in smart buildings through ultrasonic sensors network. *Intell. Build. Int.* **2020**. doi:10.1080/17508975.2020.1765723. [CrossRef]
37. Rinaldi, S.; Bellagente, P.; Camillo Ciribini, A.L.; Chiara Tagliabue, L.; Poli, T.; Giovanni Mainini, A.; Speroni, A.; Blanco Cadena, J.D.; Lupica Spagnolo, S. A cognitive-driven building renovation for improving energy effciency: The experience of the elisir project. *Electronics (Switzerland)* **2020**, *9*, 666. [CrossRef]
38. Marzouk, M.; Abdelaty, A. Monitoring thermal comfort in subways using building information modeling. *Energy Build.* **2014**, *84*, 252–257. [CrossRef]
39. Woo, J.H.; Peterson, M.A.; Gleason, B. Developing a Virtual Campus Model in an Interactive Game-Engine Environment for Building Energy Benchmarking. *J. Comput. Civ. Eng.* **2016**, *30*, C4016005. [CrossRef]
40. Solihin, W.; Eastman, C.; Lee, Y.C.; Yang, D.H. A simplified relational database schema for transformation of BIM data into a query-efficient and spatially enabled database. *Autom. Constr.* **2017**, *84*, 367–383. [CrossRef]
41. Khalili, A.; Chua, D.K.H. IFC-Based Graph Data Model for Topological Queries on Building Elements. *J. Comput. Civ. Eng.* **2015**, *29*, 04014046. [CrossRef]
42. Mazairac, W.; Beetz, J. BIMQL—An open query language for building information models. *Adv. Eng. Inform.* **2013**, *27*, 444–456. [CrossRef]
43. Alves, M.; Carreira, P.; Costa, A.A. BIMSL: A generic approach to the integration of building information models with real-time sensor data. *Autom. Constr.* **2017**, *84*, 304–314. [CrossRef]
44. Dibley, M.; Li, H.; Rezgui, Y.; Miles, J. An ontology framework for intelligent sensor-based building monitoring. *Autom. Constr.* **2012**, *28*, 1–14. [CrossRef]
45. Curry, E.; O'Donnell, J.; Corry, E.; Hasan, S.; Keane, M.; O'Riain, S. Linking building data in the cloud: Integrating cross-domain building data using linked data. *Adv. Eng. Inform.* **2013**, *27*, 206–219. [CrossRef]
46. Hu, S.; Corry, E.; Curry, E.; Turner, W.J.; O'Donnell, J. Building performance optimisation: A hybrid architecture for the integration of contextual information and time-series data. *Autom. Constr.* **2016**, *70*, 51–61. [CrossRef]
47. McGlinn, K.; Yuce, B.; Wicaksono, H.; Howell, S.; Rezgui, Y. Usability evaluation of a web-based tool for supporting holistic building energy management. *Autom. Constr.* **2017**, *84*, 154–165. [CrossRef]
48. International Organization for Standardization (ISO). *ISO 16739-1:2018—Industry Foundation Classes (IFC) for Data Sharing in the Construction and Facility Management Industries—Part 1: Data Schema*; ISO: Geneva, Switzerland, 2018.
49. International Organization for Standardization (ISO). *BS EN ISO 12006-3:2016. Building Construction—Organization of Information about Construction Works—Part 3: Framework for Object-Oriented Information*; ISO: Geneva, Switzerland, 2016.
50. Maltese, S.; Branca, G.; Re Cecconi, F.; Moretti, N. Ifc-based Maintenance Budget Allocation. *In_Bo* **2018**, *9*, 44–51. [CrossRef]
51. Dong, B.; O'Neill, Z.; Li, Z. A BIM-enabled information infrastructure for building energy Fault Detection and Diagnostics. *Autom. Constr.* **2014**, *44*, 197–211.
52. International Organization for Standardization (ISO). *BS EN ISO 29481-1:2017 Building Information Models—Information Delivery Manual. Part 1: Methodology and Format*; ISO: Geneva, Switzerland, 2017.
53. International Organization for Standardization (ISO). *ISO/PAS 16739:2005—Industry Foundation Classes, Release 2x, Platform Specification (IFC2x Platform)*; ISO: Geneva, Switzerland, 2005.

54. Becerik-Gerber, B.; Jazizadeh, F.; Li, N.; Calis, G. Application areas and data requirements for BIM-enabled facilities management. *J. Constr. Eng. Manag.* **2012**, *138*, 431–442.
55. BuildingSMART International. Technical Roadmap buildingSMART: Getting Ready for the Future. Available online: https://www.buildingsmart.org/wp-content/uploads/2020/09/20200430_buildingSMART_Technical_Roadmap.pdf (accessed on 20 November 2020).

Article

Explainable Post-Occupancy Evaluation Using a Humanoid Robot

Marina Bonomolo [1], Patrizia Ribino [2] and Gianpaolo Vitale [2,*]

1 Dipartimento di Ingegneria, Università degli Studi di Palermo, 90128 Palermo, Italy; marina.bonomolo@deim.unipa.it

2 Institute for High Performance Computing and Networking, National Research Council, 90146 Palermo, Italy; patrizia.ribino@icar.cnr.it

* Correspondence: gianpaolo.vitale@icar.cnr.it

Received: 28 September 2020; Accepted: 5 November 2020; Published: 7 November 2020

Abstract: The paper proposes a new methodological approach for evaluating the comfort condition using the concept of explainable post occupancy to make the user aware of the environmental state in which (s)he works. Such an approach was implemented on a humanoid robot with social capabilities that aims to enforce human engagement to follow recommendations. The humanoid robot helps the user to position the sensors correctly to acquire environmental measures corresponding to the temperature, humidity, noise level, and illuminance. The distribution of the last parameter due to its high variability is also retrieved by the simulation software Dialux. Using the post occupancy evaluation method, the robot also proposes a questionnaire to the user for collecting his/her preferences and sensations. In the end, the robot explains to the user the difference between the suggested values by the technical standards and the real measures comparing the results with his/her preferences and perceptions. Finally, it provides a new classification into four clusters: true positive, true negative, false positive, and false negative. This study shows that the user is able to improve her/his condition based on the explanation given by the robot.

Keywords: explainable post occupancy; humanoid robot; lighting simulation software

---

1. Introduction

The improvement of comfort conditions is one of the main goals for an optimal building design. Many standards and studies propose methods and indices to evaluate the quality of the environment [1–4]. They indicate methods and criteria for the design of indoor environments and the evaluation of the energetic performance of buildings related to the internal air quality, illuminance level, and thermal and acoustic parameters. In general, such standards are based on measurable values directly collected from the environment by employing appropriate sensors.

It is also widely recognized that comfort perception is mostly influenced by the psychological and physiological aspects of the users, as was shown in [5–11], where the Post Occupancy Evaluation (POE) approach was proposed as a suitable method for evaluating user's subjective aspects. However, traditional approaches for evaluating and reaching comfort situations show some drawbacks.

Firstly, the location of the sensors used for measuring or controlling systems can influence the reliability of the measurement [12] or the performance of the control system [13]. Indeed, the monitoring campaign of the environmental parameters in indoor spaces is commonly carried out using weather stations generally placed in the middle of the room [14–16]. Such position sometimes impede the collection of the measures close to the users as the standards claim. It can jeopardize the reliability of the measures. For example, the illuminance values should be measured on the work-plane as suggested by the EN12464 standard [17] or by the standard ISO 8995 [4]. Hence, the fixed location of the sensors may downgrade the performance of the adopted environmental control system. Moreover, because a building can have a different organization of the fixtures, the position of the station could disturb the user. Some recently published papers faced this problem by proposing different solutions based on environmental sensors [18], sensors aided by correlation and regression analysis [19], or by optimization algorithms (GA and RBF-NN) [20].

Secondly, a widely used method for reaching comfort conditions is based on Building Automation and Control Systems (BACSs). A BACS automatically gathers data related to environmental parameters, and it takes appropriate corrective actions to maintain the reference values established by the standard. It has been recently noted that the use of such systems may cause some discomfort to the building occupant [21–23]. This is generally due to the loss of control that users perceive [23], in particular when they are not aware of the reasons that induced the system to change the environmental conditions. Indeed, such systems generally do not give feedback to the users, which often do not understand the behavior adopted by the system [5,6]. Thirdly, BACSs use a dense array of sensors, placed in fixed locations of the building, for collecting data at different points of the environment. Such an approach is costly, especially when implemented in old existing buildings.

Finally, long-term measurement is the most effective and accurate way to establish databases that contain this information [24]. However, it is unfeasible to purchase and install pyranometers and illuminance meters at every orientation and tilted angle to collect all of the required data [24]. If it is not suitable to place a grid of sensors, support can be given by simulation software [25].

In this work, we propose a new methodological approach, which, addressing the drawbacks mentioned above, extends the approach proposed in [26], where a method for reaching comfort conditions was presented. The need to improve such an approach comes from retrieving information given by highly fluctuating parameters both in space and in time, such as the illuminance. Moreover, the users' consciousness and their acceptance of supporting technology were further enhanced by giving a comprehensive explanation during the interaction, transforming the traditional POE approach into Explainable POE (EPOE). Notably, the explanation is based on four kinds of possible results: (1) False Positive (FP): the environment does not comply with the standard, and the user feels good; (2) True Negative (TN): the environment does not comply with the standard, and the user does not feel good; (3) False Negative (FN): the environment complies with the standard, and the user is not feeling well; and finally (4) True Positive (TP): the environment complies with the standard, and the user feels good.

Hence, the proposed approach is founded on a new triplet of elements: (i) a humanoid robot for carrying out the application of the measurements of the environmental parameters; (ii) a POE questionnaire performed by the robot; (iii) a simulation lightning software for reproducing the trend of the light distribution both in the space and during the hours of the day. Each element was chosen to face some limitations of the traditional approaches and the particular strengths they provide.

Although the use of a robot for collecting environmental data was firstly introduced in [27–29], they did not resolve the issue of acquiring the measurement at particular positions because they are not reachable by the robot. We chose to adopt a humanoid social robot both to exploit its mobility for collecting measures, as well as to take advantage of human cooperation to locate sensors opportunely. Indeed, it has been demonstrated that both the physical and the anthropomorphic nature of humanoid social robots improves human engagement [30,31]. Humans are more pleasantly involved in their collaboration with robots that are socially competent and show humans traits. Moreover, it should be remarked that the use of humanoid robots strengthens the concept of explainability, as will be evident in the paper.

The significant contribution of this paper is a new methodological approach based on a humanoid robot along with a simulation lightning software for improving indoor comfort conditions and user awareness of his/her environment and cooperation in the control process. The robot collecting and reasoning on four kinds of information (environmental parameters, user preferences from POE, data from the simulation software, and recommendations from the standards) provides a final report. It includes the comfort conditions of the user under analysis and some suggestions to improve them.

The rest of the paper is organized as follows. Section 2 is devoted to explaining the analogies to and differences from the literature and to putting in evidence the novelties of the proposed approach. Section 3 introduces the proposed methodological approach. Section 4 illustrates the case study, and finally, Section 5 contains the conclusions.

2. Related Work

The literature shows a strong interest in the issues discussed in this paper. Indeed, several papers dealing with the topic inherent to our work have been recently published. Some of these are directly related to the Indoor Environment Quality (IEQ) [18,19,23,32–36] and the relevance of the appropriate location of sensors [12,13], while others analyzed the impact of lighting control on workers' health in-depth [20,37–40] and exploited a robot [28,29], whereas the issues related to the post occupant evaluation were studied in [5–10]. The approach proposed in this paper shows similarities and differences compared to the above-mentioned research. To focus on the novelty of this work, a detailed comparison is performed in the following.

Kallio et al. [18] proposed an approach based on relatively inexpensive sensors and exploited machine learning to assess the employees' perception of indoor environment quality, introducing a new method to classify the data. The approach proposed in [32] was based on a laboratory method. It consisted of a chamber in which thermal and luminous conditions could be varied. It was found that space, luminous, and thermal parameters positively affected the satisfaction of users; however, the influence of acoustic noise was not treated. Geng et al. [33] also considered the Energy Use Intensity (EUI) to estimate the performance of green structures, and the differences put in evidence by groups belonging to high and low EUI buildings were related to the energy used for the thermal environment according to the standards. In addition to the traditional approach based on a questionnaire, Tang et al. [19] used a correlation and regression analysis for predicting overall satisfaction. They exploited a laboratory approach with 31 prescribed sets of conditions. Different from this, Dunleavy et al. [34] investigated the difference between aboveground and underground work spaces in terms of psychological distress, and they considered a high number of samples, classified by the OFFICAIRquestionnaire, showing that different parameters influence the two categories.

Lu et al. [35] faced the problem of a personalized model for thermal comfort for people sharing the same offices by using infrared thermography to predict the thermal sensation of users. Sakellaris et al. [23] conducted a large-scale survey that raised the important role of personal control for IEQ satisfaction in office buildings. This study highlighted that personal control is a crucial aspect in obtaining a healthy, comfortable, and productive environment. Angelova and Velichkova [36] analyzed the problem of the thermo-physiological comfort of patients and surgeons in operating room where there was a conflict in the requirements for the thermal environment between the two categories of users. The analysis took into consideration factors related to clothing.

Bonomolo et al. [12] and Bellia et al. [13] explained that the position of the sensors could downgrade the performance of BACSs. As a consequence, data collection had to be done by locating the sensors at suitable points.

As in Bonomolo et al. [12] and Bellia et al. [13], in this paper, we underline the crucial aspect of the measurement points. Moreover, our paper agrees with the approach proposed by Kallio et al. [18] based on inexpensive sensors, as well as with the works proposed in [19,32–34], which collected data by questionnaire. Different from [18], we propose a robot endowed with a set of sensors. Compared to the others, our approach shows two main differences: (a) it aims to plan an interaction between the robot and the user for improving his/her conditions based on recommendations given by the robot; (b) it uses a dedicated software (aided by a few experimental values obtained by a sensor) to reproduce the illuminance. Besides, our paper describes a methodological approach; hence, the study of a great number of users and related analysis methods as in [18,19,34] can be considered as a further step. The comparison with the standard, performed in [33], is improved in our paper based on the feedback provided by users, and new classification criteria are given (i.e., TP-FP-TN-FN) with the possibility given to the users to improve their condition in real time. Different from the laboratory approach of [19], our study is focused on real operating conditions.

As concerns the role of lighting discussed in [20,37–39], it is widely recognized as crucial to workers' health; our paper is focused on this as well, and the illuminance is retrieved by software based on a measurement point. Zanon et al. [40] proposed a new index for evaluating visual comfort based on quantitative and qualitative parameters, highlighting the advantage of using such an index in a simulation software, which allowed the simulation of several parameters with good accuracy. Pragmatic open-loop procedures were explained in [37], where open loops were recognized as more competent and as an alternative to closed-loop systems. Gao et al. [20] exploited a wireless sensors network to retrieve data related to illuminance. They aimed to reduce the number of sensors using an optimization algorithm (GA and RBF-NN) obtaining a matrix with illuminance data. Bellia et al. [38] investigated the impact of Percentage Light Oscillations (PDFs) on users and related switching techniques; the shading of the sensors placed near the windows was considered, but it was evaluated as not preferable. They concluded with the need to minimize the light oscillations even if the tolerance is still under study. In contrast, De Vries et al. [39] took into consideration the influence of wall luminance on an office.

Different from [37], the proposed paper aims to make the user aware of the optimal conditions defining the comfort, and as a consequence, the user is encouraged to adjust the control system according to the recommendation of the robot. Compared to [20], the target is the same, i.e., to obtain a precise reconstruction of the I-matrix. Nevertheless, Reference [20] obtained the I-matrix by a few static sensors and optimization algorithms. On the contrary, we propose the use of the Dialux software validated by a sensor placed on the user's work-plane aided by a robot. Our approach has the drawback of using commercial software. Still, on the other hand, it exploits a single measurement point, and it can model the elements of the environment precisely. Finally, the conclusions of [38,39], in our case, can be given to the user as further advice, thanks to the user-robot interaction.

The features of a robot to improve the knowledge of the environment were exploited in [28,29,41]. In [28], a high granularity of the data distribution was obtained by a mobile sensing platform; this is very different from our approach, which adopts a humanoid able to interact with a human. In our case, the questionnaire is filled during the interaction. Yang et al. [29] used a mobile platform, as did [28], to test different algorithms for contaminant detection. In [41], R.K. Mantha et al. used a mobile robot for collecting ambient parameter data in existing houses with the ultimate aim to retrieve an optimized building retrofit decision (e.g., energy saving).

These papers confirm the approach proposed in our work, meaning that a robot easily collects data; unfortunately, they do not resolve the issue of acquiring the measurement at specific positions not reachable by the robot. In our work, we adopt a robot with an anthropomorphic nature and social abilities for our purposes. In particular, the mobility and the attitude in the interaction with a human of a social humanoid robot allow us to collect measures in appropriate locations through human cooperation as well as to make how the comfort is retrieved explainable.

Indeed, as was demonstrated by recent studies, both the physical and anthropomorphic nature of a social robot have a positive effect on human engagement. In [42], the authors investigated the user's behavior with respect to accepting advice from a physical robot against a computer agent. They showed that a humanoid robot is more valuable in giving recommendations. In [43], the authors analyzed the abilities of a robot, compared to a virtual agent, to persuade human users in a task such as following indications, showing better trust and confidence for the physical robots. In [30], the authors compared a humanoid robot with a mobile application in order to understand the most suitable system for providing recommendations. Their study demonstrated that users prefer the assistance of a social robot. Reference [31] investigated the role of the robot's appearance concerning the acceptance of recommendations. Different humanoid robots were used to provide advertisements to customers at a shopping mall. Such a study proved that people are more attracted to small robots than large ones since the interaction with the small robots was considered more straightforward.

The POE was deeply studied in [5–11]. Yu et al. [11] analyzed the subjective and objective measurements of indoor environment quality from four points of views: thermal, visual, acoustic comfort, and indoor air quality. The entire set of data related to these studies was based on the analysis of habits related to energy consumption and control of the environment when Heating, Ventilation and Air Conditioning systems (HVACs) were used. The authors found the desire of users to learn more about how environment control systems work and comfort is provided. The analysis of our paper is based on the same parameters as in [11].

Pastore et al. [6] focused their attention on the relationship between building sustainability and comfort/health conditions. This underlines that the compliance to the standards does not always fit the users' satisfaction, and it was found that the limitations in individual thermal preference degrade the comfort perception. The gender differences in the perception of IEQ (thermal discomfort in particular) were investigated in [7]. Choi and Moon [8] encompassed questionnaire surveys, environmental measurements, and building attributes; a statistical analysis of them was performed. A "quality measurement car" was devised (this is similar to the mobile platform shown in [28]); the study revealed correlations between human factors and IEQ compliance. Despite the analysis of [9], referring to outdoor spaces, the integration of spatial and temporal data and the attention to the context were interesting; it was pointed out that the users' perspective must be considered from the design phase. The study proposed in [5] showed different IEQ perceptions due to age and gender (not to season, which seems bizarre, but this is probably due to the location in Southern California). Finally, the survey [10] collected a significant part of this information in a unique paper.

Our paper agrees with the papers in which individuals were considered part of the IEQ definition process, including her/his gender, age, and preferences. Our proposed approach takes into account these results when the robot gives the report to the user, and it provides recommendations to improve his/her comfort. In the case of true negatives or false positives, the advice given by the robot can help the user make a change in his/her habits or modify the environmental conditions.

3. Explainable POE Approach

The main goal of this work is the improvement of indoor comfort conditions by making the user aware of his/her situation. The proposed explainable POE approach exploits a humanoid robot that performs a POE survey, collects data, compares them with the standards, and gives an explanation to the user. The analysis performed by the lighting simulation software, which gives information on illuminance levels at different points of the environment, was one-off validated by some measures acquired on a grid of points placed on the user work-plane to verify the correctness of the implementation of both the furniture and lighting sources; then, in normal operating conditions, it was confirmed by a single validation point acquired by the robot. Finally, the robot explained to the user the differences between his/her perception and standards, and it gave information to improve compliance with the standard. In the following, the procedure is described in detail.

3.1. *Environmental Comfort*

Although the standards define comfort, its determination represents a quite difficult task because subjective factors influence it. Indeed, it depends on the perception of the world given by our five senses, leading to an individual perception of the degree of comfort. In this work, we take into consideration three kinds of comfort [44]: thermal, visual, and acoustic comfort.

- Thermal comfort: Thermal comfort depends on our perception of the environmental temperature. Our body performs within an internal temperature range much narrower than external temperatures. In particular, due to metabolism, heat is generated and transmitted to the external environment. High external temperatures hinder this process, and a warm sensation is perceived.

Conversely, when external temperatures are low, a feeling of cold is experienced. Furthermore, the relative humidity plays a relevant role in providing the perception of thermal comfort. During frosty winters, high levels of relative humidity cause a more intense perception of cold. On the other hand, in hot environments, a greater sensation of warmth can be produced by high levels of humidity.

- Visual comfort: Visual comfort is related to the quantity and quality of light. Both insufficient and excessive intensity of light may cause visual discomfort. Visual comfort encompasses a variety of aspects, such as views of outdoor environments, the quality of light, as well as the lack of glare.
- Acoustic comfort: Acoustic comfort is achieved by minimizing noise. This improves concentration and allows for better communication.

Hence, the discomfort is experienced when the thresholds of the environmental parameters are approached. On the contrary, the sensation of comfort is mainly correlated with the ranges of the acceptability of such parameters. Several standards propose such ranges. In particular, ASHRAE 55-92 [3] is a standard that provides the recommended values of environmental parameters for thermal well-being, as is shown in Table 1.

Table 1. Comfort zones for the summer and winter seasons.

	Thresholds in SUMMER		Thresholds in WINTER	
Relative Humidity	**Low Temp**	**High Temp**	**Low Temp**	**High Temp**
30%	24 °C	27 °C	20 °C	25 °C
60%	23 °C	25 °C	20 °C	23 °C

As concerns visual comfort, the European standard EN 12464-1 [17] defines the quantity and quality of illumination as lighting requirements for indoor work spaces. It covers offices, places of public assembly, restaurants/hotels, and theaters/cinemas. Moreover, recommended light levels are provided according to the task an individual has to perform. An example is provided in Table 2. Noise Rating (NR) [45] i as standard to measure and quantify noise in buildings. Table 3 reports an excerpt of the NR levels that are recommended for different application areas. Moreover, the mapping of NR levels with the sound pressure level expressed in decibels (dB, the commonly used unit of measurement) is provided by the noise rating curves. For example, for spaces with office end-use, the maximum sound pressure level is 55 dB.

Table 2. Examples of recommended light levels for offices.

Type of Activity	Illuminance
Writing, typing, reading ...	500 lx
Technical design	750 lx
Conference and meeting	500 lx
Reception	300 lx

Table 3. Maximum recommended noise rating levels.

Type of Application	Max Noise Rating Level
Concert halls, broadcasting and recording studios, churches	NR 25
Halls, shops cloakrooms, restaurants, night clubs, offices	NR 40
Offices with business machines, typing pools	NR 50
Foundries and heavy engineering works	NR 70

3.2. Data Acquisition

As previously mentioned, the robot acquires data also from the environment and the user. The robot was equipped with wearable sensors to collect environmental data. As previously said, the location of sensors is fundamental for the reliability of measurements. In this study, the sensors were transported by the robot, and its mobility was exploited both for taking the measures close to the user and for monitoring them in the whole space. The parameters measured by the sensors were the illuminance (lx), the indoor temperature (°C), the relative humidity (%), and finally, the sound pressure level (dB). To collect the environmental parameters, we used the following sensors:

- A humidity and temperature sensor provided by the SHT31 Smart Gadget device, which is also endowed with data logging capabilities and Bluetooth Low Energy (BLE) connectivity.
- A light sensor provided by the ISO-Tech ILM350 Digital Lux Meter, which is designed to give light level readings up to 50,000 lx. It also allows light measurements away from the subject area to be collected.
- The microphone of a Huawei smartphone to obtain the sound pressure levels.

The robot, moving close to the user, gives her/him guidelines for positioning sensors in particular locations to take the validation point measurement. For example, the above-mentioned standard EN12464 recommends having 500 lx on the desk when a user performs a read task. Hence, the robot asks for the user's collaboration in positioning the sensor at a point where (s)he reads most often. Thus, it can take one measurement in a place where permanent sensors are usually not positioned.

Then, to acquire information about the user, the robot applies the POE questionnaire. Such a survey was designed to obtain user's feedback about his/her preferences and sense of well-being related to thermal, visual, and acoustic conditions. For example, some questions were about: (i) the thermal satisfaction (the expected answers were based on the ASHRAE seven point sensation scale [3] ranging from very satisfied to very dissatisfied); (ii) the illuminance conditions (i.e., "Too dark, Ok, Too bright"), and the perception of the noise in the environment.

3.3. The Lighting Simulation Software

The evaluation of the environment state depends on the knowledge of a scalar field defining the spatial distribution of a parameter. The correct knowledge would require a significant number of sensors. However, this solution is usually expensive. Moreover, it does not give a good description of the field because some points are difficult to equip with measurement sensors. In a previous work [20], a solution based on unique sensor moved to different points was revealed to be appropriate only for parameters affected by low variations. Since illuminance exhibits a greater gradient, a different approach was devised.

In this work, we improved the knowledge by employing a suitable simulation software used to reproduce the illuminance inside an environment where, due to the solar radiation in the presence of clouds or when there are artificial sources, significant variation at short distances could occur. The use of suitable software reproduces the field inside the environment with a reduced amount of information. It represents an inexpensive trade-off since it requires only a model of the environment under study and an initial validation to assure a proper position of the measurement point. We used the lighting simulation software Dialux Evo [46] to perform a set of simulations of the office selected as a case study. Hence, a single measurement station equipped with temperature, humidity, and illuminance sensors to be placed at appropriate points with the user's participation aided by the robot was employed. In principle, the illuminance software is self-consistent. It can reproduce the illuminance based on the model of the environment and lighting sources. We decided to validate the simulation model both at design time and at run time. At design time, the model was validated by some experimental points to compensate the errors due to an imprecise knowledge of both the environment (e.g., the presence of furniture or wall paintings) and the light sources (e.g., aging of artificial light). Then, at run-time, the validation was performed through the measurement acquired by the robot during the survey.

3.4. Workflow of the Explainable POE Approach

The EPOE approach is conceptually represented by the diagram depicted in Figure 1. As previously mentioned, it was implemented on a humanoid robot, which can elaborate, compare, and analyze different kinds of data. These data come from various sources: the survey submitted to the users, the values of the environmental parameters measured by the sensors, the standard opportunely formalized to be robot readable, and the simulation software results.

The process enclosed by the dotted square shown in Figure 1 was performed iteratively for each type of environmental comfort, namely thermal, acoustic, and visual comfort. This process is generically described in the following.

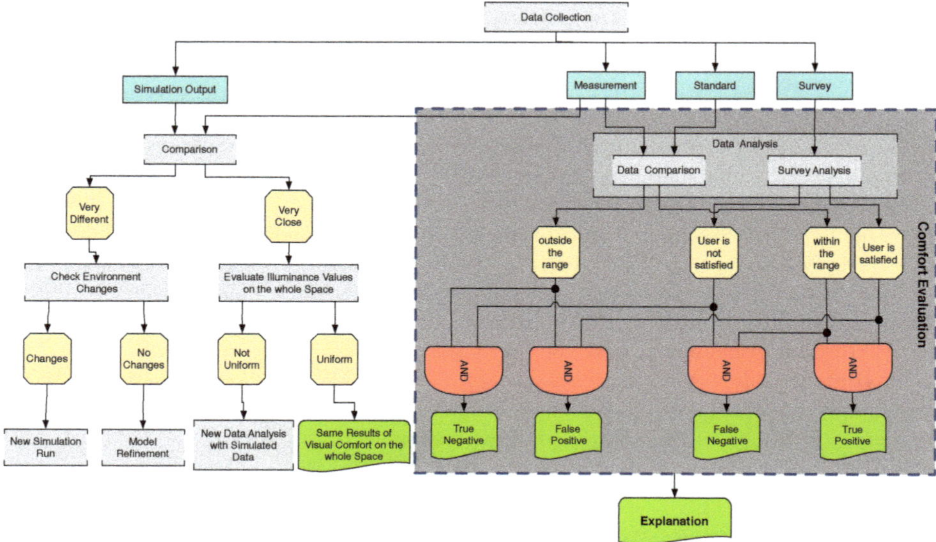

Figure 1. Conceptual flow diagram of the Explainable Post Occupancy Evaluation (EPOE) approach.

The analysis of the data collected by the survey can provide, for each kind of comfort, two outputs according to the user's answer: the user is satisfied or not (see Figure 1). The environmental parameters of the office were compared with the values suggested by the standards to evaluate the compliance. From this analysis, it is possible to know if the measured values are within or outside the range of the suggested values. These last options were compared with the user answers. Hence, four different cases can occur:

- The user feels good, but the values provided by the standard are not achieved. This is classified as a False Positive (FP) result. This implies that the robot produces a report explaining that, despite the user perception, the environmental conditions are not adequate. The robot may suggest some corrections or changes in user habits to improve user comfort, also considering the standards.
- The user does not feel good, and the values provided by the standard are not achieved. This is classified as a True Negative (TN). This implies that the robot produces a report that underlines the violated standard that causes discomfort.
- The user does not feel good, but the values provided by the standard are achieved. This is classified as a False Negative (FN). Furthermore, in this case, the user has to change something to improve the comfort. The robot may underline that some psychological or physiological conditions are the cause of this perception, and also, the robot may give some suggestions about the boundary conditions to be verified.
- The user feels good, and the standards are followed. This is classified as a True Positive (TP). In this case, the robot informs the user that her/his environment is fully compliant with the standard. No further actions are proposed.

The proposed classification allows for quantifying improvement after the robot intervention.

The second part of the methodology takes into account the output of the lighting simulation software in order to detect if the visual comfort is only a local situation or that it is reached in the whole space. Thus, firstly, the data belonging to the experimental point are compared to those obtained by the software at the same point. This step allows us to avoid errors due to some approximation of the model, both the environment and the light sources.

If the simulation is assessed (i.e., the data are very close), the simulation results are used to estimate the light distribution in the whole space. Two cases can occur (see Figure 1). The first one corresponds to the uniformity of the simulated values on the space (meaning that the standard deviation lies within a threshold). As a consequence, we can assume that the result of the visual comfort evaluation obtained by the previous step can be extended to the entire space. The second one occurs when a high variation of simulated data is noticed. In this case, a new analysis of visual comfort is performed based on the simulated data. This analysis may describe a situation characterized by different visual comfort conditions throughout the space.

Otherwise, if the comparison shows that the simulated data are very different from the validation point, it is necessary to check the cause of the problem. The model's parameters are not precise or the environmental conditions changed during the range of time from the first measurement and the simulation. Thus, the robot firstly asks the user if the environmental conditions have been changed in the last few minutes (e.g., the presence of new clouds, the actions of opening or closing the shade systems). If the environmental conditions are changed, a new simulation run is performed with the new inputs. Conversely, a model refinement is required.

The final result of the process is an explanation about the environmental conditions of the user along with some suggestions if this is the case. For example, the robot may suggest the area of the room where the best visual comfort could be achieved. In the following, a detailed case study is proposed to illustrate the proposed approach.

4. Case Study

An office located in the National Research Council building in Palermo (38°09′55.4″ N, 13°18′34.9″ E) was considered as a case study to test the proposed approach to improve the comfort conditions. Figure 2 shows the Google view of the building. The choice of the office was based on the feedback of some workers unsatisfied with their environmental conditions. The office is generally occupied by two workers on different days and different hours of the day. For this reason, in this study, only one worker was selected to conduct the questionnaire. The room selected as a case study (Figure 3) is characterized by an area of 21.41 m^2 (6.15 × 3.50 m). It includes some furniture: a table with four workstations, two book cabinets close to the door, and a small cabinet on a wall. On the shorter wall, there is a window 1 m from the floor, and it is 2.3 m large.

The real image of the office under study is shown in Figure 4. The large window is equipped with white drapes, whose transmission factor is equal to 0.70. It is exposed to 285° west. The office is equipped with: a big painted wooden desk ($r = 0.60$ where r is the reflection coefficient) with a small portion painted black located in the middle of the room; two painted wood cabinets ($r = 0.60$) equipped with glass doors ($r = 0.70$), located almost symmetrically with respect to the door; a small painted wood cabinet ($r = 0.60$) at the end of the opposite wall; and four chairs made of blue and black material ($r = 0.20$). The door of the room is made of steel painted blue ($r = 0.20$). The work-plane of the participant for the test is equipped with two desktop PCs with three big monitors. Moreover, usually, a laptop is positioned under the first monitor.

Figure 2. Google view of the National Research Council building in Palermo.

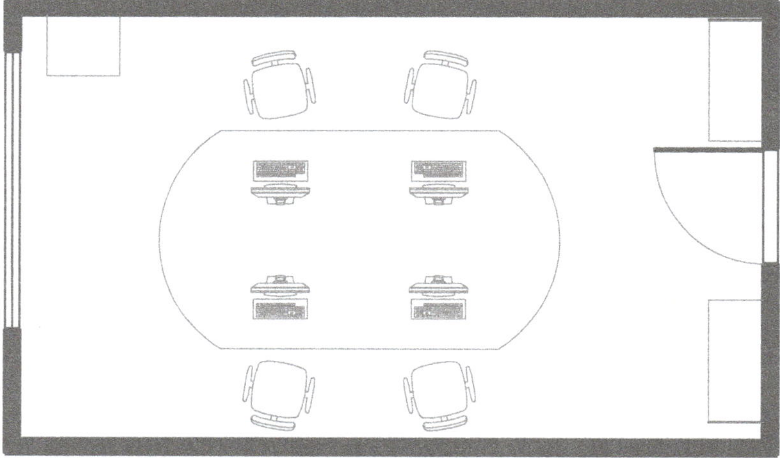

Figure 3. Floor plan of the room under study.

As already mentioned, in this study, four different environmental parameters are taken into consideration: sound level, temperature, humidity, and illuminance. To check the distribution of the measures in the office, a measurement campaign was carried out at different points. Notably, as concerns the illuminance, the measures were retrieved on a regular grid of points placed on the work-plane as a one-off condition to verify the implementation of the furniture and lighting sources. It should be remarked that in the operating conditions, a single validation point was used.

Regarding the illuminance results, in general, it is noted that the room shows some critical aspects tied to the position and exposition of the window. Indeed, the distribution of the illuminance exhibited high variations among the measurement points. Conversely, temperature and humidity showed more homogeneous values. As was confirmed by the standard deviation of the illuminance (see Table 4), there was a relevant difference between the minimum and the maximum value of the illuminance measures. On the contrary, the standard deviation of the temperature was 0.6, and the standard deviation of the humidity was 1.2. The possibility to gather values close to the user may outperform the limitations of the traditional position of the instrument. Looking at the measures of the lighting contribution in Table 4, it is possible to see that the values suggested by the standard EN12464 were not achieved on the whole work-plane. Such measurements refer only to the contribution of daylight. It was noted that artificial light did not improve the illuminance during the daylight hours.

Figure 4. Photograph of the office under study.

Table 4. Table with the maximum and minimum values.

	Illuminance [lx]	Temperature [°C]	Relative Humidity [%]
Min	52	22.95	40.89
Max	376	25	44.48
Mean	181.1	23.73	43.35
Dev.Stand	75.5	0.6	1.2

4.1. Software Output Analysis

The digital model of the office shown in Figure 5 was implemented in the software Dialux Evo. All the optical characteristics of the surfaces of the room were implemented in the model. Furthermore, a maintenance factor of 0.8 was set. To perform the daylight simulation, an hour of a day and the sky conditions (http://www.cie.co.at/publications/cie-standard-overcast-sky-and-clear-sky) were selected. Some simulations were performed at different hours of the day. A calculation surface was positioned at the height of 0.75 m from the floor (see Figure 6). This comprehends the whole area of the work-plane available for the user. The portion of the desk behind the first monitor (i.e., black semicircle on the right) was not considered due to its irregular form, which resulted in being uncomfortable for the user.

Figure 7 shows the isolines of the illuminance values calculated by the simulation software. It is a 2D representation of the values shown in Figure 6.

Figure 5. The 3D model of the office under study.

Figure 6. Calculation surface on the work-plane.

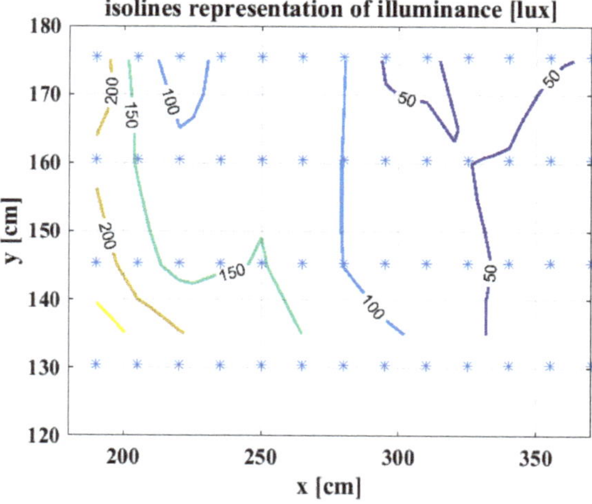

Figure 7. Simulated illuminance values on the work-plane.

As can be noticed, the obtained values decreased significantly from the points near the window towards those near the door (i.e., far from the window). This is more evident considering the 3D representation of the illuminance in Figure 8, where a variation in the **y** coordinate can be appreciated. This situation always occurs during the day since it is tied to the position of the window. Moreover, during the day, the illuminance values compliant with the technical standard EN12464 are only rarely achieved at some points.

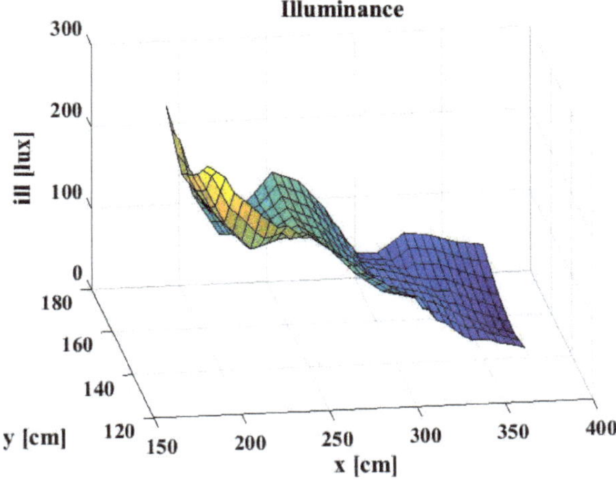

Figure 8. The 3D representation of the illuminance values on the work-plane.

Some simulations were carried out considering only the natural lighting and the simulation considering both natural and artificial light. It is noted that artificial light did not improve the illuminance during the hour of the daylight.

In the diagrams depicted in Figure 9, the simulated illuminance values are compared with some experimental values to assess the accuracy of the results. They show a variation throughout space and time. In particular, they refer to the values of the illuminance (simulated and real ones) of the office taken on 4 December 2019. Both simulations and the acquisition of the real data were performed with the artificial light off, the door closed, and the window drapes open. The first and the second data acquisition were performed at 11:00 a.m. and 12:30 a.m.when the sky was clear. Instead, during the third acquisition at 2:30 p.m., the sky was cloudy. The simulation was run by setting the same time when the measurements were taken and the same sky conditions.

The three diagrams on the top refer to a fixed time and y coordinate of the work-plane, whereas the x coordinate ranges from 170 to 360 cm. Taking into account that the origin of the reference system corresponds to the corner on the bottom left, as shown in Figure 5, these values span the whole work-plane. Conversely, the three diagrams on the bottom of Figure 9 were retrieved by fixing a point on the work-plane, varying the time from 11:00 a.m. to 2:30 p.m.

As concerns the comparison of the real and simulated values shown in the top diagrams, it can be noticed that a good agreement was maintained until the right edge of the work-plane. In this part, very low values were retrieved, making this part unsuitable throughout the hours of the day. This is confirmed by the bottom diagrams in which the discrepancy remains low at the point at coordinate $(190, 135.5)$, and it is slightly higher at the point $(280, 135.5)$, while it dramatically increases at the point $(355, 135.5)$ at the right edge of the work-plane. In general, the difference between measured and simulated values can be considered acceptable, confirming that, in operating conditions, a single validation point is sufficient to assess the reliability of the simulation.

It should be remarked that this comparison was performed at more points to show that the simulation software was quite accurate both in space and time, confirming that for practical use, a unique measurement point can be considered. Besides, a noticeable error was obtained only at points on the right edge of work-plane that were not exploited.

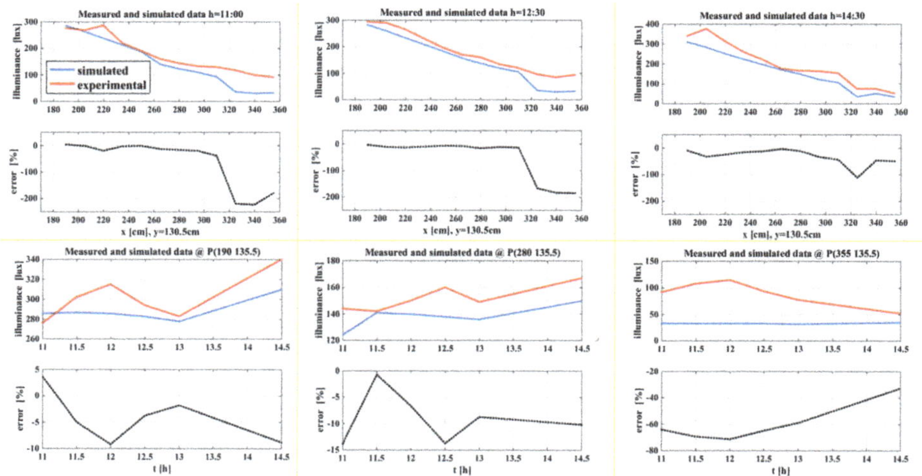

Figure 9. Representation of simulated (blue lines) and real (red lines) illuminance values versus space and time and the related percent error (dotted lines).

4.2. Survey Session and Results

In this section, a detailed description of the interaction between the user and the humanoid robot is presented. It can be considered as a novelty of the proposed approach aiming to establish a dialogue in which the user gives fruitful information to the robot to perform its analysis, and (s)he receives from the robot dedicated explanations about her/his condition and how to improve it when necessary. This is the crucial step in which POE becomes EPOE.

In this study, the experimentation was conducted by using a Nao robot. The Nao robot belongs to the class of humanoid robots (see Figure 10), and it is an autonomous and programmable robot developed by the French company Aldebaran Robotics to provide an open and accessible, high-performance robot platform for both researchers and the general public. The Nao robot is endowed with multitude of sensors, motors, and software handled by an ad-hoc operating system. It has several features to demonstrate natural interactions with humans [47]. The most important feature of the Nao robot is the possibility to be personalized utilizing Choreographe [47]. This is a user-friendly programming suite that allows us to control the robot, create behaviors, and access the data acquired by the sensors. It also allows for testing new behaviors on a simulated robot.

During the session of the POE survey, the robot recognized the user, and it moved towards her/him. Then, it provided her/him guidelines for placing the external sensors in appropriate locations to take the right measurements. After that, the robot applied the POE questionnaire to the user. Finally, by the analysis of the data obtained from the survey and the sensors, it provided the user with a report of her/his working environment quality, giving some recommendations. The interaction between the robot and the user took about 10 minutes. The whole dialogue between the user and the robot during the monitoring session at 12:30 is reported in Appendix A.

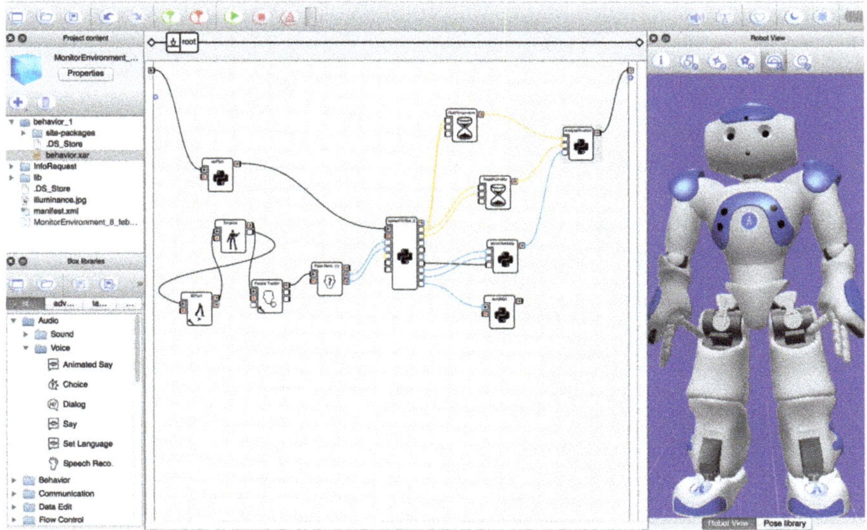

Figure 10. The robot Nao and the Choreographe environment.

During this session, the user placed the illuminance sensor at the point of coordinate (280, 135.5). From the survey, the robot acquired different information from the user. The most important information for the analysis was the following:

- The user perceives a warm working environment.
- She prefers feels a bit cooler.
- She did not like the illuminance conditions, perceiving a too dark environment.
- She prefers a warmer light.
- She considers her environment noiseless.

In addition, the robot gathered through the sensors the following values of the environmental parameters:

- Illuminance level of 160 lx.
- Sound pressure level of 35 dB.
- Temperature of 22 °C.
- Relative humidity of 43%.

Thus, analyzing such data, the robot detected the following situations:

- Thermal comfort: false negative. The environment conformed to the standard of thermal comfort. The current temperature was in the range of the value recommended by the standard in the winter months. Nevertheless, the user felt warm; she would like to have more manual control of the temperature.
- Visual comfort: true negative. The environment did not conform to the standard of visual comfort. The user did not like the illuminance, and she preferred a warmer color temperature.
- Acoustic comfort: true positive. Finally, the environment conformed to the standard of acoustic comfort. The occupant perceived a quiet environment.

Moreover, as can be noted from the simulated values reported in Figure 7, there were no locations in which the standard values of the illuminance were satisfied. For this reason, the robot did not recommend to move to a better work position. Conversely, when among the simulated values, one or more points compliant with the standard appeared, the robot may suggest that the user vary her work position by exploiting the better natural light in the office.

From the user interview, it can be noted that two situations needed a correction. Hence, the robot recommended the user to decrease the temperature up to the min value of the standard, and it also suggested to verify if she was heavily dressed. Moreover, it suggested using a table lamp with warm lights, as she preferred, and to verify if it was a more pleasant situation.

The user, according to the robot suggestions, tried to follow them to improve her comfort opportunely. In particular, she placed a table lamp with warm light (i.e., 3000 K) on her work-plane, as is possible to see from the real figure of the room in Figure 4. As concerns her perception of the temperature, she tried to set a lower operating point temperature of the fan-cooler. Thus, a new POE questionnaire was performed with the user after a suitable time interval to evaluate if the corrections effectively improved the situation. The robot detected the following new situations. Old results are reported near the most recent (old result –> new result).

- True negative –> true positive. The environment complied with the standard of visual comfort. The user perceived a better visual situation.
- False negative –> false negative. The environment conformed to the standard of thermal comfort. The current temperature was in the range of the value recommended by the standard during the winter months. Nevertheless, the user continued to feel warm.
- True positive. Finally, the environment complied with the standard of acoustic comfort. The user perceived a quiet environment.

The introduction of a table lamp effectively improved the visual comfort of the user, although this situation was obtained only close to the user's working area. However, a report recommending an improvement of the artificial lighting system was sent to the maintenance manager. Conversely, the situation of the thermal comfort remained unsolved. A further investigation put in evidence that the thermostatic control was out of operation.

4.3. Discussion

As we stated previously, the significant contribution of this paper is a new methodological approach based on a humanoid robot along with a simulation lightning software for improving indoor comfort conditions and user awareness of his/her environment and cooperation in the control process, leading to the new concept of explainable POE. Even if a measurement campaign is out of the scope of the paper (it will be performed in future work), the verification of the method on a case study allows highlighting the main strengths. The most important advantage to using a humanoid robot is to attract the attention of the users about their real environmental situation and provide them with personalized suggestions to improve their well-being. This is coherent with the literature; indeed (as was found in [26]), often, users' attention is completely absorbed by the task he/she is performing, making him/her unaware of the discomfort situation. Moreover, humans find it more pleasant to collaborate with robots that are socially competent and show human traits. Additionally, users are more inclined to follow recommendations provided by a physical anthropomorphic device than other kinds of systems. In addition, the use of a humanoid robot proved to be effective due to its mobility. Indeed, it can collect environmental data at several points of the environment without a previously installed set of sensors, which could be an expensive solution, especially when it needs to be installed in existing old buildings. The use of a robot becomes an inexpensive solution considering that the sensors represent a fixed installation that should be repeated in every environment to be monitored. The robot, on the other hand, can move from one environment to another carrying an inexpensive and small system containing a sensor for each type of parameter to be measured and can use its wireless interface system to communicate information. Hence, to control many environments, such as the dozens of offices of the National Research Council of Palermo, the use of a robot can become a more affordable solution, also considering that the same robot can be used for other applications when it is not used for monitoring. For the above-mentioned reasons, the proposed approach is less invasive compared with monitoring with a fixed sensors array. Besides, a limitation of classical approaches is the difficulty to place sensors in particular locations, such as the center of the desk where the illuminance has to be measured. Our work is based on human-robot collaboration cooperation in positioning the sensor at the measuring point. In this way, the user, for example, can take the measurement where (s)he reads most often; this type of performance cannot be achieved with fixed sensors. By endowing the robot with sensors, these can be easily substituted or increased, if necessary, to improve the measurement. Finally, when more than one occupant occupies a room, the preferences of the occupants can be considered by the robot and used to propose an eventual trade-off in the case that some occupants' preferences are different.

Along with the robot, the proposed methodological approach takes advantage of the complementary employment of simulation software that retrieves the spatial distribution of an environmental parameter. In general, the use of such software was revealed to be suitable to improve the knowledge of the parameters with a wide range variations, whereas for quantities with small spatial and temporal variations, the single sensor with which the robot is equipped is sufficient.

Conversely, a drawback in using a humanoid robot is that the space in which it has to operate needs to be robot-friendly, namely without relevant obstacles, such as steps, that can interfere with the robot's mobility. The second drawback is related to the Dialux simulation model updating, which in the current implementation of the approach requires human intervention; however, the model implementation is needed only the first time, then it can be exploited to retrieve the illuminance distribution based on a single measurement point.

5. Conclusions and Future Work

In this paper, we introduce the concept of "explainable post occupancy" (EPOE) seeking to make the user aware of his/her well-being status, taking into account both the standards and the user preferences and perceptions. Adopting the EPOE approach concerning the existing proposals in the literature, the users have a better understanding of the environmental comfort situation, thus acquiring more consciousness about their safety and health conditions. This is obtained thanks to the use of a humanoid robot. Indeed, a humanoid social robot is conceived of to reproduce more cues, to show many human characteristics (such as natural language and anthropomorphic physical aspects), and to communicate with humans in ways that resemble interpersonal relationships. Such humanization produces positive effects within human-robot interactions both at the cognitive and the emotional level, thus providing adequate support to people. The EPOE approach is different from the traditional ones in which a conventional automatic control system operates based on the difference between a reference parameter and the value measured by one or more sensors. Based on the well-known result that the illuminance is one of the most significant elements for providing comfort and that it is affected by a high variability both in space and in time, we use a lightning simulation software to reproduce the field inside the environment based on a reduced amount of information. This mitigates the problem of the spatial distribution of the real sensors, as well as reduces their installation costs. The EPOE approach is assessed on a case study allowing us to show also the interaction between the robot and the user, including the explanation given at the end. By this explanation, the user acquires consciousness about his/her real working condition, and (s)he is able to improve it in some cases. Since four different situations, for each considered parameter (i.e., illuminance, temperature, relative humidity, and noise level), are recognized, a new classification is proposed, meaning that (i) the user feels good even if the the environment does not conform to the standard; (ii) the user feels good, and the environment fits with the standard; (iii) the user is not satisfied even if the standard values are achieved; and finally, (iv) the user is not satisfied, and the environment is not compliant with the standard. These situations are respectively classified into the following four categories: false positive, true positive, false negative, and true negative.

The authors believe that their approach could be a good trade-off between the objective classification given by the standard and the subjective perception of the users.

Finally, regarding the whole comfort evaluation, we are currently working to develop a low cost embedded system endowed with sensors able to acquire air velocity as well for a better evaluation of the thermal comfort and a CO_2 sensor to evaluate the air quality within the environment as well.

Author Contributions: All authors contributed equally to the manuscript. All authors have read and agreed to the published version of the manuscript.

Funding: This work was conducted in the framework of the project AMICO—Medical Assistance In COntextual awareness (CUP B46G18000390005). Facilitation marked by the identification code ARS01 0090.

Acknowledgments: The authors are very grateful to National Research Council, Area della Ricerca di Palermo, via Ugo La Malfa 153, Palermo, for the availability of the offices where the investigation was carried out.

Conflicts of Interest: The authors declare no conflict of interest.

Appendix A

The whole dialogue between the user and the robot during the monitoring session at 12:30 is reported in the following. It is worth noting that during the dialogue, the robot uses a language with some domain specific concepts. However, the robot is able to explain such concepts that require explanations by providing definitions and figures to the user. Moreover, in some cases, we use informal terms that are more comprehensible for users. For example, we use the words "dark" and "bright", although there are scales for brightness (that has a slightly different meaning from the light level). This choice was because such terms are used in everyday language to express the perception of light levels.

```
R: Hello, I'm Nao. What's your Name?
U: My~name is Patrizia.
R: Nice~to meet you, Patrizia. I'm here for ensuring
your well-being. I~have to gather some environmental
parameters. But~I also need your collaboration. Please,
take the sensors in my belt and then place them near
to you. Tell~me ok, when~you finished.
U: Ok, finished.
```

The user placed the illuminance sensor at the point of coordinate $(280, 135.5)$. Hereafter, the robot starts the POE questionnaire.

```
R: How~old are you?
U: I'm 44 years old.
R: What~is your role in the office?
U: I'm a Researcher.
R: How~much time do you usually spend at work?
U: 8 hours.
R: Ok, the~time you usually spent at work is 8
hours. How~do you feel right now?
Choose among: hot, warm, slightly warm,
no change, slightly cool, cool, cold
U: I~feel warm
R: How~do you prefer to feel right now?
Choose among: much~warmer, warmer, a~bit
warmer, no~change, a~bit cooler, cooler,
much cooler.
U: a~bit cooler
R: Would~you like to have more manual control
in the temperature?
```

```
U: yes
R: Do~you like illuminance level condition?
Choose among: no~it is too dark, it~is ok,
no it is too bright
U: no, it~is too dark
R: Do~you like luminaries color level?
U: No
R: Do~you prefer warmer or cooler light?
U: I~prefer the warmer light
R: Ok, you~prefer warmer
Are you able to control the shading system manually,
during the day?
U: yes
R: Do~you have some eye pathologies?
U: yes
R: I'm sorry, what~kind of eye pathologies?
U: myopia
R: Is~usually your office too loud?
U: no, it~isn't
R: Do~you perceive a high humidity level?
U: No
```

Once the POE survey is finished, the robot acquires from the user illuminance and sound level that the user reads from the sensors placed on her desk.

```
R: Now, can~you read the value shown by
the illuminance sensor?
U: 160
R: Ok, Can~you read the sound pressure level
from the smartphone?
U: 35
R: Thank~you for your cooperation.
Just another request. Can~you put again
the sensor in my belt? Tell me ok,
when you finished.
U: ok
```

In the meanwhile, the robot has also acquired data coming from the Smart Gadget device about temperature and relative humidity, and it is elaborating the results.

Thus, analyzing the answers of the user and the environmental measurements, the robot detects the following situations:

```
R: Ok~Patrizia, now~I explain to you the situation
of thermal, visual and acoustic comfort
with respect to your working conditions.
I will also try to give you some suggestions
for improving your well-being if it is the case.
U: ok
R: Patrizia, the~temperature near you is 22 degree.
You feel warm in your environment even if it
respects the thermal comfort standard: ''It is
recommended that the temperature is lower than 23
```

```
degree and greater than 20 degree if the humidity
ranges between 30% and 60% if the environment
is a job area, in~the winter season.''
I suggest you to decrease the temperature up to the min
value of the standard and to verify if you are heavy
dressed. Tell~me repeat or next for going on.
U: Ok, next
R: Patrizia, your~current work-plane illuminance is 160.
You perceive bad illuminance because your environment
is not compliant with the recommendation: ``It is
recommended that the illuminance on the work-plane is
at least 500 lx if the environment is a job area.''.
Unfortunately, there~are no better illuminated
locations on your work-plane where you can move.
I suggest you to increase the brightness of your work-
plane surface. Because you prefer a warmer color
temperature, I~also suggest you to use a table lamp
with warm lights to verify if it is a more pleasant
situation for your eyes. Tell~me repeat or next for
going on.
U: Ok, next
R: Finally, Patrizia your current environment
respects the acoustic comfort standard.
I hope to be useful to you. Thank~you, bye-bye.
U: Bye.
```

Hence, the robot explains the situation to the user. In the following, the rest of the dialogue with Patrizia.

```
R: Ok~Patrizia, now~I explain to you the situation
of thermal, visual and acoustic comfort
with respect to your working conditions.
I will also try to give you some suggestions
for improving your well-being if it is the case.
U: ok
R: Patrizia, the~temperature near you is 22 degree.
You feel warm in your environment even if it
respects the thermal comfort standard: "It is
recommended that the temperature is lower than 23
degree and greater than 20 degree if the humidity
ranges between 30% and 60% if the environment
is a job area, in~the winter season."
I suggest you to decrease the temperature up to the min
value of the standard and to verify if you are heavy
dressed. Tell~me repeat or next for going on.
U: Ok, next
R: Patrizia, your~current work-plane illuminance is 160.
You perceive bad illuminance because your environment
is not compliant with the recommendation: ``It is
recommended that the illuminance on the work-plane is
at least 500 lx if the environment is a job area.''.
```

```
Unfortunately, there~are no better illuminated
locations on your work-plane where you can move.
I suggest you to increase the brightness of your work-
plane surface. Because you prefer a warmer color
temperature, I~also suggest you to use a table lamp
with warm lights to verify if it is a more pleasant
situation for your eyes. Tell~me repeat or next for
going on.
U: Ok, next
R: Finally, Patrizia your current environment
respects the acoustic comfort standard.
I hope to be useful to you. Thank~you, bye-bye.
U: Bye.
```

References

1. Comité Européen de Normalisation. *BS EN 15251:2008. Indoor Environmental Input Parameters for Design and Assessment of Energy Performance of Buildings Addressing Indoor Air Quality, Thermal Environment, Lighting and Acoustics*; Comite Europeen de Normalisation: Rome, Italy, 2008; Volume 14.
2. ISO. ISO 7730: Ergonomics of the thermal environment—Analytical determination and interpretation of thermal comfort using calculation of the PMV and PPD indices and local thermal comfort criteria. *Management* **2005**, *3*, e615.
3. Standard Ashrae. *Standard 55—2017 Thermal Environmental Conditions for Human Occupancy*; Ashrae: Atlanta, GA, USA, 2017.
4. ISO. *Lighting for Work Places Part 1: Indoor (ISO: 8995-1: 2002 E)*; International Commission on Illumination: Vienna, Austria, 2002.
5. Choi, J.H.; Lee, K. Investigation of the feasibility of POE methodology for a modern commercial office building. *Build. Environ.* **2018**, *143*, 591–604. [CrossRef]
6. Pastore, L.; Andersen, M. Building energy certification versus user satisfaction with the indoor environment: Findings from a multi-site post-occupancy evaluation (POE) in Switzerland. *Build. Environ.* **2019**, *150*, 60–74. [CrossRef]
7. Kim, J.; de Dear, R.; Candido, C.; Zhang, H.; Arens, E. Gender differences in office occupant perception of indoor environmental quality (IEQ). *Build. Environ.* **2013**, *70*, 245–256. [CrossRef]
8. Choi, J.H.; Moon, J. Impacts of human and spatial factors on user satisfaction in office environments. *Build. Environ.* **2017**, *114*, 23–35. [CrossRef]
9. Göçer, Ö.; Göçer, K.; Başol, A.M.; Kıraç, M.F.; Özbil, A.; Bakovic, M.; Siddiqui, F.P.; Özcan, B. Introduction of a spatio-temporal mapping based POE method for outdoor spaces: Suburban university campus as a case study. *Build. Environ.* **2018**, *145*, 125–139. [CrossRef]
10. Li, P.; Froese, T.M.; Brager, G. Post-occupancy evaluation: State-of-the-art analysis and state-of-the-practice review. *Build. Environ.* **2018**, *133*, 187–202. [CrossRef]
11. Yu, X.; Liu, L.; Wu, X.; Wu, X.; Wang, Z.; Liu, Q.; Shi, G. On a Post-occupancy Evaluation Study of Effects of Occupant Behavior on Indoor Environment Quality in College Buildings in Chongqing. *Procedia Eng.* **2017**, *205*, 623–627. [CrossRef]
12. Bonomolo, M.; Beccali, M.; Brano, V.L.; Zizzo, G. A set of indices to assess the real performance of daylight-linked control systems. *Energy Build.* **2017**, *149*, 235–245. [CrossRef]
13. Bellia, L.; Fragliasso, F.; Pedace, A. Lighting control systems: factors affecting energy savings' evaluation. *Energy Procedia* **2015**, *78*, 2645–2650. [CrossRef]
14. Bellia, L.; Fragliasso, F. Automated daylight-linked control systems performance with illuminance sensors for side-lit offices in the Mediterranean area. *Autom. Constr.* **2019**, *100*, 145–162. [CrossRef]
15. Gao, Y.; Cheng, Y.; Zhang, H.; Zou, N. Dynamic illuminance measurement and control used for smart lighting with LED. *Measurement* **2019**, *139*, 380–386. [CrossRef]

16. Alzoubi, H.; Bataineh, R.F. Pre-versus post-occupancy evaluation of daylight quality in hospitals. *Build. Environ.* **2010**, *45*, 2652–2665. [CrossRef]
17. EN 12464-1:2011. *Light and Lighting—Lighting of Work Places— Part 1: Indoor Work Places*; Central Bureau of the CIE: Vienna, Austria, 2011.
18. Kallio, J.; Vildjiounaite, E.; Koivusaari, J.; Räsänen, P.; Similä, H.; Kyllönen, V.; Muuraiskangas, S.; Ronkainen, J.; Rehu, J.; Vehmas, K. Assessment of perceived indoor environmental quality, stress and productivity based on environmental sensor data and personality categorization. *Build. Environ.* **2020**, *175*, 106787. [CrossRef]
19. Tang, H.; Ding, Y.; Singer, B. Interactions and comprehensive effect of indoor environmental quality factors on occupant satisfaction. *Build. Environ.* **2020**, *167*, 106462. [CrossRef]
20. Gao, Y.; Lin, Y.; Sun, Y. A wireless sensor network based on the novel concept of an I-matrix to achieve high-precision lighting control. *Build. Environ.* **2013**, *70*, 223–231. [CrossRef]
21. Huang, Y.H.; Robertson, M.M.; Chang, K.I. The role of environmental control on environmental satisfaction, communication, and psychological stress: effects of office ergonomics training. *Environ. Behav.* **2004**, *36*, 617–637. [CrossRef]
22. Lee, S.Y.; Brand, J.L. Effects of control over office workspace on perceptions of the work environment and work outcomes. *J. Environ. Psychol.* **2005**, *25*, 323–333. [CrossRef]
23. Sakellaris, I.; Saraga, D.; Mandin, C.; de Kluizenaar, Y.; Fossati, S.; Spinazzè, A.; Cattaneo, A.; Szigeti, T.; Mihucz, V.; de Oliveira Fernandes, E.; et al. Personal control of the indoor environment in offices: Relations with building characteristics, influence on occupant perception and reported symptoms related to the building—The officair project. *Appl. Sci.* **2019**, *9*, 3227. [CrossRef]
24. Li, D.H.; Lou, S. Review of solar irradiance and daylight illuminance modeling and sky classification. *Renew. Energy* **2018**, *126*, 445–453. [CrossRef]
25. Yoon, Y.; Moon, J.W.; Kim, S. Development of annual daylight simulation algorithms for prediction of indoor daylight illuminance. *Energy Build.* **2016**, *118*, 1–17. [CrossRef]
26. Ribino, P.; Bonomolo, M.; Lodato, C.; Vitale, G. A Humanoid Social Robot Based Approach for Indoor Environment Quality Monitoring and Well-Being Improvement. *Int. J. Soc. Robot.* **2020**, 1–20. [CrossRef]
27. Mantha, B.R.; Feng, C.; Menassa, C.C.; Kamat, V.R. Real-time building energy and comfort parameter data collection using mobile indoor robots. In Proceedings of the 32nd International Symposium on Automation and Robotics in Construction and Mining (ISARC 2015): Connected to the Future, Oulu, Finland, 15–18 June 2015; Vilnius Gediminas Technical University, Department of Construction Economics: Vilnius, Lithuania, 2015; Volume 32, p. 1.
28. Jin, M.; Liu, S.; Schiavon, S.; Spanos, C. Automated mobile sensing: Towards high-granularity agile indoor environmental quality monitoring. *Build. Environ.* **2018**, *127*, 268–276. [CrossRef]
29. Yang, Y.; Feng, Q.; Cai, H.; Xu, J.; Li, F.; Deng, Z.; Yan, C.; Li, X. Experimental study on three single-robot active olfaction algorithms for locating contaminant sources in indoor environments with no strong airflow. *Build. Environ.* **2019**, *155*, 320–333. [CrossRef]
30. Rossi, S.; Staffa, M.; Tamburro, A. Socially assistive robot for providing recommendations: Comparing a humanoid robot with a mobile application. *Int. J. Soc. Robot.* **2018**, *10*, 265–278. [CrossRef]
31. Shiomi, M.; Shinozawa, K.; Nakagawa, Y.; Miyashita, T.; Sakamoto, T.; Terakubo, T.; Ishiguro, H.; Hagita, N. Recommendation effects of a social robot for advertisement-use context in a shopping mall. *Int. J. Soc. Robot.* **2013**, *5*, 251–262. [CrossRef]
32. Xu, H.; Huang, Q.; Zhang, Q. A study and application of the degree of satisfaction with indoor environmental quality involving a building space factor. *Build. Environ.* **2018**, *143*, 227–239. [CrossRef]
33. Geng, Y.; Lin, B.; Zhu, Y. Comparative study on indoor environmental quality of green office buildings with different levels of energy use intensity. *Build. Environ.* **2020**, *168*, 106482. [CrossRef]
34. Dunleavy, G.; Bajpai, R.; Tonon, A.C.; Cheung, K.L.; Thach, T.Q.; Rykov, Y.; Soh, C.K.; de Vries, H.; Car, J.; Christopoulos, G. Prevalence of psychological distress and its association with perceived indoor environmental quality and workplace factors in under and aboveground workplaces. *Build. Environ.* **2020**, *175*, 106799. [CrossRef]
35. Lu, S.; Wang, W.; Wang, S.; Cochran Hameen, E. Thermal Comfort-Based Personalized Models with Non-Intrusive Sensing Technique in Office Buildings. *Appl. Sci.* **2019**, *9*, 1768. [CrossRef]

36. Angelova, R.A.; Velichkova, R. Thermophysiological Comfort of Surgeons and Patient in an Operating Room Based on PMV-PPD and PHS Indexes. *Appl. Sci.* **2020**, *10*, 1801. [CrossRef]
37. Jain, S.; Garg, V. A review of open loop control strategies for shades, blinds and integrated lighting by use of real-time daylight prediction methods. *Build. Environ.* **2018**, *135*, 352–364. [CrossRef]
38. Bellia, L.; Fragliasso, F.; Riccio, G. Daylight fluctuations effect on the functioning of different daylight-linked control systems. *Build. Environ.* **2018**, *135*, 162–193. [CrossRef]
39. De Vries, A.; Souman, J.L.; de Ruyter, B.; Heynderickx, I.; de Kort, Y.A. Lighting up the office: The effect of wall luminance on room appraisal, office workers' performance, and subjective alertness. *Build. Environ.* **2018**, *142*, 534–543. [CrossRef]
40. Zanon, S.; Callegaro, N.; Albatici, R. A Novel Approach for the Definition of an Integrated Visual Quality Index for Residential Buildings. *Appl. Sci.* **2019**, *9*, 1579. [CrossRef]
41. Mantha, B.R.; Menassa, C.C.; Kamat, V.R. Robotic data collection and simulation for evaluation of building retrofit performance. *Autom. Constr.* **2018**, *92*, 88–102. [CrossRef]
42. Powers, A.; Kiesler, S.; Fussell, S.; Fussell, S.; Torrey, C. Comparing a computer agent with a humanoid robot. In Proceedings of the ACM/IEEE International Conference on Human-Robot Interaction, Arlington, VA, USA, 10–12 March 2007; ACM: New York, NY, USA, 2007; pp. 145–152.
43. Bainbridge, W.A.; Hart, J.; Kim, E.S.; Scassellati, B. The effect of presence on human-robot interaction. In Proceedings of the RO-MAN 2008—The 17th IEEE International Symposium on Robot and Human Interactive Communication, Munich, Germany, 1–3 August 2008; pp. 701–706.
44. Frontczak, M.; Wargocki, P. Literature survey on how different factors influence human comfort in indoor environments. *Build. Environ.* **2011**, *46*, 922–937. [CrossRef]
45. Beranek, L.L.; Blazier, W.E.; Figwer, J.J. Preferred noise criterion (PNC) curves and their application to rooms. *J. Acoust. Soc. Am.* **1971**, *50*, 1223–1228. [CrossRef]
46. DIAL. DIALux Evo 8.2, Professional Lighting Design Software. 2018. Available online: https://www.dialux.com/it-IT (accessed on 27 October 2020).
47. Robotics, A. NAO Robot. 2015. Available online: https://www.softbankrobotics.com/emea/en/nao (accessed on 27 October 2020).

MDPI
St. Alban-Anlage 66
4052 Basel
Switzerland
Tel. +41 61 683 77 34
Fax +41 61 302 89 18
www.mdpi.com

Applied Sciences Editorial Office
E-mail: applsci@mdpi.com
www.mdpi.com/journal/applsci

www.ingramcontent.com/pod-product-compliance
Lightning Source LLC
LaVergne TN
LVHW070138170726
843515LV00007B/1822

* 9 7 8 3 0 3 6 5 3 9 5 2 2 *